Provisioning, Recovery,
and In-operation Planning
in Elastic Optical Networks

Provisioning, Recovery, and In-operation Planning in Elastic Optical Networks

Luis Velasco and Marc Ruiz

Registered Office
John Wiley & Sons, Inc., 111 River Street, Hoboken, NJ 07030, USA

Editorial Office
111 River Street, Hoboken, NJ 07030, USA

For details of our global editorial offices, customer services, and more information about Wiley products visit us at www.wiley.com.

Wiley also publishes its books in a variety of electronic formats and by print-on-demand. Some content that appears in standard print versions of this book may not be available in other formats.

Library of Congress Cataloging-in-Publication Data

Names: Velasco, Luis, 1966– author. | Ruiz, Marc, 1981– author.
Title: Provisioning, recovery, and in-operation planning in elastic optical networks / Luis Velasco, Marc Ruiz.
Description: Hoboken, NJ : John Wiley & Sons, 2017. |
   Includes bibliographical references and index. |
Identifiers: LCCN 2017011910 (print) | LCCN 2017027869 (ebook) |
   ISBN 9781119340416 (pdf) | ISBN 9781119340423 (epub) | ISBN 9781119338567 (cloth)
Subjects: LCSH: Optical engineering. | Computer networks. | Optics.
Classification: LCC TA1520 (ebook) | LCC TA1520 .P765 2017 (print) |
   DDC 621.382/7–dc23
LC record available at https://lccn.loc.gov/2017011910

Cover design: Wiley
Cover image: © John Lund/Gettyimages

Set in 10/12pt Warnock by SPi Global, Pondicherry, India

Printed in the United States of America

10 9 8 7 6 5 4 3 2 1

Contents

List of Contributors

Adrián Asensio
Universitat Politècnica de Catalunya
Barcelona
Spain

Alberto Castro
University California Davis
Davis, CA
USA

Jaume Comellas
Universitat Politècnica de Catalunya
Barcelona
Spain

Luis Miguel Contreras
Telefonica Investigación y Desarrollo
Madrid
Spain

Filippo Cugini
CNIT
Pisa
Italy

Lluís Gifre
Universitat Politècnica de Catalunya
Barcelona
Spain

Mirosław Klinkowski
National Institute of
Telecommunications
Warsaw
Poland

Dimitrios Klonidis
Athens Information Technology
Athens
Greece

Victor López
Telefonica Investigación y Desarrollo
Madrid
Spain

Ricardo Martínez
Centre Tecnològic de
Telecomunicacions de Catalunya
Castelldefels
Spain

Fernando Morales
Universitat Politècnica de Catalunya
Barcelona
Spain

Marc Ruiz
Universitat Politècnica de Catalunya
Barcelona
Spain

Behnam Shariati
Athens Information Technology
Athens
Greece
Universitat Politècnica de Catalunya
Barcelona
Spain

Ioannis Tomkos
Athens Information Technology
Athens
Greece

Alba Pérez Vela
Universitat Politècnica de Catalunya
Barcelona
Spain

Luis Velasco
Universitat Politècnica de Catalunya
Barcelona
Spain

1

Motivation

Luis Velasco and Marc Ruiz

Universitat Politècnica de Catalunya, Barcelona, Spain

1.1 Motivation

The huge amount of research done in the last decade in the field of optical transmission has made available a set of technologies jointly known as flexgrid, where the optical spectrum is divided into 12.5 GHz frequency slices with 6.25 GHz central frequency granularity, in contrast to the coarser 50 GHz in fixed grid Wavelength Division Multiplexing (WDM) [G694.1]. Such frequency slices can be combined in groups of contiguous slices to form frequency slots of the desired spectral width, thus increasing fiber links' capacity. To illustrate the magnitude of the capacity increment, a 40 Gb/s connection modulated using Dual-Polarization Quadrature Phase Shift Keying (DP-QPSK) can be transported on a 25 GHz slot in flexgrid, instead of 50 GHz needed with WDM [Ru14.1].

In addition to increasing network capacity, subsystems currently being developed will foster devising novel network architectures. These are as follows:

- Liquid Crystal on Silicon (LCoS)-based Wavelength Selective Switches (WSS) to build flexgrid-ready Optical Cross-Connects (OXCs) [Ji09].
- The development of advanced modulation formats to increase efficiency, which are capable of extending the reach of optical signals [Ge12].
- Sliceable Bandwidth-Variable Transponders (SBVTs) able to deal with several flows in parallel, thus adding, even more, flexibility and reducing costs [Sa15].

The resulting flexgrid networks will allow mixing optical connections of different bitrates (e.g., 10, 40, 100 Gb/s), by allocating frequency slices and using different modulation formats such as 16-State Quadrature Amplitude

Modulation (QAM16) or DP-QPSK more flexibly. Furthermore, larger bitrates (e.g., 400 Gb/s or even 1 Tb/s) can be conceived by extending the slot width beyond 50 GHz. In addition, the capability to elastically allocate frequency slices on demand and/or modify the modulation format of optical connections according to variations in the traffic of the demands allows resources to be used efficiently in response to traffic variations; this is named as Elastic Optical Networks (EONs) [Na15], [Lo16].

To understand how the finer spectrum granularity together with the flexible slot allocation of EONs can impact the network architecture, let us compare national Multi-Protocol Label Switching (MPLS) network designs when the underlying optical network is based either on a WDM network or on an EON [Ve13.1]. MPLS networks typically receive client flows from access networks and perform flow aggregation and routing. The problem of designing MPLS networks consists in finding the configurations of the whole set of routers and links to transport a given traffic matrix whilst minimizing capital expenditures (CAPEX). To minimize the number of ports a router hierarchy consisting of *metro* routers, performing client flow aggregation, and *transit* routers, providing routing flexibility, is typically created.

As a consequence of link lengths, national MPLS networks have been designed on top of WDM networks, and, thus, the design problem has been typically addressed through a multilayer MPLS-over-WDM approach where transit routers are placed alongside OXCs. Besides, multilayer MPLS-over-WDM networks take advantage of grooming to achieve high spectrum efficiency, filling the gap between users' flows and optical connections' capacity.

The advent of flexgrid technology providing a finer granularity, however, makes it possible to flatten the multilayer approach and advance toward single layer networks consisting of a number of MPLS areas connected through a core EON (Figure 1.1).

To compare the designs of national MPLS networks when such flatten network architecture is adopted, we define the overall network spectral efficiency of the interarea optical connections, as:

$$
Sp.Eff = \frac{\displaystyle\sum_{a\in A}\sum_{\substack{a'\in A\\a'\neq a}} \frac{b_{aa'}}{\Delta f \cdot B_{\mathrm{mod}}}}{\displaystyle\sum_{a\in A}\sum_{\substack{a'\in A\\a'\neq a}} \left\lceil \frac{b_{aa'}}{\Delta f \cdot B_{\mathrm{mod}}} \right\rceil}, \tag{1.1}
$$

where $b_{aa'}$ represents the bitrate between two different areas a and a' in the set of areas A, Δf is the considered spectrum granularity (i.e., 50 GHz for WDM and 12.5 GHz for EON), and B_{mod} is the spectral efficiency (b/s/Hz) of the chosen modulation format. Note that the ceiling operation computes the number

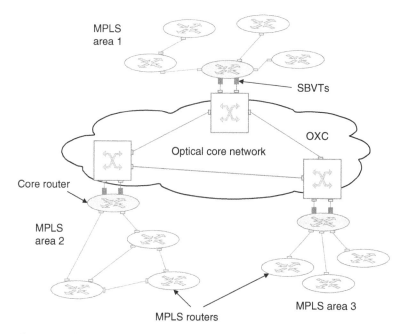

Figure 1.1 Three MPLS areas connected through a core optical network.

of slices/wavelengths needed to convey the requested data flow under the chosen technology.

Let us analyze the results obtained from solving a close-to-real problem instance consisting of 1113 locations, based on the British Telecom (BT) network. Those locations (323), with a connectivity degree of 4 or above, were selected as potential core locations. A 3.22 Pb/s traffic matrix was obtained by considering the number of residential and business premises in the proximity of each location. Locations could only be parented to a potential area if they were within a 100 km radius.

Figure 1.2 plots the amount of aggregated traffic injected to the optical core network for each solution as a function of the number of areas. The relationship between the amount of aggregated traffic and the number of areas is clearly shown; more areas entail higher aggregated traffic to be exchanged because less traffic is retained within an area since the areas are smaller. Nonetheless, the amount of aggregated traffic when all the 323 areas are opened is only 6% with respect to just opening 20 areas.

Plots in Figure 1.2 show the maximum spectral efficiency of the solutions for each technology and the number of areas selected. As illustrated, network spectral efficiency decreases sharply when the number of areas is increased since more flows with a lower amount of traffic are needed to be transported over the

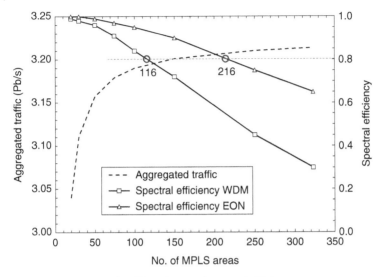

Figure 1.2 Aggregated traffic and network spectral efficiency.

core network. Note that the traffic matrix to be transported by the core network has $|A|^*(|A| - 1)$ unidirectional flows. Let us consider a target threshold for network spectral efficiency of 80% (horizontal dotted line in Figure 1.2). Then, the largest number of areas are 116 and 216 when WDM and EON, respectively, are selected. Obviously, the coarser the spectrum granularity chosen for the optical network, the larger the areas need to be for the spectral efficiency threshold selected, and, thus, the lower the number of areas to be opened.

Figure 1.3 presents the details of the solutions as a function of the number of areas, while Table 1.1 focuses on the characteristics of those solutions in the defined spectral efficiency threshold for each technology. Note that plots in Figure 1.3 are valid irrespective of the technology selected since no spectral efficiency threshold was required.

In Figure 1.3a, the size of the areas is represented. Note that in the reference interval [116, 216] the average number of locations in each area is lower than 10, being lower than 20 for the largest area. This limited area size simplifies the design of MPLS networks. Besides, switching capacity of core MPLS routers is proportional to the size of the areas as shown in Figure 1.3b. They require capacities of up to 58 Tb/s (28.5 Tb/s on average) for WDM, decreasing to 35.5 Tb/s (15.7 Tb/s on average) for EON. Hence, finer spectrum granularity might keep routers to single chassis size—far more efficient and cost effective.

Figure 1.3c surveys the size of the area internal data flows (flows where at least one end router is in a given area). The size of the internal flows is up to 72 Gb/s, 32 Gb/s on average, for WDM decreasing to 50 Gb/s, 21 Gb/s on average, for EON.

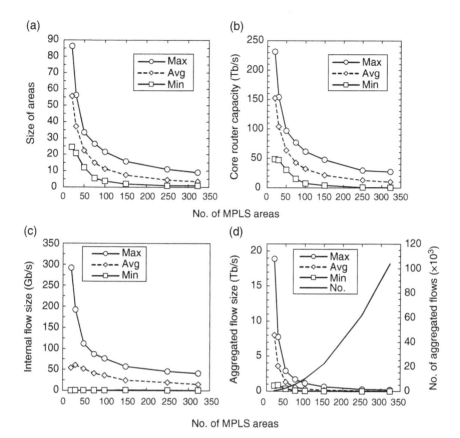

Figure 1.3 Details of the solutions against the number of MPLS areas: (a) size of the areas, (b) switching capacity of core MPLS routers, (c) size of the internal data flows, and (d) size of the aggregated data flows.

Table 1.1 Solutions details (network spectral efficiency = 0.8).

		WDM	EON
Number of IP/MPLS areas		116	216
Areas size	Max	19.66	15.04
	Avg	9.94	5.46
Core router capacity (Tb/s)	Max	58.0	35.3
	Avg	28.5	15.7
Area flows (Gb/s)	Max	72.33	50.52
	Avg	32.68	21.12
Aggregated flows (Gb/s)	Max	989.06	404.51
	Avg	264.95	82.73
	#	13,884	48,684

Finally, Figure 1.3d examines the size of the aggregated data flows in the optical core network. When the number of areas is low, the size of the aggregated flows is quite large, but the overall size of the traffic matrix is quite small: the largest flow conveys 1 Tb/s and there are 13,880 aggregated data flows when WDM is used. In this case, since the largest capacity that WDM network can convey is usually limited to 100 Gb/s, each 1 Tb/s flow has to be split into 10×100 Gb/s flows, which increases aggregated data flow count to 138,800.

In comparison, when the number of areas is larger in the case of EON, the size of the largest flow is smaller, but the size of the core network traffic matrix is very large: the largest flow conveys 400 Gb/s, but there are as many as 48,000 aggregated data flows.

As a conclusion, simpler and smaller MPLS areas containing 10–15 locations are enough to obtain good spectral efficiency when the EON-based core network are implemented. In addition, large aggregated flows need to be conveyed by the optical core network, which entails a rather limited traffic dynamicity and explains why these networks are configured and managed statically. As a result, long planning cycles are used to upgrade the network and prepare it for the next planning period. Aiming at guaranteeing that the network can support the forecast traffic and deal with failure scenarios, spare capacity is usually installed, thus increasing network CAPEX. Moreover, note that results from network capacity planning are deployed manually in the network, which limits the network agility.

However, the scenario is rapidly changing because of the maturity of the Internet of services, where cloud providers play a critical role in making services accessible to end-users by deploying platforms and services in their datacenters (DCs). The new model and the requirements derived from cloud services are contributing to the increasing amount of data that require connectivity from transport networks. Evolution toward cloud-ready transport networks entails dynamically controlling network resources, considering cloud requests in the network configuration process. Hence, that evolution is based on elastic data, control, and management planes.

In a scenario where dynamic capabilities are available, the network can be reconfigured and reoptimized in an automatic fashion in response to traffic changes. Hence, resource overprovisioning can be minimized, which reduces overall network costs; this was called as *in-operation network planning* in [Ve14.2].

In fact, network operators are deploying their own small DCs and integrating them into their networks to provide ultra-low latency new generation services, such as cloud computing or video distribution. The so-called *Telecom Cloud* infrastructure [Ve15] must support a large number of small, distributed DCs placed close to the end-users to reduce traffic in the core network and to provide services and applications that can take advantage from very low latency (Figure 1.4).

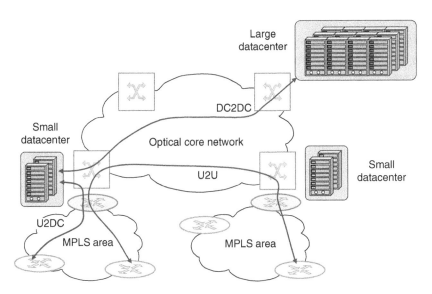

Figure 1.4 Distributed DC connected through an EON.

To support content distribution, digital contents, for example, films, need to be available in the DCs, so connectivity among DCs (DC2DC) and from users in the metro areas to the DCs (U2DC) are needed in the telecom cloud. Note that this is in addition to the connectivity among MPLS areas (denoted as U2U in Figure 1.4) considered in the previous study. In fact, the core optical network must support dynamic connectivity and time-varying traffic capacities, which might also entail changes in the traffic direction along the day.

This book is devoted to reviewing algorithms to solve provisioning problems, as well as in-operation network planning problems aiming at enhancing the performance of EONs. The feasibility of those algorithms is also validated in experimental environments.

In-operation network planning requires an architecture enabling the deployment of algorithms that must be solved in stringent times. That architecture can be based on a Path Computation Element (PCE) or a Software Defined Networks (SDN) controller. In both cases, a specialized planning tool is proposed to be responsible for solving in-operation planning problems efficiently. After the architecture to support provisioning and in-operation planning is assessed, we focus on studying a number of applications in single layer and multilayer scenarios.

The methodology followed in the book consists in proposing a problem statement and a mathematical formulation for the problems and, then, presenting algorithms for solving them. Next, performance evaluation results from the simulation are presented including a comparison of results for different scenarios and architectures. In addition, experimental assessment is carried out on realistic testbeds.

In this book we will cover the topics discussed earlier and present:

- Provisioning point to point, anycast, and multicast connections, as well as transfer-based connections for DC interconnection in single layer EON.
- Recovery connections from failures and in-operation planning in single layer EON.
- Virtual network topology (VNT) reconfiguration and recovery in multilayer MPLS-over-EON scenarios.
- New topics currently under extensive research, including high capacity optical networks based on space division multiplexing (SDM), provisioning Customer Virtual Networks (CVNs), and the use of data analytics to bring cognition to the network.

1.2   Book Outline

The remainder of this book is organized into five parts. Part I includes Chapters 2–4 and gives an overview of the concepts that will be used along the book. Part II includes Chapters 5–7 and focuses on provisioning aspects in single layer EONs. Part III includes Chapters 8–10 and centers on recovery and in-operation planning issues in single layer EONs. Part IV includes Chapters 11 and 12 and targets at recovery and in-operation planning problems in multilayer MPLS-over-EON. Finally, Part V includes Chapters 13–15 and presents future trends of optical networks.

Chapter 2 briefly introduces the main background of graph theory and optimization and their application to model a very basic routing problem following several approaches. Basic heuristic methods are introduced, and the two meta-heuristics commonly used throughout the book are presented. In addition, main concepts of the optical technology and networks' planning and operation, including provisioning and recovery, are reviewed.

Chapter 3 is devoted to the *routing and spectrum allocation* (RSA) problem. Regardless of additional constraints, the RSA problem involves two main constraints for spectrum allocation: *continuity* and the *contiguity* constraints. These two constraints highly increase the complexity of the optimization, so it is crucial that efficient methods are available to allow solving realistic problem instances in practical times. This chapter reviews several Integer Linear Programming (ILP) formulations of the RSA problem and evaluates the computational time to solve a basic offline problem. The basic RSA problem is extended to consider the capability of selecting the required modulation format among the available ones, and, therefore, the routing, modulation, and spectrum allocation (RMSA) problem is presented. This chapter is based on the work in [Ve12.1], [Ve14.1], and [Ve17].

Chapter 4 reviews control and management architectures to allow the network to be dynamically operated, as well as the capabilities of the Application-based Network Operations (ABNO) architecture proposed by the Internet Engineering Task Force (IETF). Using these dynamicity capabilities, the network can be reconfigured and reoptimized automatically in response to traffic changes. Nonetheless, dynamic connectivity is not enough to guarantee elastic DC operations and might lead to poor performance provided that not enough overprovisioning of network resources is performed. To alleviate it to some extent, the ability of EONs can be used to dynamically increase and decrease the amount of optical resources assigned to connections. This chapter is based on the work in [Ve14.2], [Gi14.1], [Ma14], and [Ve13.2].

Assuming traffic dynamicity, Chapter 5 presents basic algorithms for dynamic point-to-point (P2P) connection provisioning in EONs and extends their scope to deal with modulation formats and with dynamic spectrum adaptation, that is, elasticity. This chapter is based on the work in [Ca12.2], [Ca12.3], and [As13].

Chapter 6 presumes a DC interconnection scenario and proposes an Application Service Orchestrator (ASO) as an intermediate layer between the cloud and the network. In this architecture, DC resource managers request data transferences using an application-oriented semantic. Then, those requests are transformed into connection requests and forwarded to the ABNO in charge of the transport network. The proposed ASO can perform elastic operations on already established connections supporting transferences provided that the committed completion time is ensured. This chapter is based on the work in [Ve14.3] and [As14].

Chapter 7 explores the scenario where data needs to be conveyed from one source to multiple destinations or from one source to any destination among multiple destinations, for example, for database synchronization or content distribution. The point-to-multipoint (P2MP) RSA problem is proposed to find feasible light-trees to serve multicast demands, as well as the anycast RSA problem to find a feasible lightpath to one of the destinations. This chapter is based on the work in [Ru14.2], [Ru15], [Gi15.1], and [Gi16].

Dynamic operation of EONs might cause the optical spectrum to be divided into fragments, making it difficult to find a contiguous spectrum of the required width for incoming connection requests. To alleviate spectrum fragmentation to some extent, Chapter 8 proposes applying spectrum defragmentation, triggered before an incoming connection request is blocked, to change the spectrum allocation of already established optical connections, so as to make room for the new connection. This chapter is based on the work in [Ca12.2] and [Gi14.2].

Chapter 9 explores P2P connection recovery schemes specifically designed for EONs, in particular taking advantage of SBVTs. The bitrate squeezing and multipath restoration problem is proposed to improve restorability of large bitrate connections. Next, that problem is extended by considering modulation formats and transponder availability. Finally, recovery for anycast optical connections is studied. This chapter is based on the work in [Ca12.1], [Pa14], [Gi15.2], and [Gi16].

When a link that had failed is finally repaired, its capacity becomes available for new connections, which increases the difference between lightly and heavily loaded links and decreases the probability of finding optical paths with continuous and contiguous spectrum for future connection requests. Chapter 10 studies the effects of reoptimizing the network after a link failure has been repaired. The problem is extended to consider scenarios where bitrate squeezing and multipath restoration had been applied after the failure. This chapter is based on the work in [Zo15] and [Gi15.3].

In the context of multilayer MPLS-over-EON networks, the introduction of new services requiring large and dynamic bitrate connectivity might cause changes in the direction of the traffic along the day. This leads to large overprovisioning in statically managed VNTs designed to cope with the traffic forecast. To reduce expenses while ensuring the required grade of service, Chapter 11 proposes the VNT reconfiguration based on real traffic measurements that regularly reconfigures the VNT for the next period to adapt the topology to the expected traffic volume and direction. This chapter is based on the work in [Mo17].

In multilayer networks, an optical link failure may cause the disruption of multiple aggregated MPLS connections. Thereby, efficient recovery schemes, such as multilayer restoration and survivable VNT design, are required. Chapter 12 focuses on such issues, specifically on multilayer MPLS-over-EON operated with a Generalized Multi-Protocol Label Switching (GMPLS) control plane, where a centralized PCE is in charge of computing the route of MPLS connections. As for survivable VNT design connecting DCs, Chapter 12 focuses on both, the problem of designing VNTs to ensure that any single failure in the optical layer will affect one single virtual link (vlink) at the most and providing a VNT mechanism that reconnects the VNT in the case of vlink failure. This chapter is based on the work in [Ca13], [Ca14], and [Gi16].

Chapter 13 reviews the recent progress in the development of SDM-based optical networks and describes the enabling technologies, including fibers and switches, for the efficient realization of such networks. It is unquestionable that the efficient use of the space domain requires some form of spatial integration of network elements. Particularly, this chapter describes three switching paradigms, which are highly correlated with three fiber categories proposed for SDM networks. This chapter is based on the work in [Kh16], [Sh16.2], and [Sh16.4].

In the context of future mobile networks, Chapter 14 studies dynamic CVNs to support the Cloud Radio Access Networks (C-RAN) architecture. CVN reconfiguration needs to be supported in both metro and core network segments and must include Quality of Service (QoS) constraints to ensure specific delay requirements, as well as bitrate guarantees to avoid service interruption. This chapter is based on the work in [As16.2].

Finally, Chapter 15 explores the interesting and challenging data analytics discipline that finds patterns in heterogeneous data coming from monitoring the data plane, for example, received power and errors in the optical layer or service traffic in the MPLS layer. The output of data analytics can be used for automating network operation and reconfiguration, detecting traffic anomalies, or transmission degradation. This way of doing network management is collectively known as the *observe–analyze–act* (OAA) loop since it links together monitoring, data analytics, and operation (with or without reoptimization). Chapter 15 illustrates the OAA loop by applying it to a variety of use cases that result into a proposed architecture based on ABNO to support the OAA loop in EONs. This chapter is based on the work in [Va16], [Ru16.2], [Mo17], and [Ve17].

1.3 Book Itineraries

A quick reading of the table of contents of any chapter in this book is enough to realize the large variety of contents that they cover, including a description of new concepts and ideas, mathematical programming formulations, heuristic algorithms, simulation results, and experimental assessment details. We definitely understand that a potential reader would be interested in focusing on only part of the contents while skipping others that are far of his/her interests and skills. For this very reason, we devised the following tracks designed to cope with the expectations of three archetype readers:

- *Average reader track*: this track is conceived for the reader interested in the main concepts and ideas that each chapter tackles. Introductory subsections about the descriptions of the problems as well as a minimum notion of the algorithmic methods used are recommended. Additionally, those performance results required to fully understand subsequent key conclusions are considered in this track. We understand that this track contains the essentials of the book; therefore, it can be explored by either an uninitiated reader demanding a broad panorama of the book topics or an advanced reader for the first approach.

- *Theoretical reader track:* this track extends the contents recommended for the average reader with those related to the models and algorithms behind the proposed concepts and ideas. Mathematical programming formulations and pseudo-codes for heuristic algorithms are extensively presented within this track, as well as those results strictly related to algorithms' performance. Advanced readers looking for detailed answers about book contributions are encouraged to follow this track.

- *Practitioner reader track:* this track is intended for the reader interested in the implementation of the methods and their experimental assessment (the most singular contents of this track). It is worth noting that part of the contents recommended in the previous tracks is included in this one, for example, the basics of the presented concepts. However, only those mathematical and algorithmic methods that are finally validated by means of experimental assessment are usually included in this track.

At the beginning of each chapter, the itinerary of each track through the sections and subsections is provided together with the table of contents. With this kind suggestion, we hope that reading this book will be more comfortable and will closely fit with readers' interests and skills.

Acknowledgment

We would like to thank the following contributors for their help in preparing the book:

- Adrián Asensio, Universitat Politècnica de Catalunya (Barcelona, Spain)
- Alba Pérez Vela, Universitat Politècnica de Catalunya (Barcelona, Spain)
- Alberto Castro, University California Davis (Davis, CA, USA)
- Behnam Shariati, Athens Information Technology (Athens, Greece) and Universitat Politècnica de Catalunya (Barcelona, Spain)
- Dimitrios Klonidis, Athens Information Technology (Athens, Greece)
- Fernando Morales, Universitat Politècnica de Catalunya (Barcelona, Spain)
- Filippo Cugini, Consorzio Nazionale Interuniversitario per le Telecomunicazioni (Pisa, Italy)
- Ioannis Tomkos, Athens Information Technology (Athens, Greece)
- Jaume Comellas, Universitat Politècnica de Catalunya (Barcelona, Spain)
- Lluís Gifre, Universitat Politècnica de Catalunya (Barcelona, Spain)
- Luis Miguel Contreras, Telefonica Investigación y Desarrollo (Madrid, Spain)
- Mirosław Klinkowski, National Institute of Telecommunications (Warsaw, Poland)
- Ricardo Martínez, Centre Tecnològic de Telecomunicacions de Catalunya (Castelldefels, Spain)
- Victor López, Telefonica Investigación y Desarrollo (Madrid, Spain)

In addition, we would like to thank the following institutions for their collaboration since some of the experiments presented in this book were carried out with their collaboration:

- Telefonica Investigación y Desarrollo (Madrid, Spain)
- Consorzio Nazionale Interuniversitario per le Telecomunicazioni (Pisa, Italy)
- Centre Tecnològic de Telecomunicacions de Catalunya (Castelldefels, Spain).

Part I

Introduction

2

Background

Marc Ruiz and Luis Velasco

Universitat Politècnica de Catalunya, Barcelona, Spain

This chapter introduces the basic concepts that are needed to understand the contents of the following chapters. It is organized into five main sections. The first two sections are devoted to graph theory and optimization; basic definitions and notation are presented. The third section illustrates the application of the two previous sections to model a very basic routing problem following several approaches. Basic heuristic methods are introduced, and the two meta-heuristics commonly used through the book are presented. The last two main sections are devoted to the optical technology and how networks are planned and operated. Section 2.4 introduces the main aspects related to optical networks, starting with an introduction to opaque, translucent, and transparent networks. Then, the key features of Elastic Optical Networks (EONs) are presented. Section 2.5 briefly introduces the classical network life cycle and the essential concepts on connection provisioning and recovery.

Table of Contents and Tracks' Itineraries

Provisioning, Recovery, and In-operation Planning in Elastic Optical Networks,
First Edition. Luis Velasco and Marc Ruiz.
© 2017 John Wiley & Sons, Inc. Published 2017 by John Wiley & Sons, Inc.

2.1 Introduction to Graph Theory

In this section, we briefly introduce the graph theory and a basic algorithm for network routing.

A *graph* (also referred to as a topology or a network) $G(V, E)$ is a mathematical object that consists of a set V of nodes, or vertices in general, and a set E of edges or links connecting two nodes (see Figure 2.1). The size of a graph can be measured in terms of either its number of nodes ($|V|$) or links ($|E|$). The *average nodal degree* (δ) is defined in the next equation and can be used as a measure of the mesh degree. Note that the degree of a given node is defined as the number of edges incident to it [Ha94].

$$\delta = \frac{2 \cdot |E|}{|V|} \tag{2.1}$$

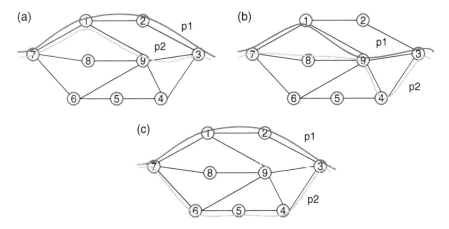

Figure 2.1 A graph and two paths that are distinct (a), link-disjoint (b), and node-disjoint (c).

For instance, the graph in Figure 2.1 has 9 nodes and 13 edges, with an average nodal degree of 2.89; node 9 has a degree of 5, node 1 has a degree of 3, and node 2 has a degree of 2.

Given a graph $G(V, E)$, a *subgraph* $G'(V', E')$ of G is a graph defined by a subset of nodes $V' \subseteq V$ and a subset of links $E' \subseteq E$, in such a way that V' includes all the endpoints of E'. If $V' = V$, G' is called *spanning subgraph* of G.

A path is a sequence of nodes in a graph that are sequentially visited without repetitions from a source to a destination following a connected set of links [Ah93]. A *cycle* is a path plus the link connecting destination and source nodes. Several relations can be defined between two different paths that have the same source and destination nodes. Two paths are *distinct* if, at least, one of their links is different (Figure 2.1a). When all their links are different, the paths are *link-disjoint* (Figure 2.1b), whereas if all the intermediate visited nodes are different, the paths are *node-disjoint* (Figure 2.1c). Note that node-disjoint paths are link-disjoint, while the inverse is not true in general.

We can define the *shortest path* between a pair of nodes as the path connecting those nodes with the minimum number of links. This definition can be generalized by considering individual weights for each link, being the shortest path with the minimum sum of weights in the visited links. The *average shortest path length* of G is defined as the mean path length of all the shortest paths in G. The *eccentricity* of a node v is defined as the maximum shortest path length between the node and any other in the network. The maximum eccentricity in a graph is called *diameter* (the longest shortest path) whereas the minimum eccentricity is called *radius*.

Among the characteristics that allow classifying graphs, we focus on the definition of *directionality, connectivity,* and *planarity*. A graph is *directed* (also known as *digraph*) if links $e = (n_1, n_2)$ and $e' = (n_2, n_1)$ defined between any pair of nodes are different; on the contrary, the graph is *undirected*. A graph is said to be *connected* if there exists at least one path connecting every pair of nodes; otherwise the graph is *disconnected*. For instance, graphs in Figure 2.2a and b are examples of connected graphs, while the graph in Figure 2.2c is disconnected. A disconnected graph consists of a number of *connected* subgraphs (or *components*) with one or more nodes (e.g., the graph in Figure 2.2c consists of two connected components). Following with connectivity, a graph is k-connected if there exist at least k link-disjoint paths between every pair of nodes (e.g., the graph in Figure 2.2a is 2-connected, while the graph in Figure 2.2b is 1-connected or just connected). Finally, a graph is *planar* if and only if the graph can be drawn on a 2D plane in such a way that no edges cross each other. All graphs in Figure 2.2 are planar.

A *tree* T is an undirected, connected, acyclic graph. Equivalently, T is a tree if any two nodes are connected by exactly one path. A subgraph T of G is called *spanning tree* if it is a tree and a spanning subgraph of G; such tree satisfies the condition $|E'| = |V| - 1$. The example in Figure 2.3 shows a graph G in

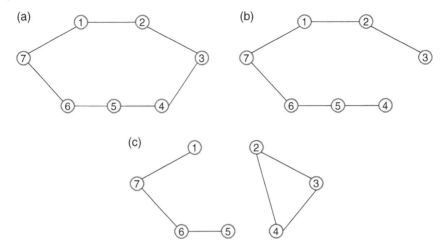

Figure 2.2 Connected graphs and connected components: (a) 2-connected, (b) connected, and (c) disconnected graphs.

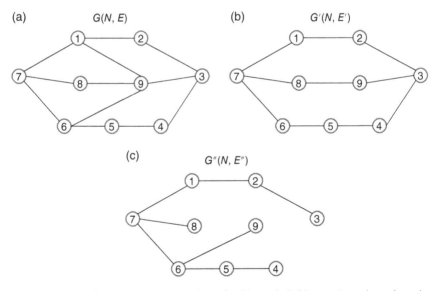

Figure 2.3 A graph and two spanning subgraphs: (a) graph G, (b) spanning subgraph, and (c) spanning tree.

Figure 2.3a and two spanning subgraphs in Figure 2.3b and c, where graph G'' in Figure 2.3c is a spanning tree of graph G in Figure 2.3a whereas the subgraph in Figure 2.3b is not.

Once the main concepts of graph theory have been reviewed, let us now focus on the algorithms for computing the shortest paths. We assume that

Table 2.1 Dijkstra's algorithm.

INPUT: $G(V, E)$, o
OUTPUT: $dist, pred$

```
 1: Q ← V
 2: for each i ∈ V do
 3:    dist[i] ← ∞
 4:    pred[i] ← ∅
 5: dist[o] ← 0
 6: while Q ≠ ∅ do
 7:    i ← arg min {dist[j] | j ∈ Q}
 8:    Q ← Q \ {i}
 9:    for each j ∈ Q do
10:       if e = {i, j} ∈ E then
11:          aux ← dist[i] + e.weigth
12:          if aux < dist[j] then
13:             dist[j] ← aux
14:             pred[j] ← i
15: return dist, pred
```

---

every link has its own weight, so the shortest path between two nodes is that where the sum of the weights of its links is the minimum. We focus on the particular case where all the links' weights are nonnegative.

Several algorithms for solving the shortest path problem considering different characteristics of the graph, such as directionality, can be found in the literature. Among them, Dijkstra's algorithm [Di59] is the most commonly used; several variants and extensions can be found in the literature [Bh99]. Table 2.1 presents the pseudo-code of Dijkstra's algorithm, which returns a tree with the shortest paths from the source node to all the rest of nodes. The algorithm receives a connected graph (or component) G and a source node o and returns two vectors, one with the shortest path distances and another with the predecessor for each of the nodes. The algorithm starts initializing the set of unvisited nodes Q and both distances and predecessor vectors (lines 1–6 in Table 2.1). Then, while Q is not empty, the node with the minimum distance in Q is selected and removed from Q, and the shortest path distance to each of its adjacent nodes is updated if and only if it is decreased by traversing the current link (lines 7–15).

As already stated, a solution of the Dijkstra's algorithm is a tree with the shortest paths from o to every other node in the graph. To construct the shortest path from o to a given destination t, we have to get track from the target t in the vector of predecessors until reaching the source o (see Table 2.2). Note that the length of the path is stored in position t in the distances vector.

Table 2.2 Getting the shortest path from the shortest path tree.

INPUT: $pred, o, t$
OUTPUT: p

```
1:  p ← ∅
2:  i ← t
3:  while pred[i] ≠ ∅ do
4:    p.push_front(i)
5:    i ← pred[i]
6:  p.push_front(o)
7:  return p
```

An interesting extension to the shortest path problem is its generalization to obtain k distinct shortest paths. To solve this problem efficiently, Yen's algorithm [Ye71] can be used; it computes k-shortest paths in a graph with nonnegative link weights.

For an extended review of the concepts introduced here, we refer the interested reader to [Ha94], [Ah93], [Bh99], and [Gr04].

2.2 Introduction to Optimization

Mathematical programming or *optimization* is a mathematical method to find an *optimal* point x^* that results into the minimum (or maximum) value of a function $f(x)$ while satisfying a set of constraints [Ch83]; such point x^* is said to be optimum. More formally, an optimization problem can be defined as follows:

$$\min \quad z = f(x) \tag{2.2}$$

subject to:

$$g_i(x) \geq b_i, \quad \forall i \in R \tag{2.3}$$

where x represents a vector of variables, $f(x)$ is the objective function, and R represents the set of constraints. A constraint is an inequality defined by a function $g_i(x)$ and a constant b_i.

Let us define X as the set of all possible x vectors. Then, we can define the set S of feasible solutions of the problem as follows:

$$S = \left\{ x' \in X \mid g_i(x') \geq b_i, \forall i \in R \right\}, \tag{2.4}$$

that is, S contains all elements in X satisfying the whole set of constraints. Note that a problem could have alternative optimal solutions, that is, several x^* with the same z^* value. Therefore, we can define the set of optimal solutions X^* as:

$$X^* = \left\{ x' \in S \mid f(x') \leq f(x''), \forall x'' \in S \right\} \tag{2.5}$$

The problem, however, could have no feasible solution, that is, $S = \emptyset$; in such case the problem is unfeasible. Finally, when $f(x^*) = -\infty$ the problem is unbounded.

A *Linear Programming* (LP) problem is a special case of mathematical programming, where $f(x)$ and $g_i(x)$ are linear functions of real variables. When variables are restricted to be integer, the problem is called *Integer Linear Programming* (ILP), whereas if the problem combines integer and real variables, the problem is defined as *Mixed Integer Linear Programming* (MILP). Finally, *Nonlinear Programming* (NLP) entails at least one nonlinear function [Ch83].

Regarding time complexity, a problem is said to be Nondeterministically Polynomial (NP) if the size of the candidate solution set is exponential with respect to the input size and if it takes polynomial time to verify the correctness of any candidate solution; therefore it is faster to verify a candidate than find a solution from scratch. Among NP problems, NP-complete are the hardest computational problems [Ga79]. As an example, ILP belongs to NP-complete.

Exact procedures have been developed to solve mathematical programming problems. For example, the *simplex algorithm* is used for solving LPs, whereas *Branch&Bound* and *Branch&Cut* algorithms are used to solve ILPs [Ch83]. However, when the problem is large-scale, its size in terms of number of variables and number of constraints makes impractical to solve it with the previous algorithms. In such case, decomposition methods, for example, *column generation*, in combination with exact procedures can be applied to efficiently reduce the number of integer decision variables [De05]. In the case of NLPs, the use of interior point methods such as the *Newton barrier method* has been exploited for nonlinear constrained problems [Fo02].

Although the output of these exact methods is the optimal solution, the required computation time tends to be too high for practical purposes when real-life instances need to be solved, even in the case of using powerful solver engines, such as CPLEX [CPLEX] or Gurobi [Gur]. Thus, some relaxation methods such as *Lagrangian relaxation* [Ah93] or *randomized rounding* [Ra87] provide near optimal solutions to ILP problems by relaxing some integer constraints, thus decreasing the problem's complexity.

As an alternative to relaxation methods, heuristic algorithms have also been proposed. A heuristic is an algorithm to obtain feasible solutions, where the guarantee of finding an optimal solution is sacrificed for the sake of getting good solutions in a significantly reduced amount of time.

2.3 ILP Models and Heuristics for Routing Problems

In this section, we review some ILP formulations and heuristics for routing problems. To that end, we propose a very basic problem, where a set of demands needs to be routed over a capacitated network. The problem statement reads as follows:

Given:
- a network, represented by a connected graph $G(V, E)$, V being the set of nodes and E the set of links connecting two nodes in V. Each link e in E with a capacity q_e;
- a set D of demands to be routed. Each demand d is represented by a tuple $<o_d, t_d, b_d>$, where o_d and t_d are the source and the destination nodes, respectively, and b_d is the requested amount of capacity.

Output: The route over the network for each of the demands, ensuring that demands are served using one single path with enough capacity or, on the contrary, they are rejected.

Objective: Minimize the total capacity that cannot be served (primary objective) and the amount of capacity used in the links (secondary objective).

2.3.1 ILP Formulations

Since the topology is given in the problem, one possible modeling approach consists in precomputing a set of k distinct paths for every demand and assign one of them to the demand provided that there is enough capacity in the links that the path traverses. This formulation is usually known as *arc-path* (AP).

The following sets and parameters have been defined.

V	Set of nodes, index v.
E	Set of links, index e.
q_e	Available capacity in link $e \in E$.
D	Set of demands, index d.
$<o_d, t_d>$	Source and target nodes of demand $d \in D$.
b_d	Amount of capacity requested by demand $d \in D$.
P	Set of precomputed paths, index p.
$P(d)$	Subset of precomputed (distinct) paths for demand d. $|P(d)| \leq k \ \forall d \in D$.
h_p	Length of path p in number of hops.
r_{pe}	Equal to 1 if path p uses link e.
α	Cost function multiplier.

The decision variables are:

w_d	Binary, equal to 1 if demand d cannot be served; 0 otherwise.
x_p	Binary, equal to 1 if path p is selected; 0 otherwise.

Using this notation, the AP formulation is as follows:

$$(AP) \quad \min \quad \alpha \cdot \sum_{d \in D} b_d \cdot w_d + \sum_{d \in D} \sum_{p \in P(d)} h_p \cdot b_d \cdot x_p \tag{2.6}$$

subject to:

$$\sum_{p \in P(d)} x_p + w_d = 1 \quad \forall d \in D \tag{2.7}$$

$$\sum_{d \in D} \sum_{p \in P(d)} r_{pe} \cdot b_d \cdot x_p \leq q_e \quad \forall e \in E \tag{2.8}$$

The objective function (2.6) minimizes the amount of demand capacity that is rejected and the total amount of used capacity. The former objective becomes the primary one if parameter α takes a large value, for example, the sum of b_d for all demands.

Constraint (2.7) ensures that either a path is selected for each demand provided that the demand is served or the demand cannot be served, and therefore it is rejected. Constraint (2.8) guarantees that the total amount of capacity allocated to a link does not exceed the available capacity of such link.

The AP formulation requires from a large number of x_p variables for the whole set of possible paths. However, only a few of them will be nonzero in the optimal solution. Based on this fact, path generation solving techniques to generate reduced sets of variables providing high-quality solutions can be easily applied when the AP formulation is used (e.g., see [Pi04]).

An alternative to AP formulation is to perform routing computation by means of specific flow conservation constraints typically used in network flow problems [Ah93]. This formulation is named as *node-arc* (NA) and presents two versions for directed and undirected graphs.

For directed graphs, we extend the previous notation with the following additional subsets and parameters:

$E^+(v)$ Subset of E with links leaving from node v.
$E^-(v)$ Subset of E with links arriving at leaving from node v.

The routing decision variable x needs to be redefined as follows:

x_{de} Binary, equal to 1 if demand d is routed through link e; 0 otherwise.

The NA formulation for the directed graph case (NA-D) is as follows:

$$(NA-D) \quad \min \quad \alpha \cdot \sum_{d \in D} b_d \cdot w_d + \sum_{d \in D} \sum_{e \in E} b_d \cdot x_{de} \tag{2.9}$$

subject to:

$$\sum_{e \in E^+(v)} x_{de} - \sum_{e \in E^-(v)} x_{de} = \begin{cases} (1 - w_d) & \forall d \in D, \ v = o_d \\ 0 & \forall d \in D, \ v \in V - \{o_d, t_d\} \\ -(1 - w_d) & \forall d \in D, \ v = t_d \end{cases} \quad (2.10)$$

$$\sum_{d \in D} \sum_{e \in E} b_d \cdot x_{de} \leq q_e \quad \forall e \in E \quad (2.11)$$

The objective function (2.9) only differs from that in (2.6) in the way the total amount of used capacity is computed, according to NA variables.

Constraint (2.10) finds a path for every demand. Specifically, if the demand is served it ensures that (i) one path for each demand leaves from the source node, (ii) the path enters in the target node, (iii) every path entering an intermediate node for a demand leaves that node. Note that solutions with cycles can be produced since the length of paths is not strictly minimized and thus, a postprocessing phase needs to be applied to remove them (if any) with no impact on feasibility and optimality of solution. Constraint (2.11) is similar to constraint (2.8) in the AP formulation, and it guarantees that the capacity of links is not exceeded.

The NA formulation for the undirected graph case (NA-UD) adapts constraint (2.10) of NA-D to deal with undirected links and uses constraint (2.11) to prevent link capacity violation. The formulation is as follows:

$$(NA - UD) \quad \min \quad \alpha \cdot \sum_{d \in D} b_d \cdot w_d + \sum_{d \in D} \sum_{e \in E} b_d \cdot x_{de} \quad (2.12)$$

subject to:

$$\sum_{e \in E(v)} x_{de} + w_d = 1 \quad \forall d \in D, \ v \in \{o_d, t_d\} \quad (2.13)$$

$$\sum_{\substack{e' \in E(v) \\ e' \neq e}} x_{de'} \geq x_{de} \quad \forall d \in D, v \in V \setminus \{o_d, t_d\}, e \in E(v) \quad (2.14)$$

$$\sum_{e \in E(v)} x_{de} \leq 2 \quad \forall d \in D, v \in V \setminus \{o_d, t_d\} \quad (2.15)$$

Constraint (2.11)

Constraint (2.13) guarantees that the path of a demand starts and terminates at source and destination nodes provided that the demand is served. To guarantee flow routing, constraint (2.14) ensures that traffic cannot end in an intermediate node. This is enough to guarantee a connected flow between source and destination nodes. Constraint (2.15) forces that intermediate nodes cannot be incident with more than two path link hops and therefore loopless paths are guaranteed.

Further details on mathematical formulations for basic routing problems can be found in [Gr04] and [Pi04].

2.3.2 Heuristics

Among heuristics, constructive methods and local search methods can be distinguished. Constructive algorithms generate solutions from scratch by adding (to an initially empty partial solution) components until a solution is complete. Two main constructive approaches can be distinguished, namely, *greedy* and *randomized*. To illustrate both approaches, Tables 2.3 and 2.4 present the pseudo-code of greedy and randomized algorithms, respectively, for solving the routing problem stated and formulated previously.

The greedy algorithm (Table 2.3) is an iterative procedure where the component that provides the best *greedy cost* is added to the solution at each iteration. In our problem, while the set of demands to route is nonempty, a greedy cost function ($q(·)$) is evaluated for every remaining demand to obtain the greedy cost (lines 3–5). The greedy cost function is usually defined as an incremental cost of adding to the partial solution one given component. In our problem, to avoid rejecting large demands at the end of the constructive procedure, the

Table 2.3 Greedy algorithm.

INPUT: $G(V, E)$, D
OUTPUT: sol, $f(sol)$

```
 1: sol ← ∅
 2: cost ← 0
 3: while D ≠ ∅ do
 4:   for each d ∈ D do
 5:     d.q ← q(d)
 6:   sort(D, d.q, ASC)
 7:   d ← D.pop_first()
 8:   sol ← sol U {d}
 9:   if d.q < ∞ then
10:     allocate(G, d)
11:     cost ← cost + |d.p|
12:   else
13:     cost ← cost + α·d.b
14: return sol, cost
```

Table 2.4 Randomized algorithm.

INPUT: $G(V, E)$, D
OUTPUT: sol, $f(sol)$

```
 1: sol ← ∅;
 2: cost ← 0
 3: D ← shuffle(D)
 4: while D ≠ ∅ do
 5:     d ← D.pop_first()
 6:     sol ← sol ∪ {d}
 7:     if allocate(G, d) then
 8:         cost ← cost + |d.p|
 9:     else
10:         cost ← cost + α·d.b
11: return sol, cost
```

greedy cost function can provide a smaller cost for those demands that have higher b_d. Note that every greedy cost function evaluation entails computing routes over the network with residual capacity, for example, using the Dijkstra algorithm in Table 2.2.

Once the greedy cost is computed for all remaining demands, the one with the minimum cost is returned and added to the partial solution (lines 6–8). The demand can be either rejected or served, in which case the required capacity needs to be allocated (lines 9–13). Once all demands have been processed, the solution is returned. It is worth noting that the greedy algorithm is deterministic, so it returns the same solution every time the algorithm is run with the same input data.

To add some sort of randomization in the hope of finding better solutions, the randomized algorithm (Table 2.4) uses a purely random sort of the demands instead of using a greedy criterion. An advantage of the randomized algorithm with respect to the greedy one is the simplicity and lower complexity since no sorting re-evaluation is required during solution construction. Note that since the randomized constructive algorithm returns a different solution every time it is run, a number of iterations of the randomized constructive algorithm can be performed.

Due to the fact that a feasible solution *sol* resulting from the constructive algorithms defined earlier has no guarantee of being locally optimal, a local search procedure can be applied starting at *sol* to find such local optimal (Table 2.5). Local search algorithms start then from an initial solution *sol* and iteratively try to replace the current solution with a better solution in an appropriately defined neighborhood (*N*(*sol*)) of *sol*. *N*(*sol*) contains those feasible

Table 2.5 Local search.

INPUT: *sol*
OUTPUT: *sol*

1: **while** *sol* is not locally optimal **do**
2:   find *s* ∈ *N*(*sol*) with *f*(*s*) < *f*(*sol*)
3:   *sol* ← *s*
4: **return** *sol*

---

solutions that can be reached from *sol* by a move. The local search procedure ends when no feasible movement can be done.

In our example, since the sequential constructive procedure could lead to rejection of some demands that were accepted with a different demand sorting, a neighborhood of a solution could consist in removing one accepted demand from the solution and trying to route a rejected one (element exchange). Note that every possible exchange could potentially improve the solution; in this regard, several improvement strategies can be followed such as updating the solution as soon as the *first improvement* is found or evaluating all exchanges in order to find the one leading to the *best improvement*.

2.3.3 Meta-Heuristics

Constructive and local search algorithms can be subordinated to other high-level procedures called *meta-heuristics*. A meta-heuristic can be defined as an iterative master process that guides and modifies the operations of subordinate heuristics to produce high-quality solutions efficiently. Meta-heuristics are generic algorithm frameworks that can be applied to different optimization problems with relatively few modifications. Examples of meta-heuristics are Tabu Search [Gl97], Variable Neighborhood Search, Ant Colony, and Simulated Annealing [Ge10]. In the following, we introduce the details of two meta-heuristics that are used for solving several optimization problems through this book; these are the Greedy Randomized Adaptive Search Procedure (GRASP) and the Biased Random Key Genetic Algorithm (BRKGA).

GRASP is an iterative two-phase meta-heuristic method based on a multistart randomized search technique with proven effectiveness in solving hard combinatorial optimization problems [Fe95]. It has been used to solve a wide range of problems (see, e.g., [Pa10], [Pe10]) with many varied applications in real life such as the design of communication networks, collection and delivery operations, and computational biology [Re10]. Table 2.6 presents the general GRASP algorithm adapted to our routing problem.

Table 2.6 GRASP algorithm.

INPUT: $G(V, E)$, D, α, maxIter
OUTPUT: sol

1: $bestSol \leftarrow \emptyset$
2: **for** 1…$maxIter$ **do**
3:   $sol \leftarrow$ constructivePhase(G, D, α)
4:   $sol \leftarrow$ localSearch (sol)
5:   **if** $bestSol = \emptyset$ OR $f(sol) > f(bestSol)$ **then**
6:     $bestSol \leftarrow sol$
7: **return** $bestSol$

Table 2.7 GRASP constructive phase.

INPUT: $G(V, E)$, D, α
OUTPUT: sol

1: $sol \leftarrow \emptyset$; $Q \leftarrow D$
2: **while** $Q \neq \emptyset$ **do**
3:   **for each** $d \in Q$ **do** $d.q \leftarrow q(d)$
4:   $q^{min} \leftarrow \min\{d.q \mid d \in Q\}$
5:   $q^{max} \leftarrow \max\{d.q \mid d \in Q\}$
6:   $RCL \leftarrow \{d \in Q \mid d.q \geq q^{max} - \alpha(q^{max} - q^{min})\}$
7:   Select an element d from RCL at random
8:   $Q \leftarrow Q \setminus \{d\}$
9:   $sol \leftarrow sol \cup \{d\}$
10: **return** sol

In the first phase of the multistart GRASP procedure, a greedy randomized feasible solution of the problem is built by means of a construction procedure. Then, in the second phase, a local search technique to explore an appropriately defined neighborhood is applied in the hope of improving the current solution. The fitness of the solution, that is, the value of the objective function $f(\cdot)$, is computed and the solution is stored if it improves the best solution obtained so far. These two phases are repeated until a stopping criterion is met and, when the procedure finishes, the best solution is returned.

The GRASP constructive algorithm is reproduced in Table 2.7. In contrast to the greedy algorithm in Table 2.3 that selects the best demands at every iteration, a demand is randomly selected from a restricted candidate list (RCL) containing those demands with the best greedy cost $q(\cdot)$; parameter α in the real interval $[0, 1]$ determines the quality of the demands in RCL.

Note that with the basic GRASP methodology, iterations are independent of each other as previous solutions of the algorithm do not have any influence on the current iteration, which opens the opportunity to parallelize the algorithm thus decreasing its computation times.

The second meta-heuristic that it is used along the book is BRKGA, a class of genetic algorithms that has been proposed in the last years to solve optimization problems effectively; in particular, network-related problems such as routing in IP networks and routing and wavelength assignment in Wavelength Division Multiplexing (WDM) optical networks [Go11], [Re11]. Compared to other meta-heuristics, BRKGA has provided better solutions in shorter running times.

As in genetic algorithms, each individual solution is represented by an array of n genes (chromosome), and each gene can take any value in the real interval $[0, 1]$. Each chromosome encodes a solution to the problem and a fitness value. A set of p individuals, called a population, evolves over a number of generations. At each generation, individuals of the current generation are selected to mate and produce offspring, making up the next generation. In BRKGA, individuals of the population are classified into two sets: the elite set with those individuals with the best fitness values and the nonelite set. Elite individuals are copied unchanged from one generation to the next, thus keeping track of good solutions. The majority of new individuals are generated by crossover combining two elements, one elite and another nonelite, selected at random. An inheritance probability is defined as the probability that an offspring inherits the gene from its elite parent (>0.5). Finally, to escape from local optima a small number of mutant individuals (randomly generated) to complete a population are introduced at each generation. A deterministic algorithm, named decoder, transforms any input chromosome into a feasible solution of the optimization problem and computes its fitness value.

In the BRKGA framework, the only problem-dependent parts are the chromosome internal structure and the decoder, and thus one only needs to define them to specify a BRKGA heuristic completely. Note that this notably facilitates heuristic development. In addition, the decoder algorithm can be easily derived from a greedy algorithm by, for instance, considering genes as a multiplier that modulates the greedy cost. An example of such a simple decoder is presented in Table 2.8, where the algorithm takes as input a chromosome of size the number of demands and uses the value of each gene to modulate the real greedy cost in line 5. Of course, this is just a first approximation and not a general rule for decoders.

Table 2.8 Example of decoder algorithm.

INPUT: $G(V, E)$, D, chr
OUTPUT: sol, $f(sol)$

```
 1: sol ← ∅
 2: cost ← 0
 3: while D ≠ ∅ do
 4:   for each d ∈ D do
 5:     d.q ← q(d) * chr[d]
 6:   sort(D, d.q, ASC)
 7:   d ← D.pop_first()
 8:   sol ← sol U {d}
 9:   if d.q < ∞ then
10:     allocate(G, d)
11:     cost ← cost + |d.p|
12:   else
13:     cost ← cost + α·d.b
14: return sol, cost
```

2.4 Introduction to the Optical Technology

An optical network is a network composed of optical nodes, which are connected using optical fibers. In these networks, data flows are transmitted from a source to a destination node through an optical connection, also known as a *lightpath.*

WDM technology allows transmitting different data flows on different optical wavelengths. Most WDM systems use the frequency region around 1530 nm because this is one of the frequency regions where the signal attenuation reaches a local minimum. Figure 2.4 shows an example of the WDM technology.

WDM systems with channel spacing ranging from 12.5 to 100 GHz have been specified [G694.1]. With that technology, the number of wavelengths being multiplexed onto a single-fiber ranges 50–400. When such a large number of channels can be transported by the WDM system, the term Dense Wavelength Division Multiplexing (DWDM) is used, in contrast to Coarse Wavelength Division Multiplexing (CWDM), which was typically considered for the metropolitan network and multiplexes a limited number of wavelengths onto a single fiber.

Light emitters (usually semiconductor lasers) are key components in any optical network. They convert the electrical signal into a corresponding light signal, on a single wavelength, that can be injected into the fiber.

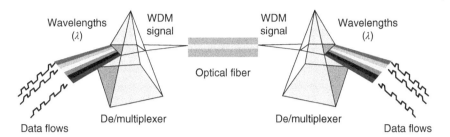

Figure 2.4 WDM technology.

Besides, a WDM system uses a multiplexer at the transmitter to multiplex the different wavelengths together in a bundle and a demultiplexer at the receiver to split them apart. An optical fiber transmits optical signal through long distances. The receiver sensitivity indicates the minimum power required to detect the incoming signal since the power of the signal is reduced when it propagates over distance (this is called *attenuation*). In order to compensate for the effect of attenuation, the optical signal can be amplified within the optical domain.

The first WDM deployments were point-to-point systems. However, the introduction of Optical Add/Drop Multiplexer (OADM) allows dropping a specific wavelength out of the bundle of WDM-multiplexed signals and adding another channel on the same wavelength; this enables the configuration of ring-based topologies. Finally, with the development of complex optical devices, such as Wavelength Selective Switches (WSSs) [Ts06], optical cross-connects (OXCs), the key element to build optical mesh-based networks, are available [Ro08].

2.4.1 From Opaque to Transparent Optical Networks

With OXCs, optical networks have evolved from *opaque* infrastructures, where reamplifying, reshaping, and retiming (3R) regeneration is performed for every optical connection at every intermediate node, toward *transparent* infrastructures, where the signal remains within the optical domain from the source to the destination node [Ra99]. Thereby, transparency eliminates the optical–electronic–optical (OEO) conversions or 3R regenerators at the intermediate nodes, which yields important benefits such as freedom on the bitrates and data formats of the transported signals, elimination of well-known electronic bottlenecks, and reduction of both network cost and power consumption [Po08]. The key enablers to deploy transparent all-optical networks are the advances on relevant optical signal functions such as amplification, filtering, and dispersion compensation and the availability of node architectures capable of routing and switching in the optical domain, for example, OXCs.

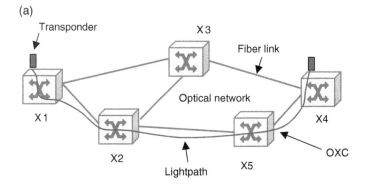

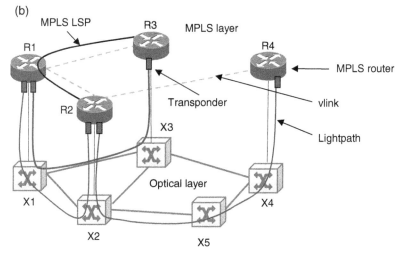

Figure 2.5 (a) Single-layer and (b) multilayer networks.

In the evolution from opaque toward transparent networks, *translucent* networks are considered as an intermediate step [Ya05.1]. This infrastructure uses a set of sparsely but strategically 3R regenerators placed throughout the network to do signal regeneration [Sh07]. Therefore, a translucent network is a cost-efficient infrastructure that aims at attaining an adequate trade-off between network cost, mainly due to the high cost of 3R regenerators, and the service provisioning performance.

2.4.2 Single-Layer and Multilayer Networks

Figure 2.5a shows a single-layer optical network where a transparent connection between OXCs X1 and X4 is established. Due to the lack of electronic regeneration, two constraints impose fundamental limitations to transparent single-layer optical networks based on WDM technology: the wavelength continuity constraint and the physical layer impairments [Ya05.1].

The wavelength continuity constraint implies that a lightpath is subject to use the same wavelength on each link along the route. This restriction raises important challenges at the time of computing lightpaths, specifically to achieve an acceptable network performance in terms of connection *blocking probability*, which holds for the probability to refuse a connection request at a random time.

Regarding physical layer impairments, such as amplification noise, cross-talk, dispersion (polarization mode dispersion (PMD), chromatic dispersion (CD)), and fiber nonlinearity, they accumulate while the optical signal travels from the source to the destination node. In consequence, the maximum optical transmission reach is limited, since the received signal quality may render unacceptable bit error rate (BER) and the lightpath becomes disconnected [Po08], [Hu05]. Although BER is the main performance parameter to measure the signal quality, it can only be measured at the receiver in transparent optical networks. Therefore, efficient substitutes to estimate the BER, such as the Optical Signal-to-Noise Ratio (OSNR) [Hu05] and the Q factor [Po08], are needed.

Optical networks can support Multi-Protocol Label Switching (MPLS) [RFC3031] networks. A demand in the MPLS layer is served by an MPLS Label-Switched Path (LSP) routed on a virtual network topology (VNT) that consists of virtual links (vlinks) supported by lightpaths in the underlying optical network.

Figure 2.5b illustrates a multilayer MPLS-over-optical network with an MPLS LSP established between routers R2 and R3. In the figure, three vlinks can be identified in the MPLS layer: R1–R3, R1–R2, and R2–R4. For illustrative purposes, the lightpaths supporting those vlinks are also depicted. Since routers operate in the electrical domain, transponders need to be used to convert signals from the electrical to the optical domain.

Currently, WDM-based networks provide support to optical connections of up to 100 Gb/s. However, new services and applications are demanding higher bitrates and flexibility, as well as the stringent quality of service requirements, which is pushing telecom operators to upgrade their WDM networks to a more flexible technology for the more dynamic and variable traffic that is expected to be conveyed. Lastly, academy- and industry-driven research on EONs has turned out to be a mature enough technology ready to gradually upgrade WDM-based networks. The next section reviews key technologies behind EON.

2.4.3   EON Key Technologies

Among the technologies facilitating EON, *flexgrid* [Ji09], [Ji10], [Ge12] highlights as a potential candidate; flexgrid uses the flexible grid defined in [G694.1], where the optical spectrum is divided into 12.5 GHz frequency *slices* and the

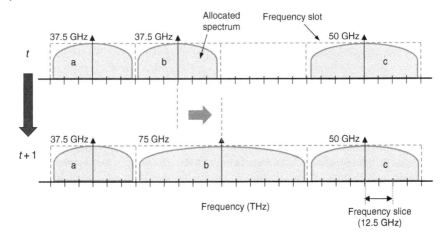

Figure 2.6 Example of flexible and elastic spectrum allocation.

nominal *central frequency* granularity, that is, the spacing between neighboring central frequencies, is equal to 6.25 GHz.

Interestingly, by using a variable number of consecutive frequency slices, arbitrary-sized frequency *slots* can be created and allocated to lightpaths, thus allowing one to elastically adapt the allocated resources to the connections' capacity requirements.

Figure 2.6 illustrates flexible and elastic spectrum allocation (SA), where resources to support three connections requiring 37.5 GHz (connection a), 37.5 GHz (b), and 50 GHz (c) are allocated at a given time t. At time $t + 1$, the resources allocated to support connection b have been increased, from 37.5 to 75 GHz, to satisfy the demanded capacity at that time.

The finer spectrum granularity is only part of EONs; dynamic spectrum allocation and elasticity are their flagship features that require several additional key technologies.

One of these requirements is the availability of bandwidth-variable (BV) WSSs to build BV OXCs, in comparison with traditional OXCs. OXCs can be assembled using existing devices like the *WaveShaper* programmable optical processor [Finisar].

The development of bandwidth-variable transponders (BVTs) that can be tuned in both bitrate and spectrum width to the values required to support a specific connection is essential for EON deployment. Therefore, one single BVT, for example, 400 Gb/s, can be used for the entire connection, which can be specifically configured to cope with the actual needs. However, when BVTs need to transmit at lower bitrates (e.g., 100 Gb/s), part of their capacity is unused. That is the reason why the sliceable bandwidth-variable transponder (SBVT), also known as a multiflow transponder, was proposed in [Ji12] and [Lo14].

SBVTs support the configuration of multiple subcarriers, which can be either corouted or, thanks to the sliceable functionality, transmitted along different routes. Each subcarrier is created using one modulator and configured with ad hoc transmission parameters, such as bitrate, baud rate, modulation format, coding, and allocated to a frequency slot. In addition, SBVTs can provide even higher levels of elasticity and efficiency to the network by enabling transmitting from one point to multiple destinations, changing the bitrate to each destination and the number of destinations on demand.

Last but not the least, the development of new advanced modulation formats and techniques allows increasing spectral efficiency and extending the reach of optical signals, thus avoiding expensive electronic 3R regeneration. Advances on both single-carrier formats, such as Quadrature Phase Shift Keying (QPSK) and Quadrature Amplitude Modulation (QAM), and multicarrier formats, such as Optical Orthogonal Frequency-Division Multiplexing (O-OFDM), have been proposed [Ji10].

Thanks to this flexible technology, an EON can adjust to varying traffic conditions over time, space, and bandwidth, thereby creating a network scenario where frequency slots are both switched and dimensioned (bitrate/reach/signal bandwidth) according to temporary traffic requirements.

Scenarios, where EON features provide advantages over fixed optical networks while the network is in operation, include connectivity provisioning and recovery, both introduced in the following sections. For a complete review of the EON technology, architecture, and applications, we refer the reader to [Lo16].

2.5 Network Life Cycle

Figure 2.7 illustrates the classical network life cycle, where a number of steps are performed sequentially [Ve14.2]. The life cycle begins by planning and forecasting the network based on the service layer requirements, the state of the resources in the already deployed network, and inventory information. The output of this first step is a set of recommendations to satisfy the forecast traffic for a period of time, the duration of which depends on factors involving the operator and the traffic type. A set of recommendations is used in the architecture and design step to define the changes to be applied to the already deployed network; these changes are later implemented in the network.

During network operation, connection requests arrive at the network, and they are provisioned using the available capacity. Such capacity is continuously monitored, and the collected data is used as input for the next planning cycle. In addition, failures might affect the network, so it is of paramount importance to define recovery strategies. The planning cycle may be restarted in case of network changes or unexpected increases in demands.

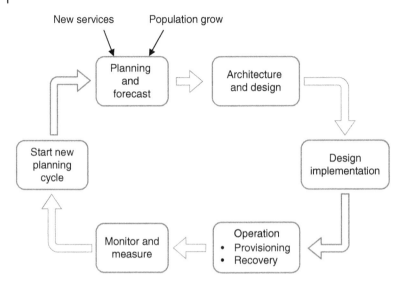

Figure 2.7 Classical network life cycle.

The next subsections focus on provisioning and recovery aspects.

2.5.1 Connection Provisioning

Connection provisioning is the process of establishing new optical connections in the network to serve incoming demands. In single-layer networks, it requires establishing a dedicated lightpath from the source to the destination OXCs. In multilayer networks, connections are established on the VNT created on top of the optical network. However, such MPLS connection provisioning might entail adding new vlinks or adding more capacity to existing ones in the VNT, which requires establishing new lightpaths to support them.

Specifically for EONs, connection provisioning consists in finding a physical route and a set of frequency slices that, in absence of spectrum converters, must be the same along the links in the route (spectrum continuity constraint, similar to the wavelength continuity constraint in WDM networks) and contiguous in the spectrum (contiguity constraint), that is, the slices must define one single-frequency slot.

To find the optimal route and frequency slot, the *Routing and Spectrum Allocation* (RSA) problem [Ve14.1] must be solved for each incoming demand. Both the central frequency and the slot width parameters are determined according to the requested bitrate and the selected modulation format. Figure 2.8 depicts three lightpaths set up on a network. Lightpath P1 (from X4 to X1) has been set up through a four-slice slot, P2 (from X1 to X2) has been set up through an eight-slice slot, and P3 (from X3 to X4) has been set up through a six-slice slot.

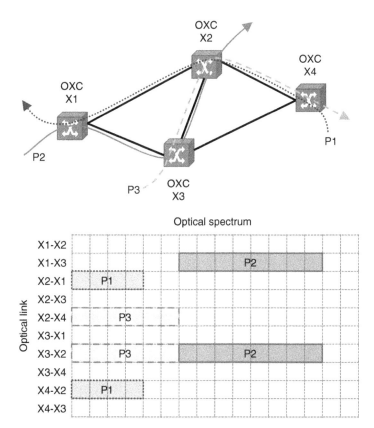

Figure 2.8 Example with three lightpaths.

Lightpaths allow connecting a source node to a single destination node, and they are also named as *point-to-point* (p2p) connections. In contrast, when a source OXC needs to transfer the same data to a number of destination OXCs, *point-to-multipoint* (p2mp) connections, also known as *light-trees*, can be set up (Figure 2.9). Note that light-trees reduce the total amount of optical resources required to serve all the destinations; for instance, establishing a p2mp connection from X1 to X2 and X4 using two connections requires in total 32 frequency slices (8 slices along 4 hops), whereas establishing a light-tree requires in total 24 frequency slices (8 slices along 3 hops).

2.5.2 Connection Recovery

Owing to the huge bitrate associated with each established path in the network, *recovery* schemes need to be used to guarantee that the associated client demand continue being served even in the case of failures [Gr04]. As in WDM networks, recovery in EONs can be provided by either *protection*, where the

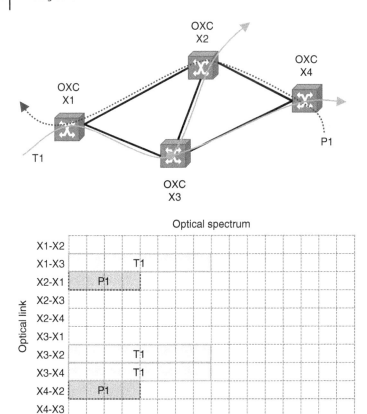

Figure 2.9 A lightpath and a light-tree set up.

failed working path is substituted by a preassigned backup one, or *restoration*, which is based on rerouting the working path without prior resource reservation. Optical resources, that is, each of the frequency slices in a fiber link, used by backup paths can be *dedicated* to protect a single working path, or they can be *shared* to provide protection to multiple working paths. As a consequence, the former scheme is called dedicated path protection (DPP) and the latter shared path protection (SPP) (see Figure 2.10).

Protection schemes reserve resources to guarantee that all protected paths are recovered in case of any single failure. Among the schemes, SPP provides better resource utilization than DPP due to spare resources are shared among several working paths. In comparison, restoration is a more efficient scheme since resources are only allocated after a failure impacts a working path. However, recovery times are usually much shorter when protection is used. In the case of DPP, since spare resources are already in use, recovery times are

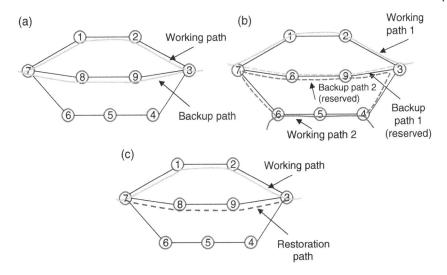

Figure 2.10 DPP (a), SPP (b), and path restoration (c).

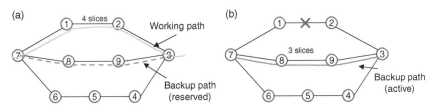

Figure 2.11 Partial SPP in action. (a) Working path is active and backup path is reserved. (b) After a failure in affecting the working path, the backup path is activated.

really short, being slightly longer for SPP since spare resources are reserved beforehand and activated in case of failure.

In flexgrid optical networks, the bitrate requested by client demands is converted to a number of frequency slices to be allocated. Therefore, service level agreements between network operators and clients can now include specific terms so that all or *only part* of the requested bitrate is recovered in case of failure [So11].

For illustrative purposes, Figure 2.11 presents how partial SPP works. In Figure 2.11a one client demand is requesting some amount of bitrate that also translates into four slices; the working path serves all the requested bitrate, and there are three slices reserved along the backup route. After a failure has affected the working path, the backup path is activated in Figure 2.11b. Note that the recovery path guarantees that at least 75% of the requested bitrate is served in the case of failure. The same effect can be achieved if restoration is applied.

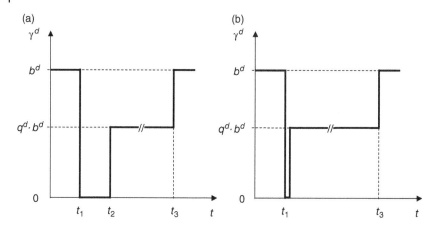

Figure 2.12 Time to recover when partial path restoration (a) and protection (b) are implemented.

Let us now generalize this concept of partial protection. Let q^d be a percentage of bitrate to be assured to demand d in the case of failure and B_{mod} the spectral efficiency of the selected modulation format; hence the number of slices, $n^d_{re\,cov\,ery}$, that must be used to recover demand d can be computed to ensure:

$$12.5 \cdot n^d_{re\,cov\,ery} \cdot B_{mod} > q^d \cdot B(d) \qquad (2.16)$$

However, protection and restoration provide different recovery times. Figure 2.12 presents the evolution of the bitrate actually served (γ^d) for a demand d as a function of time when restoration (Figure 2.12a) and protection (Figure 2.12b) are used for recovery. In the period $t < t_1$ all the requested bitrate is served. In $t = t_1$ a failure impacts the working path of demand d, and in $t = t_3$ the failure is repaired. When restoration is used for recovery in Figure 2.12a, no bitrate is served until the restoration path is established for the demand in $t = t_2$. On the contrary, when protection is used for recovery, the minimum bitrate is served just after the failure is detected in $t = t_1$, as shown in Figure 2.12b. In $t = t_3$ the failure is eventually repaired, and all the requested bitrate is served again in both schemes.

2.6 Conclusions

Relevant background related to the main topics of this book has been presented in this chapter. Specifically, basic notation and concepts of graphs and the Dijkstra's shortest path algorithm have been introduced first. Next, basic concepts and notation of optimization have been presented.

Several ILP formulations have been presented to solve an illustrative routing problem. Taking advantage of such problem, basic constructive and local search heuristic algorithms have been proposed. Finally, GRASP and BRKGA meta-heuristics have been introduced due to their frequent use through the rest of the book.

Finally, a brief description of the evolution of optical networks, from point-to-point WDM systems to both single and multilayer networks has been presented, as an introduction to EONs. The main key technological features supporting EONs, such as flexgrid technology for switching, advanced modulation formats for transmission, and bandwidth-variable equipment, have been presented as enablers of the most important EON capabilities, such as spectrum dynamic allocation and elasticity. Those capabilities are specifically exploited for connection provisioning and recovery (the main topics of this book), and more generally during the whole network life cycle.

3

The Routing and Spectrum Allocation Problem

Luis Velasco[1], Marc Ruiz[1] and Mirosław Klinkowski[2]

[1] *Universitat Politècnica de Catalunya, Barcelona, Spain*
[2] *National Institute of Telecommunications, Warsaw, Poland*

To properly design, operate, and reconfigure Elastic Optical Networks (EONs), the *routing and spectrum allocation* (RSA) problem must be solved. Regardless of additional constraints, the RSA problem involves two main constraints for spectrum allocation: the *continuity* constraint to ensure that the allocated spectral resources are the same along the links in the route and the *contiguity* constraint to guarantee that those resources are contiguous in the spectrum. These two constraints highly increase the complexity of the optimization, so it is crucial that efficient methods are available to allow solving realistic problem instances in practical times. In this chapter, we review several Integer Linear Programming (ILP) formulations of the RSA problem and evaluate their performance regarding the computational time to solve a basic offline problem.

From the basic RSA problem, additional variables and constraints can be added to consider various EON key features. Among them, the capability of selecting the required modulation format among the available ones becomes essential for a vast set of EON optimization problems. Therefore, the *routing, modulation, and spectrum allocation* (RMSA) problem is eventually presented as the most relevant basic optimization problem, and, consequently, some efficient formulations for connection provisioning and topology design problems are reviewed.

Provisioning, Recovery, and In-operation Planning in Elastic Optical Networks,
First Edition. Luis Velasco and Marc Ruiz.

Table of Contents and Tracks' Itineraries

3.1 Introduction

The *Routing and Spectrum Allocation* (RSA) problem involves finding feasible routes and spectrum allocations for a set of traffic demands. As in the case of the Routing and Wavelength Assignment (RWA) problem in fixed-grid Wavelength Division Multiplexing (WDM) networks, the spectrum continuity constraint must be enforced. In the case of flexgrid, the spectrum allocation is represented by a frequency *slot*, and thus in the absence of spectrum converters the same slot must be used along the links of a given routing path. This also entails that the allocated frequency *slices* must be contiguous in the spectrum; this is known as spectrum contiguity constraint. The RSA problem was proved to be *NP*-complete in [Ch11] and [Wa11.3]. As a consequence, it is crucial that efficient methods are available to allow solving realistic problem instances in practical times.

Due to the spectrum contiguity constraint, RWA problem formulations developed for WDM networks are not applicable for RSA in flexgrid optical networks, and they need to be adapted to include such constraint. Several works can be found in the literature presenting Integer Linear Programming (ILP) formulations of RSA [Ch11], [Wa11.3], [Kl11], [Ve12.1]. In [Ch11] the authors address the problem of planning a flexgrid optical network, where a traffic matrix containing the requested bitrate of a set of demands is given. To solve the problem, an ILP formulation that aims at minimizing the spectrum used to serve the traffic matrix is proposed. The RSA formulation cannot be solved in practical times and therefore the authors present a decomposition method that breaks the previous formulation into two subproblems: a demand routing subproblem and a spectrum allocate subproblem. Since both subproblems are solved sequentially, global optimality cannot be guaranteed. In [Wa11.3] the authors study the RSA problem by providing a different ILP formulation but with the same objective as in [Ch11]. Authors in [Kl11] formulate the RSA problem and propose an effective heuristic algorithm to obtain near optimal solutions. Finally, a formulation based on precomputing frequency slots, thus moving the complexity of the contiguity constraint to the input data generation phase, is presented in [Ve12.1] and extended in [Ve14.1].

Furthermore, the capability of Elastic Optical Networks (EONs) to support different modulation formats providing different spectral efficiencies (which translate into different slot widths) needs to be considered. For instance, an optical connection requesting 400 Gb/s would need 50 GHz if the Dual Polarization (DP)–Quadrature Amplitude Modulation 16 (QAM16) modulation format is selected in contrast to 200 GHz needed if the DP–Quadrature Phase-Shift-Keying (QPSK) modulation format is used. However, the transmission reach of more efficient modulation formats is shorter than that of those less efficient and therefore the selection of a modulation format depends on the length of the selected routing path.

The remainder of the chapter is organized as follows. Firstly, a comparative analysis between several approaches to model the RSA problem is presented, including the ILP formulations and their performance in terms of required solving time to meet the optimal solution. Next, an extension of the most efficient RSA formulation to model the RMSA problem is presented. Basic RMSA-based offline problems are presented and proposed as essential optimization problems to be solved for designing, operating, and reoptimizing EONs [Ve17].

3.2 The RSA Problem

3.2.1 Basic Offline Problem Statement

Let us define a very basic RSA problem that will be used to compare the different ILP formulations that will be reviewed in this chapter. The problem involves finding a lightpath for every demand in a given traffic matrix with the objective of minimizing or maximizing some utility function. Several alternatives for this problem may exist, for instance, we can assume that all the traffic matrix needs to be served, or alternatively some demands cannot be served, that is, they are blocked.

The offline RSA problem is formally stated as follows:

Given:

- an EON, represented by a directed connected graph $G(V, E)$, V being the set of optical nodes and E the set of fiber links connecting two nodes in V;
- an ordered set S of frequency slices in each link in E; $S = \{s_1, s_2, ..., s_{|S|}\}$. A guard band B (number of slices) is required between two spectrum allocations;
- a set D of demands to be transported. Each demand d is represented by a tuple $<o_d, t_d, b_d, n_d>$, where o_d and t_d are the source and the destination nodes, respectively, b_d is the requested bitrate, and n_d is the requested number of slices.

Output: The route over the network and the spectrum allocation of every transported demand.

Objective: Minimize the amount of rejected bitrate.

Note that from the problem definition, demands can be served or alternatively rejected. We selected this objective to avoid infeasibility that may appear when trying to serve large sets of demands over a capacitated network.

3.2.2 Notation

The following sets and parameters have been defined for the formulations presented in this chapter:

Topology

V	Set of nodes, index v.
E	Set of fiber links, index e.
$E(v)$	Subset of fiber links incidents to location v.
$E^+(v)$	Subset of E with links leaving from node v.
$E^-(v)$	Subset of E with links arriving at node v.
g_{ve}	Equal to 1 if link e is incident to node v, 0 otherwise.

Demands and Paths

D	Set of demands, index d.
$\{o_d, t_d\}$	Subset of nodes with source and destination nodes of demand d.
b_d	Bitrate of demand d in Gb/s.
$P(d)$	Subset with the predefined candidate paths for demand d. Each path p consists of a set of links $e \in E$ so that nodes o_d and t_d are connected.
P	Set of precomputed paths for all the demands, that is, $P = \bigcup_{d \in D} P(d)$, index p.
δ_{pe}	Equal to 1 if path p uses link e, 0 otherwise.

Spectrum

S Set of frequency slices, index s.

C Set of frequency slots, index c. Each slot contains a subset of contiguous frequency slices.

B Guard band in number of slices required between two spectrum allocations.

γ_{cs} Equal to 1 if slot c includes frequency slice s, 0 otherwise.

$C(d)$ Subset of frequency slots for demand d.

N Set of different slot sizes to be requested by the traffic demands, index n.

n_d Number of slices to transport the requested bitrate of demand d.

The decision variables are:

x_d Binary. Equal to 1 if demand d is rejected, 0 otherwise.

y_p Binary. Equal to 1 if path p is selected, 0 otherwise.

f_d Positive integer containing the starting slice index for demand d.

$f_{dd'}$ Binary. Equal to 1 if the starting slice index of demand d is smaller than that of d', that is, $f_d < f_{d'}$, 0 otherwise.

y_{ps} Binary. Equal to 1 if slice s is assigned to path p, 0 otherwise.

y_{pc} Binary. Equal to 1 if slot c is assigned to path p, 0 otherwise.

z_e Binary. Equal to 1 if link e is installed, 0 otherwise.

w_{dec} Binary. Equal to 1 if demand d uses slot c in link e, 0 otherwise.

w_{dc} Binary. Equal to 1 if demand d uses slot c, 0 otherwise.

Using this notation, the objective function of every subsequent formulation of the RSA problem computes the amount of unserved bitrate:

$$\Phi = \sum_{d \in D} x_d \cdot b_d \tag{3.1}$$

3.3   ILP Formulations Based On Slice Assignment

Aiming at solving the basic RSA problem described in the previous section, we propose an arc-path formulation (see Chapter 2). Since the topology is given, we can precompute a set of k distinct paths for every demand in D.

Next subsections present two different formulations with specific constraints to allocate continuous and contiguous sets of slices to those selected paths.

3.3.1   Starting Slice Assignment RSA (SSA-RSA) Formulation

This ILP formulation, adapted from [Ch11], involves assigning the starting slice to every demand to be transported while avoiding slice overlapping of demands in which paths share at least one common link. Note that intermediate slices are not explicitly assigned in this formulation. The formulation is as follows:

$$(SSA - RSA) \quad \min \quad \Phi \tag{3.2}$$

subject to:

$$\sum_{p \in P(d)} y_p + x_d = 1 \qquad \forall d \in D \tag{3.3}$$

$$f_d + n_d \cdot \left(1 - x_d\right) \leq |S| \qquad \forall d \in D \tag{3.4}$$

$$f_{dd'} + f_{d'd} = 1 \quad \forall d, d' \in D : \exists p \in P(d) \cap \exists p' \in P(d') \cap \left(p \cap p' \neq \varnothing\right) \tag{3.5}$$

$$f_{d'} - f_d < |S| \cdot f_{dd'} \\ \forall d, d' \in D : \exists p \in P(d) \cap \exists p' \in P(d') \cap \left(p \cap p' \neq \varnothing\right) \tag{3.6}$$

$$f_d - f_{d'} < |S| \cdot f_{d'd} \\ \forall d, d' \in D : \exists p \in P(d) \cap \exists p' \in P(d') \cap \left(p \cap p' \neq \varnothing\right) \tag{3.7}$$

$$f_d + n_d \cdot y_p + B - f_{d'} \leq \left(|S| + B\right) \cdot \left(1 - f_{dd'} + 2 - y_p - y_{p'}\right) \\ \forall d, d' \in D \cap \forall p \in P(d) \cap \forall p' \in P(d') : p \cap p' \neq \varnothing \tag{3.8}$$

$$f_{d'} + n_{d'} \cdot y_{p'} + B - f_d \leq \left(|S| + B\right) \cdot \left(1 - f_{d'd} + 2 - y_p - y_{p'}\right) \\ \forall d, d' \in D \cap \forall p \in P(d) \cap \forall p' \in P(d') : p \cap p' \neq \varnothing \tag{3.9}$$

The objective of the SSA-RSA formulation (3.2) is to minimize the amount of unserved bitrate given by equation (3.1). Constraint (3.3) either assigns one feasible path or blocks the demand. Constraint (3.4) guarantees that the starting slice for each demand leaves enough room for the number of slices that that demand requests. Constraints (3.5)–(3.7) manage starting slice ordering for all demand pairs such that any of their paths share at least one common link. They compute whether the starting slice of one of the demands in the pair is lower than the starting slice of the other demand. Constraints (3.8)–(3.9) perform spectrum continuity and non-overlapping spectrum allocation. They ensure that the spectrum assigned to all pairs of demands and all pairs of paths of the demands share at least one common link and that the paths are activated and do not overlap.

3.3.2 Slice Assignment RSA (SA-RSA) Formulation

This ILP formulation, adapted from [Wa11.3], involves explicitly assigning slices to demands ensuring the spectrum contiguity constraint. In contrast to

[Ch11], it makes use of variables and constraints that depend on the frequency resources. A similar approach can be found in [Kl11].

The formulation uses constraint (3.3) and defines new constraints:

$$(SA - RSA) \quad \min \quad \Phi \tag{3.10}$$

subject to:
Constraint (3.3)

$$\sum_{s \in S} y_{ps} = n_d \cdot y_p \qquad \forall d \in D, p \in P(d) \tag{3.11}$$

$$\sum_{d \in D} \sum_{p \in P(d)} \delta_{pe} \cdot y_{ps} \leq 1 \qquad \forall e \in E, s \in S \tag{3.12}$$

$$\sum_{\substack{d' \neq d : p' \in P(d') \cap (p \cap p' \neq \varnothing) \\ s' \in [\max(0, s-B), \min(|S|, s+B)]}} y_{p's'} \leq \left(1 - y_{ps}\right) \cdot |P| \cdot |S| \quad \forall d \in D, p \in P(d), s \in S \tag{3.13}$$

$$-|S| \cdot \left(y_{ps} - y_{p(s+1)} - 1\right) \geq \sum_{s' \in [s+2, |S|]} y_{ps'} \quad \forall d \in D, p \in P(d), s \in \left[1, |S| - 1\right] \tag{3.14}$$

In the SA-RSA formulation, constraint (3.11) assigns the requested number of slices to each demand. Constraint (3.12) ensures that each slice is allocated to one path at the most. Constraint (3.13) reserves a guard band between spectrum allocations belonging to two different demands. Finally, constraint (3.14) focuses on spectrum contiguity and ensures that if a slice is allocated to a demand and the next slice is not, all the remaining slices of the spectrum will not be allocated to that demand.

The formulations presented in this section require a set of dedicated variables and constraints so as to deal with the spectrum contiguity constraint. The next section presents an ILP formulation that uses precomputed frequency slots and, as a consequence, spectrum contiguity is ensured in the input data.

3.4   ILP Formulations Based On Slot Assignment

In this section, we present ILP formulations based on frequency slot assignment that remove the spectrum contiguity constraint from the formulation; a precomputed set of candidate slots defining a set of contiguous frequency slices is given as an input parameter to the formulation [Ve12.1], [Ve14.1]. Additionally, we define a relaxed ILP formulation that provides tight lower bounds to the problem in really short computation times.

3.4.1   Slot Precomputation

The definition of slots can be mathematically formulated as follows. Let γ_{cs} be a coincidence coefficient that is equal to 1 whenever slot $c \in C$ uses slice $s \in S$

and 0 otherwise. Let us assume that a set of slots $C(d)$ is predefined for each demand d, which requests n_d slices. Then, $\forall c \in C(d)$ the spectrum contiguity constraint is implicitly imposed by the proper definition of γ_{cs} such that $\forall s_i, s_j : \gamma_{csi} = \gamma_{csj} = 1, s_i < s_j \Rightarrow \gamma_{csk} = 1, \forall s_k \in \left\{ s_i, ..., s_j \right\}, \sum_{s \in S} \gamma_{cs} = n_d$. We consider that each set $C(d)$ consists of all possible slots of the size requested by d that can be defined in S.

The algorithm in Table 3.1 computes $C(d)$.

Since $|C(n)| = |S| - (n - 1)$, the size of the complete set of slots C that needs to be defined is $|C| = \sum_{n \in N} \left[|S| - n + 1 \right] < |N| \cdot |S|$ Note that computation of slots is trivial, and thus no additional complexity is added to the precomputation phase.

To account for guard bands, without loss of generality, we consider that they are included as a part of the requested spectrum (n_d).

Therefore, we can define the Routing and Slot Assignment problem as the problem that finds a route and assigns a proper slot to a set of input demands, so that the number of active slices in the slot guarantees that the bitrate requested by each demand can be transported. Therefore, the Routing and Slot Assignment and the RSA problems are in fact the same problem; a route and a set of contiguous slices are allocated for each input demand. However, the complexity of the former is much lower than that of the latter since the slices in the slot are already contiguous in the spectrum.

3.4.2 Slot Assignment RSA (CA-RSA) Formulation

The proposed CA-RSA formulation is as follows:

$$(CA - RSA) \quad \min \quad \Phi \tag{3.15}$$

subject to:

$$\sum_{p \in P(d)} \sum_{c \in C(d)} y_{pc} + x_d = 1 \qquad \forall d \in D \tag{3.16}$$

Table 3.1 $C(d)$ precomputation.

INPUT: S, d
OUTPUT: $C(d)$

```
1: Initialize: C(d) ← 0⌊(|S|-nd+1) × |S|⌋
2: for each i in [0, |S|-n_d] do
3:     for each s in [i, i+n_d - 1] do
4:         C(d)[s] = 1
5: return C(d)
```

$$\sum_{d \in D} \sum_{p \in P(d)} \sum_{c \in C(d)} \gamma_{cs} \cdot \delta_{pe} \cdot y_{pc} \leq 1 \qquad \forall e \in E, \, s \in S \tag{3.17}$$

Constraint (3.16), similar to constraint (3.3) in the SSA-RSA formulation, either assigns one feasible path and slot or blocks the demand. In addition, constraint (3.17), similar to constraint (3.12) in the SA-RSA formulation, ensures that each slice in a link is assigned to one demand at the most.

In line with [Ch11], a lower bound on the objective function can be introduced aiming at accelerating solving times. The bound is given by the solution of the RSA problem with relaxed spectrum contiguity and continuity constraints.

The relaxed CA-RSA model is as follows, where the constraint (3.17) in the CA formulation has been redefined as constraint (3.19) to ensure that the capacity of each link is not exceeded.

$$(relaxed \ CA - RSA) \quad \min \quad \Phi \tag{3.18}$$

subject to:
Constraint (3.3)

$$\sum_{d \in D} \sum_{p \in P(d) \cap P(e)} y_p \, n_d \leq |S| \qquad \forall e \in E \tag{3.19}$$

3.5 Evaluation of the ILP Formulations

In this section, we first analyze and compare the complexity of the slice-based and slot-based ILP models. Note that they are exact RSA formulations and the solutions that they provide are equivalent; therefore, we focus on analyzing the size of the formulation and the time to return optimal solutions. Next, in view that the CA-RSA formulation solves problem instances in short computation times, real problem instances are generated and solved using such ILP formulation.

3.5.1 Model Size Analysis

Table 3.2 presents expressions to estimate the number of variables and constraints of the considered ILP formulations for the RSA problem. In contrast to the number of variables, the number of constraints greatly differs among the formulations; in the SSA-RSA formulation it grows quadratically also with k, in SA-RSA it depends also on the size of the sets of slices and links, whereas in the CA-RSA formulation it is proportional to the number of demands set and to the size of E times the size of S. As an example, the number of constraints in a network with 14 nodes and 23 links (named as the DT network in the next subsection), considering $|S| = 30$, $|D| = 36$, and $k = 5$, is 3.2×10^4, 1.15×10^4, and 7.3×10^2 for the SSA-RSA, SA-RSA, and CA-RSA formulations, respectively. In contrast, the relaxed RSA model has only 180 variables and 56 constraints.

3.5.2 Performance Comparison

The performance of the previous ILP formulations has been evaluated using four different optical topologies from [Ch11], [Pv10], and [SND] covering a wide range of sizes and mesh degrees: a 9-node ring topology, the 10-node Brazilian topology, the 12-node Abilene topology, and the 14-node DT topology.

It is clear in view of Table 3.2 that the value of k and the size of the demands and slice sets highly impact on the size of the models. Consequently, scenarios consisting of 36 demands randomly generated and two different slice sets ($|S| \in \{30, 60\}$) were tested. The number of slices requested by demands was uniformly distributed ranging 1–4 slices. Finally, different sizes for the set of precomputed paths, k, were generated for each demand, ranging from 1 to 5, except for the ring topology; the range of k was limited as a consequence of the resulting computation times as explained here. All the formulations were implemented in iLOG-OPL and solved by the CPLEX v.12.5 optimizer [CPLEX] on a 2.4 GHz Quad-Core machine with 8 GB RAM memory running Linux.

Table 3.3 shows the obtained computations times for the three ILP formulations. We can see that computation times increase considerably with k for both SSA-RSA and SA-RSA and, in particular, these formulations are clearly impractical for higher values of k. Note that in many problems, $k \geq 10$ needs to be used to ensure optimality [Pi04].

Note also that as soon the size of the considered spectrum increases from 30 to 60, the effective load decreases and thus the problem becomes easier to be solved since it is easier to find free spectrum resources in the network. In contrast, as observed in Table 3.3, the CA-RSA formulation provides computation times in the order of seconds for all tested scenarios. The computational efficiency is at least three orders of magnitude higher than that of SSA-RSA or SA-RSA formulations; it is a result of representing the spectrum contiguity constraints by means of the precomputed data. In the CA-RSA formulation,

Table 3.2 Size of the models.

Formulation	Variables	Constraints
SSA-RSA	$O(|D|^2 + k{\cdot}|D|)$	$O(k^2{\cdot}|D|^2)$
SA-RSA	$O(k{\cdot}|D|{\cdot}|S|)$	$O(2{\cdot}k{\cdot}|D|{\cdot}|S| + |E|{\cdot}|S|)$
CA-RSA	$O(k{\cdot}|D|{\cdot}|C|)$	$O(|D| + |E|{\cdot}|S|)$
relaxed CA-RSA	$O(k{\cdot}|D|)$	$O(|D| + |E|)$

Table 3.3 Computation times (seconds).

Network (|V|, |E|)	Formulation	|S|=30					|S|=60				
		$k=1$	$k=2$	$k=3$	$k=4$	$k=5$	$k=1$	$k=2$	$k=3$	$k=4$	$k=5$
RING (9, 9)	SSA-RSA	7.70	1325.94	—	—	—	4.67	68.64	—	—	—
	SA-RSA	9.14	1695.54	—	—	—	6.20	56.11	—	—	—
	CA-RSA	0.07	0.08	—	—	—	0.19	0.32	—	—	—
BRASIL (10, 12)	SSA-RSA	10.42	2358.26	$>10^5$	$>10^5$	$>10^5$	9.27	189.44	$>10^5$	$>10^5$	$>10^5$
	SA-RSA	3.16	1242.66	$>10^5$	$>10^5$	$>10^5$	3.28	84.04	$>10^5$	$>10^5$	$>10^5$
	CA-RSA	0.24	1.94	2.65	5.69	9.18	0.24	0.36	0.41	0.52	0.91
ABILENE (12, 15)	SSA-RSA	9.84	53.26	3198.35	$>10^5$	$>10^5$	4.37	27.49	142.31	$>10^5$	$>10^5$
	SA-RSA	6.27	42.01	2798.58	$>10^5$	$>10^5$	1.47	19.44	88.32	5856.34	$>10^5$
	CA-RSA	0.25	0.26	0.29	4.90	19.71	0.14	0.24	0.33	0.52	0.58
DT (14, 23)	SSA-RSA	2.94	9.87	15.92	31.73	123.34	3.07	5.87	19.31	50.24	92.31
	SA-RSA	1.80	4.35	6.48	13.51	27.65	2.60	4.86	16.48	30.28	64.28
	CA-RSA	0.09	0.10	0.17	0.37	0.42	0.10	0.19	0.28	0.41	0.45

thus, the complexity of the RSA problem is comparable to that of the RWA since only spectrum continuity needs to be considered.

3.5.3 Evaluation in Real Scenarios

In order to carry out experiments under more realistic scenarios, we consider the 21-node and 35-link Spanish Telefonica network topology shown in Figure 3.1. We consider an optical spectrum of 800 GHz and $|S| = 64$ slices of 12.5 GHz. For the demands, the requested bitrate is 10, 40, or 100 Gb/s (requiring 1, 2, or 4 slices). Both bitrate and traffic distribution follow the uniform distribution.

Seven increasing traffic loads and 10 randomly generated traffic matrices per load were generated ranging from 5 to 11 Tb/s in steps of 1 Tb/s. All traffic matrices but those with the highest load could be completely served, that is, none of the demands was blocked. In contrast, some of the matrices with total bitrate equals to 11 Tb/s could not be completely served. This kind of studies is designed to find the highest amount of traffic that a network can serve.

For each demand in each traffic matrix, $k = 10$, 15, and 20 paths were precomputed, and each resulting problem instance was solved using the CA-RSA formulation in the scenario described earlier. The results in terms of

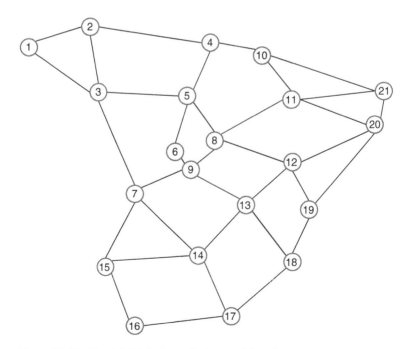

Figure 3.1 The Spanish Telefonica optical network topology.

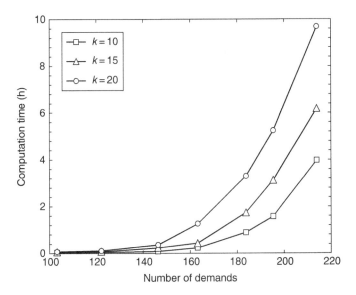

Figure 3.2 Computation time as a function of the number of demands for k values.

Table 3.4 Exact versus relaxed formulations.

Formulation	Computation time	Optimal value (rejected bitrate)
CA-RSA	$>10^5$ s	60 Gb/s (0.48%)
relaxed CA-RSA	0.61 s	10 Gb/s (0.08%)

computation time as a function of the number of demands are shown in Figure 3.2, where each point represents the average value of 10 instances with the same total bitrate. Note that although the figure shows a clear exponential behavior, computation times are still practical (few hours).

Finally, we tested the goodness to the proposed relaxed RSA formulation against the exact CA-RSA formulation in terms of goodness of the solutions and computation times. We generated a traffic matrix with 250 demands and 12.5 Tb/s of total bitrate and solved both the relaxed CA-RSA and the CA-RSA formulations. Results in Table 3.4 show that lower bound provided by the relaxed formulation is really tight to the optimal value obtained by solving the exact one; the relaxed formulation blocked one 10 Gb/s demand whereas the exact formulation blocked three demands (two 10 Gb/s demands and one 40 Gb/s demand). Rejected bitrate and percentages are also shown in Table 3.4.

In light of the performance results of the CA-RSA formulation, we can conclude that such formulation can be used for RSA problems involving realistic optical networks since they can be solved in practical times, much shorter than using slice-based RSA formulations.

In the next section, we extend the CA-RSA formulation (hereafter simply referred to as RSA formulation) for solving the RMSA problem.

3.6 The RMSA Problem

3.6.1 Notation Extensions

As previously introduced, the RMSA adds the selection of modulation format to the RSA problem. To allow this, the following sets and parameters need to be defined in addition to the previous notation:

$len(e)$ Length in kilometers of link $e \in E$.
$len(p)$ Length in kilometers of path $p \in P$. Note that $len(p) = \sum_{e \in E} \delta_{pe} \cdot len(e)$
M Set of modulation formats, index m.
$len(m)$ Reach in kilometers of modulation format m.
q_{cm} Equal to 1 if slot $c \in C$ is computed for modulation format m.

Regarding the definition of slots detailed in Section 3.4.1, the notion of the modulation format needs to be added. Let us assume that a set of slots $C(d)$ is predefined for each demand d that requests b_d bitrate. Depending on the modulation format m selected among the available ones in set M, the size of the slot would be different, so $C(d) \cup_{m \in M} C(d,m)$, where every slot in $C(d, m)$ contains the same number n_{dm} of slices. For convenience, we define q_{cm} equal to 1 if $c \in C(d, m)$. We consider that each set $C(d, m)$ consists of all possible slots of size n_{dm} that can be defined in S. Since $|C(n_{(\cdot)})| = |S| - (n_{(\cdot)} - 1)$, the size of the complete set of slots C that needs to be defined is $|C| = \sum_{d \in D} \sum_{m \in M} [|S| - n_{dm} + 1] < |D| \cdot |M| \cdot |S|$.

3.6.2 Basic Offline Problem

A very basic RMSA problem involves finding a lightpath represented by the tuple <p, m, c>, containing a path p, a modulation format m, and a frequency slot c, for every demand in a given traffic matrix with the objective of minimizing or maximizing some utility function. As for the RSA example, we aim at minimizing the total amount of rejected bitrate.

The problem can be formally stated as follows.

Given:
- a directed connected graph $G(V, E)$,
- the optical spectrum as an ordered set of frequency slices S,

- the set of modulation formats M,
- a set of demands D with the amount of bitrate exchanged between every pair of locations in V.

Output: the lightpath $<p, m, c>$ for each demand in D.
Objective: Minimize the amount of rejected bitrate.

We follow an arc-path approach since the topology is given. Moreover, because of the use of precomputed modulation format–aware slots for each demand, we call this formulation as arc-path routing modulation and slot assignment (AP-RMSA).
The AP-RMSA formulation is as follows:

$$\left(AP - RMSA\right) \quad \min \quad \Phi \tag{3.20}$$

subject to:
Constraint (3.16)
Constraint (3.17)

$$\sum_{p\in P(d)}\sum_{c\in C(d)} q_{cm} \cdot len(p) \cdot y_{pc} \leq len(m), \quad \forall d \in D, m \in M \tag{3.21}$$

The objective function (3.20) minimizes the amount of bitrate that cannot be served (rejected). This formulation extends the CA-RSA one, so it inherits two constraints from the latter. Firstly, constraint (3.16) is required to ensure that a lightpath is selected for each demand provided that the demand is served; otherwise, the demand cannot be served and is therefore rejected. Secondly, constraint (3.17) is used to guarantee that every slice in every link is assigned to one demand at the most. The main difference between RSA and RMSA formulations resides in the new constraint (3.21), which guarantees the selection of a modulation format with enough reachability for the selected path.
The size of the AP-RMSA formulation is $O(k \cdot |D| \cdot |C|)$ variables and $O(|E| \cdot |S| + |D| \cdot |M|)$ constraints. Despite an apparent similarity with respect to RSA size, the inclusion of modulation formats within slot precomputation increases the number of variables significantly. For instance, in the DT network assuming $M = |3|$, the number of variables and constraints is 1.6×10^4 and 8.1×10^2, respectively.

3.6.3 Topology Design Problem as an RMSA Problem

In the previous RMSA problem, the network topology was given. Let us now consider that the problem consists of designing the network topology, so as to serve all the demands in the given traffic matrix. Since each installed link increases network capital expenditures (CAPEX) as a result of optical interfaces, including amplifiers, to be installed in the end nodes and some intermediate locations, minimizing the number of links in the resulting network topology would reduce the total CAPEX cost.

The problem can be formally stated as follows:

Given:
- a directed connected graph $G(V, E)$,
- the optical spectrum S and the set of modulation formats M,
- a demand set D.

Output: the lightpath $<p, m, c>$ for each demand in D and the links to be equipped.

Objective: Minimize the number of links to be equipped to transport the given traffic matrix.

Note that we could precompute k distinct paths for each demand in the traffic matrix, as we did when the topology was given. However, since only part of the links will be installed eventually, the number of routes k to be precomputed for each demand would need to be highly increased to counteract the fact that some of the routes would become useless. For that very reason, we present a *node-arc* ILP formulation (see Chapter 2) that performs routing computation within the optimization. Similarly, as before, since we precompute modulation format–aware slots for each demand, we call this formulation as node-arc modulation format and slot assignment (NA-RMSA).

The NA-RMSA formulation is as follows:

$$\left(NA-RMSA\right) \quad \min \quad \sum_{e\in E} z_e \tag{3.22}$$

subject to:

$$\sum_{e\in E^+(v)}\sum_{c\in C(d)} w_{dec} - \sum_{e\in E^-(v)}\sum_{c\in C(d)} w_{dec} = \begin{cases} 1 & \forall d \in D, v = o_d \\ 0 & \forall d \in D, v \in V - \{o_d, t_d\} \\ -1 & \forall d \in D, v = t_d \end{cases} \tag{3.23}$$

$$\sum_{e\in E} w_{dec} \leq |E| \cdot w_{dc} \quad \forall d \in D, c \in C \tag{3.24}$$

$$\sum_{c\in C} w_{dc} = 1 \quad \forall d \in D \tag{3.25}$$

$$\sum_{e\in E}\sum_{c\in C(d)} q_{cm} \cdot len(e) \cdot w_{dec} \leq len(m) \quad \forall d \in D, m \in M \tag{3.26}$$

$$\sum_{d\in D}\sum_{c\in C(d)} q_{cs} \cdot w_{dec} \leq z_e \quad \forall e \in E, s \in S \tag{3.27}$$

The objective function (3.22) minimizes the number of links to be installed. Constraint (3.23) finds a lightpath for every demand. Specifically, it ensures that one lightpath for each demand leaves from the source node and enters in the target node of every demand, whereas it guarantees that every lightpath entering an intermediate node for a demand leaves that node. As introduced in Chapter 2, solutions with cycles can be produced since the length of paths is not strictly minimized and thus a postprocessing phase needs to be applied to remove them (if any) with no impact on feasibility and optimality of solution.

Constraints (3.24) and (3.25) guarantee that every lightpath is assigned one and only one frequency slot along its route; constraint (3.24) collects the slots that are assigned to every lightpath whereas constraint (3.25) limits that number to one. Like constraint (3.21) in the AP-RMSA formulation, constraint (3.26) selects a modulation format with enough reachability for the selected path. Finally, constraint (3.27) prevents that any frequency slice in any link is used by more than one demand while installing the link when any slice is used.

The size of the NA-RMSA formulation is $O(|D|\cdot|E|\cdot|C|)$ variables and $O(|E|$ $\cdot|S| + |D|\cdot(|N| + |M| + |C|))$ constraints. The size of this formulation for the national DT network previously considered is 6.45×10^4 variables and 4.5×10^3 constraints, higher than for the AP-RMSA formulation.

It is obvious that minimizing the number of links to be installed can be different than minimizing CAPEX since some other costs need to be considered. For this very reason, the previous problem needs to be extended to take into account all the costs and to calibrate every equipment in the network.

Then, a more specific network dimensioning problem can be formally stated as follows.

Given:
- a connected graph $G(V, E)$,
- the optical spectrum S and the set of modulation formats M,
- a set of demands D,
- the cost of every component, such as optical cross-connects (OXCs), transponder types, and regenerators specifying its capacity and reach. The cost of installing each link is also specified.

Output:
- the lightpath $<p, m, c>$ for each demand in D,
- the links that need to be equipped and the network dimensioning including the type of OXC, TPs, and regenerators in each location;

Objective: Minimize the total CAPEX to transport the given traffic matrix.

Although we do not present any specific ILP model for this problem, it could be easily derived from the NA-RMSA formulation together with a CAPEX model (see, e.g., [Pe12], [Ve13.1]).

3.7 Conclusions

Among the variety of EON features, the capability to create optical connections with different bitrates and beyond 100 Gb/s and to adapt the spectrum allocation are of great interest for network operators, since it opens the opportunity to provide new services and to improve the way transport networks are currently designed and operated.

This chapter reviewed basic optimization problems based on solving the RSA problem. Several alternatives to deal with the spectrum-related constraints (continuity and contiguity) are presented. From the comparative analysis, we concluded that the formulation based on slot assignment (CA-RSA) outperforms the other proposed formulations based on slice assignment in terms of required solving time to reach an optimal solution.

To take advantage of the selection of the most proper modulation format for each of the demands, the RSA problem was extended to include modulation formats. Thus, the routing, modulation, and spectrum allocation (RMSA) problem was presented, and two different ILP formulations (the arc-path and the node-arc) were detailed.

The remaining chapters in this book that target complex problems for designing, operating and reoptimizing EONs will be based on the formulations presented in this chapter.

4

Architectures for Provisioning and In-operation Planning

Luis Velasco and Marc Ruiz

Universitat Politècnica de Catalunya, Barcelona, Spain

Current transport networks are statically configured with big static *fat pipes* based on capacity overprovisioning. The rationality behind that is guaranteeing traffic demand and Quality of Service. As a result, long planning cycles are used to upgrade the network and prepare it for the next planning period (see the network life cycle in Chapter 2). Aiming at guaranteeing that the network can support the forecast traffic and deal with failure scenarios, spare capacity is usually installed, thus increasing network expenditures. Moreover, results from network capacity planning are manually deployed in the network, which limits the network agility. In this chapter, we review control and management architectures to allow the network to be dynamically operated. Employing those dynamic capabilities, the network can be reconfigured and reoptimized in response to traffic changes in an automatic fashion. Hence, resource over-provisioning can be minimized and overall network costs reduced.

Because of the large capacity of optical connections, optical networks are commonly used to support virtual topologies on which client demands are served. Therefore, each optical connection is shared by a number of client connections, which balances the changes in the traffic that they have to support over time (e.g., during day and night hours). However, the benefits of capacity sharing among clients are canceled if some of them require large capacity to transport traffic that highly varies over time. This is the case of datacenters (DCs) interconnection, where optical connections are used to perform virtual machine (VM) migration and database (DB) synchronization allowing elastic DC operations. Such static connection configuration entails large bitrate overprovisioning, and thus high operational costs for DC operators, so dynamic inter-DC connectivity is needed to improve resource utilization and save costs.

Provisioning, Recovery, and In-operation Planning in Elastic Optical Networks,
First Edition. Luis Velasco and Marc Ruiz.
© 2017 John Wiley & Sons, Inc. Published 2017 by John Wiley & Sons, Inc.

Nonetheless, dynamic connectivity is not enough to guarantee elastic DC operations and might lead to poor performance provided that not enough overprovisioning of network resources is performed. To alleviate that to some extent, the ability of Elastic Optical Networks (EONs) can be used to dynamically increase and decrease the amount of optical resources assigned to connections.

Table of Contents and Tracks' Itineraries

	Average	Theoretical	Practitioner
4.1 Introduction	☑	☑	☑
4.2 Architectures for Dynamic Network Operation	☑	☑	☑
4.2.1 Static versus Dynamic Network Operation	☑	☑	☑
4.2.2 Migration toward In-operation Network Planning	☑	☑	☑
4.2.3 Required Functionalities			☑
4.2.4 The Front-end/Back-end PCE Architecture			☑
4.3 In-operation Planning: Use Cases			
4.3 1 VNT Reconfiguration after a Failure			☑
4.3.2 Reoptimization			☑
4.4 Toward Cloud-Ready Transport Networks	☑	☑	☑
4.5 Conclusions	☑	☑	☑

4.1 Introduction

Network capacity planning focuses on the placement of network resources to satisfy expected traffic demands and network failure scenarios. Today, the network capacity planning process is typically an offline process, and it is based on very long planning cycles (e.g., yearly, quarterly) (see Chapter 2). Generally, this is due to the static and inflexible nature of current networks. This is true for both the transport layer (optical and Ethernet) and the Multi-Protocol Label Switching (MPLS) layer. However, the latter should be inherently more dynamic compared to the underlying transport infrastructure. The MPLS layer might use automated *Traffic Engineering* (TE) techniques to place MPLS traffic where the network resources are.

Increasing transport capacity to meet predicted MPLS traffic changes and failures provides limited network flexibility. However, due to the fixed and rigid nature of provisioning in the transport layer, network planning and TE still require significant human intervention, which entails high Operational Expenditures (OPEX). In addition, to ensure that the network can support the forecast traffic and all the failure scenarios that need to be protected against, operators add spare capacity (overprovision) in different parts of the network to address likely future scenarios. This entails the inefficient use of network resources and significantly increases network Capital Expenditures (CAPEX).

Notwithstanding, optical transport platforms are designed to facilitate setting up and tearing down optical connections (*lightpaths*) within minutes or even seconds [ASON]. Combining remotely configurable optical cross-connects (OXCs) with a control plane provides the capability of automated lightpath set up for regular provisioning and real-time reaction to failures, thus reducing OPEX. However, to exploit existing capacity, increase dynamicity, and provide automation in future networks, current management architectures, utilizing legacy Network Management Systems (NMSs), need to be radically transformed.

In a scenario where lightpath provisioning can be automated, network resources can be made available by reconfiguring and/or reoptimizing the network on demand and in real time. We call this *in-operation network planning*. We propose to take advantage of novel network reconfiguration capabilities and new network management architectures to perform in-operation planning, aiming at reducing network CAPEX by minimizing the overprovisioning required in today's static network environments.

Significant research and standardization effort has assisted in defining control plane architectures and protocols to automate connection provisioning. Starting from a distributed paradigm, the control plane has lately moved toward a centralized one led by the development of the software-defined networking (SDN) concept with the introduction of the OpenFlow protocol [ONF]. The Internet Engineering Task Force (IETF) is also moving in that direction with the definition of the Application-Based Network Operations (ABNO) architecture [ABNO] based on functional elements defined by the IETF, like the active stateful Path Computation Element (PCE) [Cr16.1].

In this chapter, we highlight current standardization work and propose a control and management architecture based on the ABNO model, which is capable of performing in-operation planning. Finally, a distributed *orchestrator* to coordinate cloud and the interconnection Elastic Optical Networks (EONs) is presented. The orchestrator can manage inter-datacenter (inter-DC) connectivity not only dynamically, but also elastically by increasing or reducing the amount of bandwidth allocated to already established optical connections. The orchestrator interfaces a network controller based on the ABNO architecture.

4.2 Architectures for Dynamic Network Operation

4.2.1 Static versus Dynamic Network Operation

The operation of the currently deployed carriers' transport networks is very complex since multiple manual configuration actions are needed for provisioning purposes (e.g., hundreds of thousands of node configurations per year in a midsize network [Wi01]). In fact, transport networks are currently configured with big static *fat pipes* based on capacity overprovisioning, since they are needed for guaranteeing traffic demand and Quality of Service. Furthermore, network solutions from different vendors typically include a centralized service provisioning platform, using vendor-specific NMS implementations along with an operator-tailored umbrella provisioning system, which may include a technology-specific Operations Support System (OSS). Such complicated architectures (Figure 4.1) generate complex and long workflows for network provisioning: up to 2 weeks for customer service provisioning and more than 6 weeks for core router connectivity services over the optical core.

Figure 4.2 illustrates the fact that such static networks are designed to cope with the requirements of several failure scenarios and predicted short-term increases in bandwidth usage, thus requiring capacity overprovisioning and significantly increasing CAPEX. The figure shows a simple network consisting of three internet protocol (IP) routers connected to a central one through a set of lightpaths established on an optical network. Two different scenarios are considered, although the same amount of IP traffic is conveyed in each of them. In scenario A, router R3 needs three lightpaths to be established to transport its IP traffic toward R4, whereas R1 and R2 need one single lightpath each.

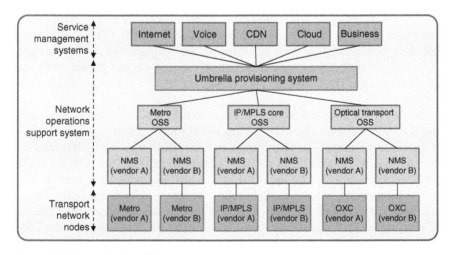

Figure 4.1 Current static architecture.

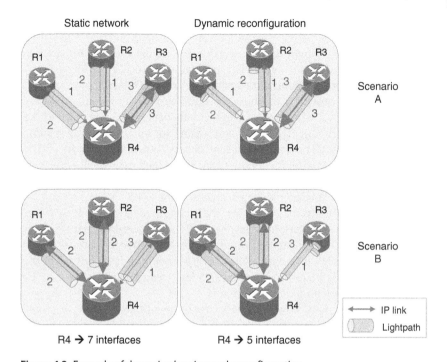

Static network   Dynamic reconfiguration

Scenario A

Scenario B

IP link

Lightpath

R4 ➔ 7 interfaces   R4 ➔ 5 interfaces

Figure 4.2 Example of dynamic planning and reconfiguration.

In contrast, R1 and R2 need two lightpaths in the scenario B, whilst R3 needs just one lightpath. In static networks, where lightpaths in the optical network are statically established, each pair of routers has to be equipped with the number of interfaces for the worst case of every IP link, resulting in R4 being equipped with seven interfaces (two to support IP link R1–R4, two for link R2–R4, and three for link R3–R4). However, if the optical network could be dynamically reconfigured by setting up and tearing down lightpaths on demand, each router could be dimensioned separately for the worst case scenario, regardless of the peering routers. As a result, R4 would need to be equipped with only five interfaces, thus saving 28.5% of interfaces.

4.2.2 Migration toward In-operation Network Planning

As introduced in Chapter 2, the classical network planning life cycle typically consists of several steps that are performed sequentially. The initial step receives inputs from the service layer and from the state of the resources in the already deployed network and configures the network to be capable of dealing with the forecast traffic for a period of time. That period is not fixed and actual time length usually depends on many factors, which are operator- and traffic-type-specific. Once the planning phase produces recommendations, the next step is to design, verify, and manually implement the network changes. While in operation, the network capacity is continuously monitored, and that data is used as input for the next planning cycle. In the case of unexpected increases in demand or network changes, nonetheless, the planning process may be restarted.

As technologies are developed to allow the network to become more agile, it may be possible to provide a response to traffic changes by reconfiguring the network near real time. In fact, some operators have deployed Generalized MPLS (GMPLS) control planes or SDN controllers, mainly for service setup automation and recovery purposes. However, those control only parts of the network and do not support holistic network reconfiguration. This functionality will require an in-operation planning tool that interacts directly with the data and control planes and operator policies via OSS platforms, including the NMS.

Assuming the benefits of operating the network in a dynamic way are proven, the classical network life cycle has to be augmented to include a new step focused on reconfiguring and reoptimizing the network, as represented in Figure 4.3. We call that step as *in-operation planning* and, in contrast to the traditional network planning, the results and recommendations can be immediately implemented on the network.

To support dynamicity, however, the current network architecture depicted in Figure 4.1 will need to evolve and include a functional block between the service layer and the network elements to support multiservice provisioning in multivendor and multitechnology scenarios; two standard interfaces are also required. Firstly, the *northbound* interface that, among other tasks, gives an

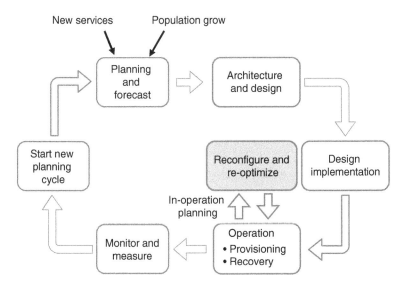

Figure 4.3 Network life cycle.

abstracted view of the network, enabling a common entry point to provision multiple services and to provision the planned configuration for the network. Moreover, this interface allows coordinating network and service layer according to service requirements. Secondly, the *southbound interface* covering provisioning, monitoring, and information retrieval.

Finally, operators will typically require human–machine interaction to ensure new configurations, and network impact are reviewed and acknowledged before being implemented in the network.

4.2.3 Required Functionalities

Standardization bodies, especially the IETF, have been working to address all these requirements, and, as a result, the ABNO architecture is now being proposed as a candidate solution. The ABNO architecture consists of a number of standard components and interfaces which, when combined together, provide a solution for controlling and operating the network.

A simplified view of the ABNO architecture is represented in Figure 4.4; it includes:

- The ABNO controller as the entrance point to the network for NMS/OSS and the service layer for provisioning and advanced network coordination. It acts as a system orchestrator invoking its inner components according to specific workflows.
- The PCE [RFC4655] serves paths computation requests. The PCE Communication Protocol (PCEP) is used to carry path computation requests and responses. Requests can be processed independently of each other or in groups, utilizing a view of the network topology stored in the TE Database (TED) (*stateless PCE*), or considering as well information regarding Label Switched Paths (LSPs) that have been set up in the network and are stored in

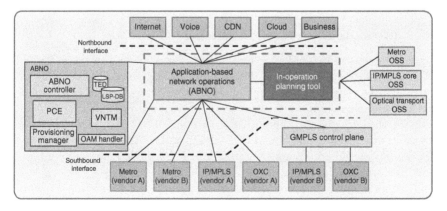

Figure 4.4 Dynamic architecture based on ABNO enabling in-operation planning.

the LSP Database (LSP-DB) (*stateful PCE*). Finally, a PCE is said to be *active* if it can modify in-place LSPs based on network trends.

- The Virtual Network Topology Manager (VNTM) [RFC5623] coordinates Virtual Network Topology (VNT) configuration by setting up or tearing down lower-layer LSPs supporting virtual links, and advertising the changes to higher-layer network entities.
- The Provisioning Manager (PM) is responsible for the establishment of LSPs. This can be done by interfacing the GMPLS control plane [Cr16.1], [Cr16.2] or by directly programming the data path on individual network nodes using Network Configuration Protocol (NetConf) or acting as an SDN controller using OpenFlow.
- The Operations, Administration, and Maintenance (OAM) handler is responsible for detecting faults and taking actions to react to problems in the network. It interacts with the nodes to initiate OAM actions, such as monitoring and testing new links and services.

Directly connected to the ABNO architecture, the in-operation planning tool can be deployed as a dedicated back-end PCE (bPCE) for performance improvements and optimizations. The bPCE is accessible via a PCEP interface, so the ABNO components can forward requests to the planning tool.

Furthermore, in-operation network planning can only be achievable if planning tools are synchronized with the state of network resources, so new configurations can be computed with updated information, and those configurations can be easily deployed in the network. In the proposed architecture, the bPCE gathers network topology and the current state of network resources, via the ABNO components, using protocols such as Border Gateway Protocol (BGP) with extensions to distribute link-state information [RFC7752].

There are several architectures utilizing IETF components that are suitable for providing in-operation network planning. These are described in Table 4.1 with the corresponding strengths and weaknesses.

4.2.4 The Front-end/Back-end PCE Architecture

Figure 4.5 presents two different architectures for deploying an active stateful PCE. In the classical centralized PCE architecture (Figure 4.5a), the PCE executes all tasks to serve connection requests. However, specialized PCEs can be devised, for instance, but not restricted, to run reoptimization algorithms that might require intensive computation (Figure 4.5b). Note the replacement of the centralized PCE by front-end PCE (fPCE) and bPCE modules. While the fPCE is responsible for simple computations (e.g., single path computation), intensive computations are delegated to the bPCE.

When an intensive task needs to be computed, the fPCE forwards the set of path computation requests to the bPCE to solve the task. To this end, PCEP can be used; a single Path Computation Request (*PCReq*) message can contain the whole set of path computations, and a Synchronization Vector (SVEC) object

Table 4.1 Strengths and weaknesses of several architectures supporting provisioning and in-operation planning.

Architecture	Features	Strengths	Weaknesses
Stateless PCE	• It performs path computation using topology and the state of the network resources in the TED.	• Path computation can be offloaded into a dedicated entity capable of complex computations with customized algorithms and functions. • It has a standard and mature interface and protocol. • It supports simple optimization, such as bulk path computation [RFC5557].	• It is unaware of existing LSPs and has no view of the current network resource utilization and key choke points. • It cannot configure by itself any LSP in the network. • Delays need to be introduced to sequence LSP setup [Ca14].
Stateful PCE	• It maintains the LSP-DB with the LSPs that are active in the network.	• Requests can be more efficiently placed optimizing network resources. • Supports optimization involving already established LSPs.	• It is more complex than a stateless PCE and requires an additional database and synchronization. • No existing LSPs can be modified, for example, for network reconfiguration purposes.
Active stateful PCE	• It is capable of responding to changes in network resource availability and predicted demands and reroute existing LSPs for increased network resource efficiency [Cs13].	• It supports complex reconfiguration and re-optimization, even in multilayer networks.	• No new LSPs can be created, for example, for VNT reoptimization purposes. • It requires protocol extensions to modify and/or instantiate LSPs (if the capability is available).
ABNO	• It provides a network control system for coordinating OSS and NMS requests to compute paths, enforce policies, and manage network resources for the benefit of the applications that use the network.	• New LSPs can be created for in-operation planning. VNTM in charge of VNT reconfiguration. • It supports deployment of solutions in multitechnology scenarios (NetConf, OpenFlow, control plane, etc.)	• It requires implementation of a number of key components in addition to the PCE function. • Some interfaces still need to be defined and standardized.

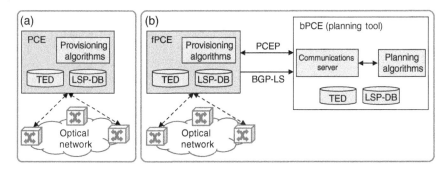

Figure 4.5 Centralized (a) versus fPCE/bPCE architecture (b).

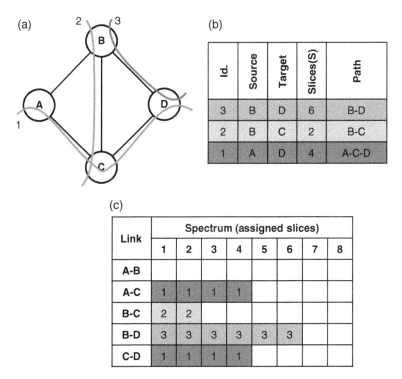

Figure 4.6 Example of RSA for two heuristic iterations: (a) network topology and (b) the routed demands. (c) Spectrum allocation in each link.

is included to group together several path requests, so that the computation is performed for all those requests (bulk), thus attaining a global optimum.

For illustrative purposes, Figure 4.6 presents an example of a bulk consisting of three demands to be computed on a 4-node and 5-link network

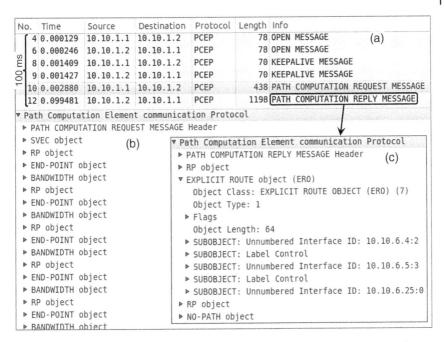

Figure 4.7 Exchanged messages and details: (a) message list, (b) PCReq message, and (c) PCRep message.

topology. Figure 4.6a shows the network topology and the routed demands, Figure 4.6b gives an insight of the order and the route of each demand, whereas Figure 4.6c depicts the spectrum allocation in each link (each slice identifies the demand using it).

Let us now validate the fPCE/bPCE architecture experimentally. The setup deployed consists of an fPCE (IP: 10.10.1.1) and a bPCE (IP: 10.10.1.2) developed in C++ that communicate through PCEP interfaces. The data plane consists of a set of emulated OXCs (IP subnetwork 10.10.6.x) to create the 30-node and 56-link network topology.

When the fPCE delegates the computation to the bPCE (Figure 4.7a), the former opens a new PCEP session, creates a PCReq message (Figure 4.7b) containing every path request and a SVEC object, and forwards the message to the bPCE. Upon receiving a PCReq message, the bPCE solves the Routing and Spectrum Allocation (RSA) problem (see Chapter 3) for the set of received requests. The bPCE sends a Path Computation Reply (*PCRep*) message (Figure 4.7c) containing either the RSA for each request or the object NO_PATH if no feasible solution was found for a request.

Another issue of the fPCE/bPCE architecture is that of DB synchronization. As discussed earlier, the fPCE stores the current state of the network resources

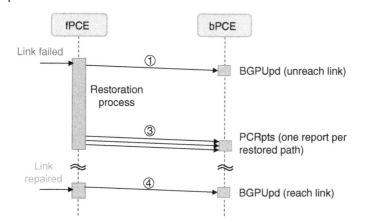

Figure 4.8 Exchanged PCEP and BGP-LS messages for DB synchronization.

and LSPs on its local copies of the TED and the LSP-DB, respectively. However, the bPCE needs also to access those data since they are input parameters for in-operation planning algorithms. This requires deploying a local TED and LSP-DB in the bPCE and using protocols to synchronize them with the fPCE. For network topology synchronization between ABNO components, BGP-LS is proposed in RFC7491 [ABNO]. Therefore, BGP speakers can be placed at both fPCE and bPCE to perform TED synchronization; BGP Update (*BGPUpd*) messages are used to notify the availability (*BGPUpd*—Reach) of new nodes and links and its unavailability (*BGPUpd*—Un-Reach). In addition, the state of the LSPs can be synchronized by means of PCEP with stateful extensions [Cr16.2]; the Path Computation Report (*PCRpt*) message can be used for LSP-DB synchronization.

To illustrate DB synchronization between fPCE and bPCE, let us use the example of a link failure and link repair conditions. The fPCE/bPCE sequence diagram for TEDs and LSP-DBs synchronization is illustrated in Figure 4.8.

When a link failure is received in the fPCE, it sends a BGP-LS link unreach message (labeled as 1 in Figure 4.8) to the bPCE to notify such condition. The fPCE continues with its restoration process that can be delegated to the bPCE (not shown in Figure 4.8). When the restoration paths have been established in the network, the fPCE notifies the bPCE with the new routes by means of PCRpt messages (3). Later, once the link is repaired, the fPCE notifies the bPCE using a BGP-LS link reach message (4).

The workflow discussed was demonstrated in an experimental test bed and the capture in Figure 4.9 shows relevant BGP-LS and PCEP messages exchanged between fPCE and bPCE. When the fPCE is notified about the link failure, it notifies that condition to the bPCE (1 in Figure 4.9) and requests the computation of the new routes, for example, by issuing a bulk PCReq to the bPCE and waits for

Source	Destin	Proto	Info
CTRL	fPCE	PCEP	Path Computation Request (PCReq)
fPCE	bPCE	BGP	UPDATE Message
fPCE	bPCE	BGP	UPDATE Message
fPCE	bPCE	PCEP	Path Computation Request (PCReq)
bPCE	fPCE	PCEP	Path Computation Reply (PCRep)
fPCE	bPCE	PCEP	Path Computation LSP State Report (PCRpt)
fPCE	bPCE	PCEP	Path Computation LSP State Report (PCRpt)
fPCE	bPCE	PCEP	Path Computation LSP State Report (PCRpt)
fPCE	CTRL	PCEP	Path Computation Reply (PCRep)
fPCE	bPCE	BGP	UPDATE Message
fPCE	bPCE	BGP	UPDATE Message

Figure 4.9 Exchanged BGP-LS and PCEP messages.

the bPCE reply (2). When the bPCE replies, the fPCE sets up the new LSPs and issues report messages to the bPCE with the new routes (3). Later, when the link is repaired, the fPCE notifies the bPCE the new state of the links (4).

4.3 In-operation Planning: Use Cases

Aiming at illustrating the applicability and benefits of in-operation network planning, two use cases regarding transport networks are described: (i) VNT reconfiguration as a consequence of network failures and (ii) reoptimization to improve network resource efficiency and utilization.

4.3.1 VNT Reconfiguration after a Failure

In our multilayer network scenario, the VNT is supported by a set of optical connections, so the MPLS layer can operate independently from the optical one. However, in some cases, the VNT needs to be reconfigured based on changing network conditions and demands. Many network operators prefer that such reconfiguration is supervised and approved by a person in charge before it is implemented in the network. This human intervention allows the application of additional policies and considerations and the integration with existing business policy and the OSS, for example, new services not completely integrated with the tool.

Figure 4.10a represents a multilayer network consisting of four OXCs in the optical layer and three routers in the MPLS layer. MPLS routers are connected through 10 Gb/s lightpaths, thus creating a VNT. Three bidirectional MPLS LSPs have been set up on the virtual topology. After a failure occurs, either in the optical or in the MPLS layer, Fast Reroute (FRR) [RFC4090] can be used to

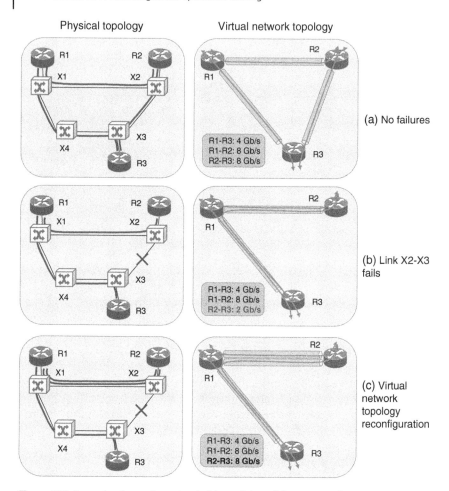

Figure 4.10 Example of reconfiguration. In (a) one lightpath between each router is set up. In (b), a failure affects an optical link, and FRR recovers part of the MPLS traffic. VNT reconfiguration is activated in (c).

recover part of the affected MPLS LSPs immediately after the failure. In addition, the state of the network after the failure can be updated in the control plane also within seconds. However, the capacity of some MPLS LSPs might be reduced or even remain disconnected as a consequence of high congestion in some links in the VNT, and VNT reconfiguration needs to be performed to aggregate traffic and distribute traffic away from choke points and heavily utilized resources. An example is presented in Figure 4.10b, where the LSP R2–R3 gets its capacity reduced from 8 Gb/s to only 2 Gb/s. To cope with VNT reconfigurations, our approach relies on the ABNO architecture presented in Figure 4.11, where the process is also represented.

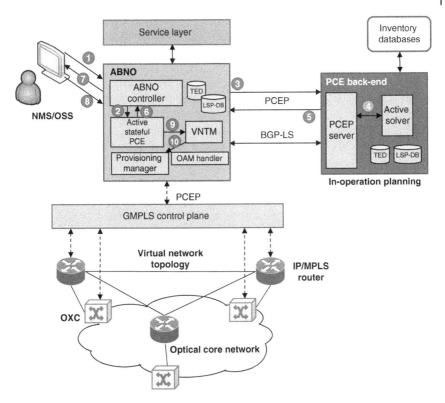

Figure 4.11 VNT reconfiguration process.

When the network operator wishes to perform a network-wide VNT reconfiguration, a request is sent from the NMS/OSS to the ABNO controller (1), who then forwards it to the PCE via the VNTM (2). To alleviate PCE workload, a bPCE, which contains an active solver, is responsible for computing new VNT layout taking into account the current state of the network. Therefore, the PCE sends a request toward the in-operation planning tool running in that bPCE to compute the new VNT (3). The tool considers all the surviving resources, which may include router interfaces and transponders connected to failed optical connections, spare interfaces that typically exist in the network for normal growth, and possibly some spare routers that have been installed ahead of time; all those resources can be stored in an inventory DB. The tool must consider how to implement the desired MPLS connectivity over the optical layer. To this end, it needs to know which optical links and nodes are up and which connections are optically feasible, considering physical layer impairments. When a result is obtained (4), the set of lightpaths is replied in a PCRep message (5) toward the originating PCE. In case an operator needs to approve the new (virtual) layout, the PCE forwards it to the ABNO controller (6). The computed layout is then presented to the operator for final approval (7). When the operator acknowledges the new optimized layout (8), it is passed to the VNTM (9) that computes the sequence of operations to carry out in terms of rerouting existing new LSPs to minimize traffic disruption. The sorted sequence of actions on existing LSPs is passed to the PM (10), which is able to interact with each head-end node. The provisioning interface used by the PM to suggest rerouting of LSPs is based on *PCUpd* PCEP messages [Cr16.2]. The newly allocated resources are reported back to the PM and ultimately the VNTM, using PCRpt messages. Note that the LSP-DB is updated accordingly after a successful reoptimization. In our example in Figure 4.10c, a new lightpath is created between R1 and R2, and as a result, the MPLS LSP R2–R3 can be rerouted and its initial capacity restored.

4.3.2 Reoptimization

Algorithms in the control or management planes compute routes and find feasible spectrum allocation for connection requests taking into account the state of network resources at the time each connection is requested. Nonetheless, as a consequence of network dynamics, some resources may not be released so that better routes could be computed and thus reoptimization could not be applied to improve network efficiency. For example, imagine an optical connection that due to network congestion is required to circumnavigate optimal nodes and links so that the end-to-end connection requires intermediate regenerators; at some point additional paths become available, and the service could be rerouted to use the shorter route and eliminate regeneration. Additionally, other existing services could be rerouted to remove the bottlenecks and avoid network congestion, or even allow some connections to increase their capacity when needed.

In this use case, we study a specific problem that arises in EONs and where reoptimization could bring clear benefits; as a consequence of the unavailability of spectrum converters, spectrum fragmentation appears, increasing the blocking probability of connection requests, thus making worse the network grade of service.

An example is shown in Figure 4.12a where the optical spectrum of a link is represented. Three already established lightpaths share that link; each lightpath uses a different frequency slot width. If a new lightpath needing 37.5 GHz is requested, it would be blocked as a consequence of the lack of spectrum contiguity. In such scenario, reoptimization could be applied to the network before a connection request is blocked, by reallocating already established lightpaths in the spectrum (Figure 4.12b) to make enough room for the triggering connection requested (Figure 4.12c). Defragmentation algorithms to efficiently compute the set of connections to be reallocated will be presented in Chapter 8.

Figure 4.13 illustrates the proposed control plane architecture to support EON reoptimization, which also facilitates human verification and acknowledgment

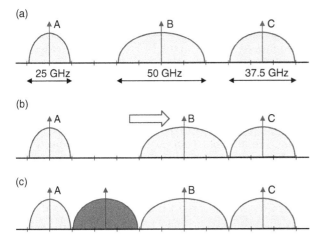

Figure 4.12 Example of reoptimization. (a) Initial spectrum, (b) reallocation, and (c) connection setup.

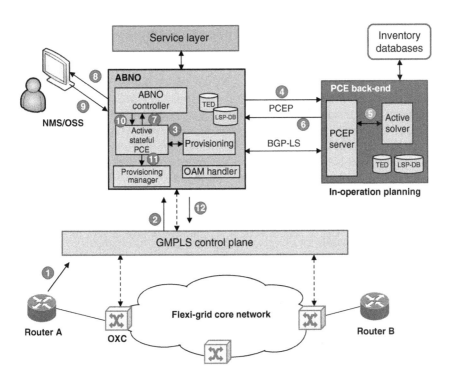

Figure 4.13 Reoptimization process.

of network changes. When Router A needs a new connection to Router B, it sends a request to the control plane of the optical network (1). After checking admission policies, a PCReq message is sent to the PCE (2), which invokes its local provisioning algorithm (3). In the event of insufficient resources being available due to spectrum fragmentation, the active PCE recommends the defragmentation of relevant connections, utilizing the right algorithm to provide such reoptimizations. Similarly, as in the previous use case, let us assume that the bPCE providing such algorithm will perform the computation upon receipt of a request (4). When a result is obtained (5), it is sent back to the fPCE (6). In case an operator needs to approve implementing the computed solution in the network, a request is sent to the NMS/OSS via the ABNO controller (7, 8). When the solution has been verified and acknowledged by the operator, the NMS/OSS informs the PCE via the ABNO controller (9, 10), which forwards the solution toward the PM. Existing connection reallocations are requested using PCUpd messages (11). Once the dependent connections have been set up, the responsible PCE will invoke the local provisioning algorithm for the original connection request between routers A and B and sends a PCRep message to the originating control plane node (12).

4.4 Toward Cloud-Ready Transport Networks

A different use case is when EONs are used to support DC interconnection. The capacity of each inter-DC optical connection is dimensioned in advance based on some volume of foreseen data to transfer. Once in operation, scheduling algorithms inside cloud management run periodically trying to optimize some cost function, such as energy costs, and organize VM migration and DB synchronization as a function of the available bitrate. To avoid transference overlapping that happens when some VM migrations or DB synchronization could not be performed in the current period, which may eventually lead to performance degradation, some overdimensioning is needed. Obviously, this static connectivity configuration adds high costs since large connectivity capacity remains unused in low periods.

Therefore, demands for cloud services require new mechanisms to provide reconfiguration and adaptability of the transport network to reduce the amount of overprovisioned bandwidth; the efficient integration of cloud-based services among distributed DCs, including the interconnecting network, then becomes a challenge. The *cloud-ready transport network* was introduced as an architecture to handle this dynamic cloud and network interaction, allowing on-demand connectivity provisioning [Co12].

Evolution toward cloud-ready transport networks entails dynamically controlling network resources, considering cloud requests in the network configuration process. Hence, that evolution is based on elastic data and control planes, which can interact with multiple network technologies and cloud services. An orchestrator between the cloud and interconnection network is eventually required to coordinate resources in both strata in a coherent manner. The considered reference architecture to support cloud-ready transport networks is shown in Figure 4.14.

In the scope of this book, the architecture of cloud-ready transport network is used to interconnect DCs in different locations to provide bandwidth-on-demand. To support these huge adaptive connections, EONs are the best-positioned network technology, since they can support optical connections created using the required spectral bandwidth based on the users' requirements. By deploying EONs in transport networks, network providers can improve spectrum utilization, thus achieving a cost-effective solution to support their services.

As for the control plane to dynamically set up and tear down connections, the entry point from applications to the network is ABNO that uses an active stateful PCE to dynamically set up and tear down connections based on the requests from the application layer (DCs). Similarly, the PCE can change the bitrate of established connections when needed.

In contrast to static connectivity, *dynamic* connectivity allows DCs to manage optical connections to remote DCs, requesting connections as they are really needed to perform data transferences and releasing them when all data has been transferred. Furthermore, the fine spectral granularity and the wide range of bitrates in EONs make the actual bitrate of the optical connection to fit connectivity needs closely. After requesting a connection and negotiating its capacity as a function of the current network availability, the resulting bandwidth can be used by scheduling algorithms to organize transferences.

Nonetheless, the availability of resources is not guaranteed and the lack of network resources at requesting time may result in long transference times and even in transference period overlapping. Note that the connection's bitrate cannot be renegotiated and remains constant along the connection's holding time. To reduce the impact of the unavailability of required connectivity resources, connection elasticity can be used, allowing an increase/decrease in the amount of spectral resources assigned to each connection, and thus its capacity. This adaptation is made if there are not enough resources at request time, requesting for more bandwidth at any time after the connection has been set up. We call this type of connectivity *dynamic elastic*.

To implement the earlier-mentioned dynamic connectivity models, an orchestrator module is used to coordinate cloud and network (Figure 4.15). In addition to the cloud and network management for local DC resources, the orchestrator implements new components to facilitate cloud operations: (i) an IT resources coordination and synchronization module in charge of coordinating VM migrations and DB synchronizations among DCs, and (ii) a connection manager and a virtual topology DB.

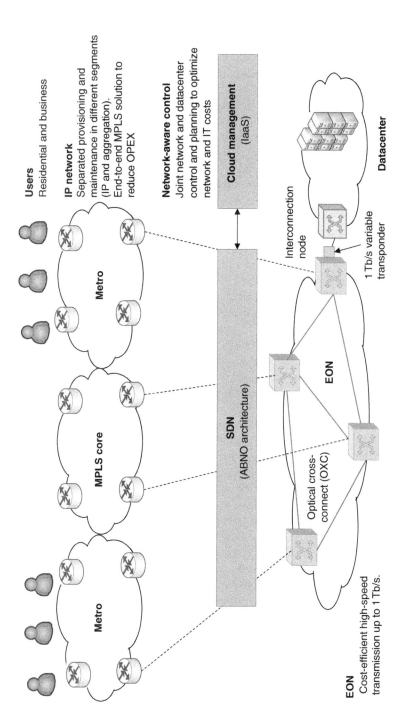

Users
Residential and business

IP network
Separated provisioning and
maintenance in different segments
(IP and aggregation).
End-to-end MPLS solution to
reduce OPEX

Network-aware control
Joint network and datacenter
control and planning to optimize
network and IT costs

Cloud management
(IaaS)

Metro

MPLS core

Metro

SDN
(ABNO architecture)

Interconnection
node

Datacenter

1 Tb/s variable
transponder

EON

Optical cross-
connect (OXC)

EON
Cost-efficient high-speed
transmission up to 1 Tb/s.

Figure 4.14 Architecture to support cloud-ready transport networks.

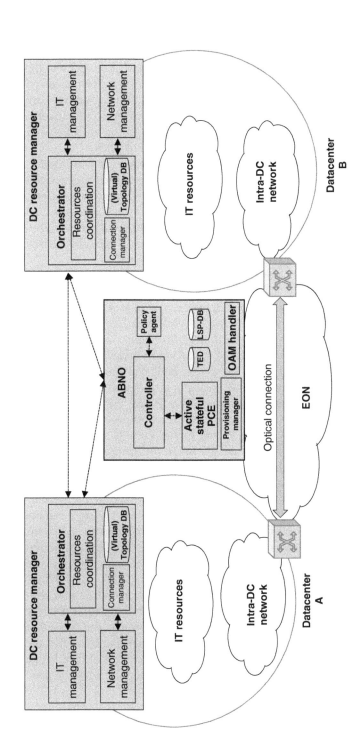

Figure 4.15 Orchestrated architecture.

The set of VMs to migrate and the set of DBs to synchronize are computed by the scheduler inside the cloud manager, which takes into account the availability of servers in the rest of DCs and the high-level performance and availability goals of the workloads hosted in the DCs. Once those sets are computed, the orchestrator module coordinates intra- and inter-DC networks to perform the transferences. The orchestrator interfaces the local network controller, the remote orchestrators, and the inter-DC controller.

The inter-DC controller implements the ABNO architecture with the following modules: the controller, a policy agent that enforces the set of policies received from an NMS, an active stateful PCE to perform path computations, a PM in charge of implementing connections in the network elements, and an OAM handler to receive notifications. Note that although a centralized approach is assumed with a single ABNO, other approaches could be devised to reduce control plane latency, such as creating subdomains.

Once a request is received from an orchestrator, the controller in the ABNO verifies rights asking the policy agent for checking maximum bandwidth, origin and destination nodes, and so on. Then, the controller requests the PCE for a path between both locations. Once the PCE finds a path for the requested capacity, it delegates its implementation to the PM. The PM creates the path using some interface; PCEP is used in our scenario to forward the request to the source node in the underlying EON so that node can start signaling the connection [Al16]. After the connection has been properly set up, the source node notifies the PM, which in turn updates TED and LSP-DB. A notification is sent back to the originating orchestrator after the whole process ends.

Using the architecture in Figure 4.16, the orchestrator is able to request optical connections dynamically to the ABNO negotiating its bitrate. The sequence diagram in Figure 4.16 for dynamic connectivity illustrates messages exchanged between the originating orchestrator and ABNO to set up and tear down an optical connection. Once the orchestrator has computed a transference to be performed, it requests an 80 Gb/s optical connection to a remote DC (time t_0); the policy agent inside ABNO verifies local policies and performs internal operations (i.e., it forwards the message to the PCE) to find a route and spectrum resources for the request. Assuming that not enough resources are currently available for the bitrate requested, an algorithm inside ABNO finds the maximum bitrate and sends a response to the originating orchestrator with such information. On receiving the maximum available bitrate, 40 Gb/s in the example, the orchestrator recomputes the transference, reduces the amount of data to transfer, and requests a connection with the available bitrate. When the transference ends (t_1), the orchestrator sends a message to ABNO to tear down the connection and the used resources are released, so that they can be assigned to any other connection.

In the dynamic elastic connectivity model, the orchestrator is able to request increments in the bitrate of already established connections. In the example in

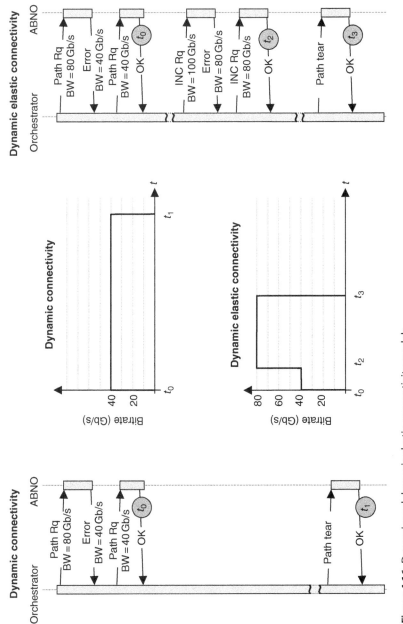

Figure 4.16 Dynamic and dynamic elastic connectivity models.

Figure 4.16, after the connection has been established in t_0 with half of the initially requested bitrate, the orchestrator sends periodical retrials to increment its bitrate. In the example, some resources have been released after the connection has been established and, after a request is received in t_2, they can be assigned to increment the bitrate of the already established connection; the assigned bitrate then increases to 80 Gb/s, which reduces total transference time. Note that this is beneficial for both the cloud operator, since better performance could be achieved, and the network operator, since unused resources are immediately occupied.

4.5 Conclusions

A control and management architecture of transport networks has been reviewed to support provisioning and in-operation network planning. The architecture is based on ABNO and allows carriers to operate the network in a dynamic way and to reconfigure and reoptimize the network near real time in response to changes, like traffic changes or failures. Networks' life cycle is extended to achieve better resource utilization, thus reducing network CAPEX, while process automation reduces manual interventions and, consequently, OPEX. To illustrate how in-operation planning can be applied, two use cases were presented: the first use case focused on reconfiguring the VNT in a multilayer scenario and the second one aimed at performing spectrum reallocation to improve the grade of service.

Even with dynamic connectivity, some resource overprovisioning still needs to be done to guarantee that resources are available when needed. Aiming at improving the performance of dynamic connectivity, elastic operations supported by EONs can be performed on already established optical connections to allow cloud orchestrators doing retries when not enough resources are available at connection's setup time.

Part II

Provisioning in Single Layer Networks

5

Dynamic Provisioning of p2p Demands

Alberto Castro[1], Adrián Asensio[2], Marc Ruiz[2] and Luis Velasco[2]

[1] University California Davis, Davis, CA, USA
[2] Universitat Politècnica de Catalunya, Barcelona, Spain

In the presence of traffic dynamicity, dynamic connectivity control and management allows fitting the allocated capacity resources with the actual traffic bitrate requirements. Mature wavelength division multiplexing (WDM) optical technology enables dynamic (online) provisioning of point-to-point (p2p) transparent optical connections. However, the algorithms developed for WDM cannot be used in elastic optical networks (EONs) because of the spectrum contiguity constraint. In this chapter, we present basic algorithms for dynamic p2p connection provisioning in EONs and extend their scope to deal with modulation formats, as well as propose new algorithms for spectrum adaptation. Finally, when the transparent transmission is not always possible due to reach limitations, impairment-aware provisioning is proposed to guarantee a target Quality of Transmission (QoT) in translucent optical networks.

Table of Contents and Tracks' Itineraries

Provisioning, Recovery, and In-operation Planning in Elastic Optical Networks,
First Edition. Luis Velasco and Marc Ruiz.
© 2017 John Wiley & Sons, Inc. Published 2017 by John Wiley & Sons, Inc.

5.1 Introduction

Many works in the literature have addressed the problem of provisioning point-to-point (p2p) demands in transparent wavelength division multiplexing (WDM) optical networks, which entails solving the routing and wavelength assignment (RWA) problem subject to the wavelength-continuity constraint (WCC). Authors in [Le96] propose to use a dedicated graph for each of the wavelengths available in the spectrum, that is, assuming 50 GHz of spectrum granularity, the number of wavelengths in every link is 100, which is the number of parallel graphs that need to be considered. That algorithm has also been used to compute link disjoint path pairs for path protection in [Yu05] and [Ve09]. In such an RWA algorithm, a shortest path algorithm, for example, Dijkstra's algorithm [Bh99] (see Chapter 2), is used to compute the shortest path in each parallel graph. Wavelength assignment is performed by choosing the wavelength that provides the shortest of all these paths; a number of heuristics have been proposed to decide the final wavelength assignment including the first fit (FF) one [Zh00].

However, in elastic optical networks (EONs) based on flexgrid technology with finer spectrum granularity (12.5 GHz), the number of parallel graphs that would need to be considered to represent the spectrum slices increases noticeably and thus the parallel graphs solution becomes impractical. In view of that, the authors in [Ji10] proposed a heuristic algorithm to be used in rings consisting of an RWA-based algorithm based on FF under the spectrum-continuity constraint. Authors in [Wa11.2] proposed three different routing and spectrum allocation (RSA) algorithms for mesh networks. However, they explicitly omit different modulation formats, physical layer constraints like distance as a function of the requested bitrate, slice width, and guard bands between adjacent connections to avoid spectral clipping caused by the cascaded optical filters in the nodes.

Authors in [Wa11.1] studied the influence of network traffic with a mix of bitrate requests on the optimal width of frequency slices. In fact, the optimal frequency slice width needed to be investigated since a large slice width wastes spectrum and reduces network capacity, whereas a small slice width may improve the spectrum utilization but imposes higher hardware requirements.

They concluded that the optimal slice width could be computed as the greatest common factor of the spectrum width requested by the optical connections. In a similar work, the authors in [Sa11.1] studied the network performance as a function of the slice width. They showed that a slice width of 12.5 GHz results in the best trade-off between grid granularity and blocking probability when all connections requested 400 Gb/s. However, such statements should be somehow clarified, weighting better the number of connections per bitrate.

All these works assumed that transparent connections could be established subject to feasible end-to-end optical transmission. However, when transparent connectivity is not possible, reamplifying, reshaping, and retiming (3R) regenerators need to be used to guarantee the Quality of Transmission (QoT). In such translucent networks, the Impairment-Aware RSA (IA-RSA) becomes a key online provisioning algorithm to be solved. The objective of the IA-RSA problem is to find lightpaths for connection requests fulfilling the requirement of adequate optical signal quality, for example, measured in terms of Optical Signal to Noise Ratio (OSNR) while achieving optimal network resource utilization.

It is worth recalling that one of the main advantages of EONs is the capability to allocate spectrum resources elastically according to traffic demands. Therefore, an alternative to deal with dynamic traffic is to manage allocated spectrum resources of existing connections in an elastic way, so the study of EONs to support connections that vary in time is of particular interest. Indeed, the resources may be used efficiently, firstly, because of the finer granularity of flexgrid, which allows that the allocated spectrum can fit the required bitrate closely and, secondly, due to the elastic (adaptive) spectrum allocation (SA) in response to traffic variations.

This chapter concentrates on dynamic p2p connection provisioning and presents:

1) efficient algorithms to solve the dynamic p2p provisioning in EON, based on the RSA and the routing, modulation, and spectrum allocation (RMSA) problems introduced in Chapter 3;
2) an IA-RSA algorithm relying on the OSNR computation to evaluate the quality of the signal conveyed by optical connections; OSNR is used to decide whether regenerators need to be used, a key decision when 3R are scarce and sparsely allocated;
3) dynamic elastic connections to facilitate elastic SA in dynamic scenarios by extending existing static SA policies, so as to explore the ability to request elastic operations on currently established connections dynamically.

The optimization problems tackled in this chapter are based on the basic offline provisioning problems introduced in Chapter 3, and, therefore, description of the mathematical formulation is avoided. The notation of the algorithms is mainly based on that proposed for the slot assignment arc-path RSA and RMSA formulations described in Chapter 3.

5.2 Provisioning in Transparent Networks

In this section, the basic dynamic provisioning algorithm is first stated, which is similar to that presented in Chapter 3. Algorithms based on both the RSA and RSMA problems are detailed to solve such a problem and simulation results are provided to validate its performance. An analysis of the slice width for a wide range of traffic profiles (TPs) is eventually presented.

5.2.1 Problem Statement

The problem of provisioning a set (bulk) of demands (or connection requests) can be stated as follows:

Given:

- an EON, represented by a directed connected graph $G(N, E)$, N being the set of optical nodes and E the set of fiber links connecting two nodes in N.
- the availability of every frequency slice in the optical spectrum of every link in E;
- set of demands D to be served, each demand $d \in D$ defined by the tuple $<o_d, t_d, b_d>$, o_d and t_d being the origin and destination nodes, respectively, and b_d the requested bitrate.

Output: the lightpath $l_d = <p, c>$ for each of demand d, where p represents the route and c the SA (frequency slot).

Objective: to minimize the amount of rejected bitrate, while minimizing the used spectrum resources.

The problem needs to be solved in stringent times (e.g., tens to hundreds of milliseconds); as a consequence, the Integer Linear Programming (ILP) formulations presented in Chapter 3 cannot be used, and some heuristic algorithm needs to be devised. Before entering into the details of the heuristic algorithm, however, algorithms for solving the RSA and the RMSA for a single demand are needed.

5.2.2 Dynamic RSA Algorithm

An approach to solve the RSA problem for a single demand consists in facing both RSA problems separately: for the routing problem, a shortest path algorithm, for example, k-shortest paths (kSPs) [Ye71], can be adapted to include spectrum availability, whereas for SA, any existing heuristic can be implemented, for example, FF or random selection.

The kSP algorithm needs to be slightly changed to deal with spectrum availability. Specifically, the underlying Dijkstra's shortest path algorithm in the kSP should extend the distance it considers to label nodes. Recall that each node i is labeled with the aggregated distance $dist(i)$ from the source node o and with its predecessor $pred(i)$. Consequently, the route o-i, defined by the subset of links $E(o, i) \subseteq E$, can be computed repetitively visiting the predecessor node starting from i until the source node o is reached (see Chapter 2). At this point, let η_{es} be equal to 1 if slice s in link e is available, 0 otherwise, and let $S(c)$ be the set of contiguous slices in slot c. Then, labels can be augmented with $\eta_s(i)$, the aggregated state of slice s ($\eta_s(i) = \prod_{e \in E(o,i)} \eta_{es}$) for each $s \in S$. The downstream node j of node i updates the label only if at least one slot is available as described in equation (5.1), that is, only if $\sigma(i, j) = 1$.

$$\sigma(e = (i, j)) = \begin{cases} 1 & \exists c \in C(d) : \eta_s(i) \cdot \eta_{es} = 1 \quad \forall s \in S(c) \\ 0 & \text{otherwise} \end{cases} \tag{5.1}$$

The extended kSP algorithm returns now up to kSPs and the width of the smaller frequency slot available between source and destination nodes, such that every path has at least one available frequency slot of the required width. Table 5.1 presents the algorithm to solve the RSA problem for a single demand. A set of kSPs is first computed between the end nodes of the demand (line 1 in Table 5.1). If no path satisfies the previous constraint, the connection request is blocked (line 2); otherwise, the shortest path is chosen (line 3) and a slot of the proper width is selected (line 4). The computed lightpath is eventually returned (line 5).

The complexity of the RSA algorithm strongly depends on that of the selected kSP algorithm; in this regard, Yen's kSP algorithm [Ma03] with complexity $O(k \cdot |N| \cdot (|E| + |N| \cdot \log |N|))$ can be used.

5.2.3 Dynamic RMSA Algorithm

The previous provisioning algorithm can be extended to include modulation format selection. To this aim, a set M of modulation formats, where M is ordered by its spectral efficiency, must be provided. The output includes the selected modulation format m as part of the solution.

Table 5.2 presents the algorithm to solve the RMSA problem for a single demand. A set of shortest paths is first computed between end nodes (line 1 in

Table 5.1 RSA algorithm.

```
INPUT:  G(N, E), <o, t, b>
OUTPUT: <p, c>
```
```
1: Q = {<p, n>} ← kSP(G, o, t, width(b))
2: if Q = ∅ then return ∅
3: p ← first(Q).p
4: c ← selectSlot(p, width(b))
5: return <p, c>
```

Table 5.2 RMSA algorithm.

INPUT: $G(N, E)$, $<o, t, b>$, M
OUTPUT: $<p, m, c>$

1: $Q = \{<p, n>\} \leftarrow$ kSP(G, o, t, width(b, first(M)))
2: $Q' \leftarrow \emptyset$
3: **for each** q **in** Q **do**
4:     $q.m \leftarrow 0$
5:     **for each** m **in** M **do**
6:         **if** len($q.p$) $\leq$ len(m) AND width(b, m) $\leq q.n$ **then**
7:             $q.m \leftarrow m$
8:             **break**
9:     **if** $q.m \neq 0$ **then** $Q' \leftarrow Q'$ U $\{q\}$
10: **if** $Q' = \emptyset$ **then return** $\emptyset$
11: sort(Q', $|q.p|$, ASC)
12: $q \leftarrow$ first(Q')
13: $q.c \leftarrow$ selectSlot($q.p$, width(b, $q.m$))
14: **return** q

Table 5.2); each path includes its physical route p and the width of the largest continuous slot in that route, n, such that n is at least the width of the slot required if the most efficient modulation format is used. Next, the best modulation format is selected from set M provided that the reach works for the length of the route and the available slot width in the route supports that modulation format (lines 3–9).

If no path satisfies the previous constraints, the connection request is blocked (line 10); otherwise, the set of paths found is sorted by the length of its route, and the best path is selected (lines 11–12). A slot of the proper width is selected (line 13) and the computed lightpath is eventually returned (line 14).

Note that modulation format selection does not entail the addition of significant complexity to the previous RSA algorithm, and consequently we can assume that both algorithms present comparable complexity.

5.2.4 Bulk RSA Algorithm

Once the RSA and RMSA algorithms have been defined, we can review the bulk RSA algorithm that computes a lightpath for every demand in a given set. Although meta-heuristics can be used for such algorithm, because of the stringent times in which a solution is needed, a simple bulk path computation algorithm based on the random heuristic presented in Chapter 2 can be used.

The algorithm (Table 5.3) performs a number of iterations that depends on the configured limit *maxIter* and the number of demands (denoted as $|D|$). If the number of distinct permutations of elements in D (i.e., $|D|!$) is lower than

Table 5.3 Bulk RSA heuristic algorithm.

INPUT: $G(N, E)$, D, $maxIter$
OUTPUT: $bestS$

```
 1: bestS ← ∅; random ← True
 2: if |D|! < maxIter then
 3:   maxIter ← |D|!
 4:   random ← False
 5: for i = 1..maxIter do
 6:   sol ← ∅; G' ← G
 7:   if not random then permute(D, i)
 8:   else shuffle(D)
 9:   for each d ∈ D do
10:     rsa = <k, cₖ> ← getRSA(G', d.o, d.t, d.b)
11:     if rsa = ∅ then continue
12:     allocate(G', d, rsa)
13:     sol ← sol ∪ {d}
14:   if Φ(sol) > Φ(bestS) then
15:     bestS ← sol
16: return bestS
```

---

maxIter, demands are sorted according to the *i*th permutation at iteration *i*; otherwise, demands in the bulk are randomly sorted at each iteration (lines 2–8 in Table 5.3). Following the resulting sequence, the RSA problem is solved for each demand and the resources are allocated (lines 9–13). At the end of each iteration, the objective function is computed, and the best solution is updated (lines 14–15). Finally, the best solution is returned (line 16).

Note that the algorithm discussed earlier can be upgraded to consider modulation formats by simply substituting the RSA algorithm in line 6 by the RMSA one.

5.2.5 Illustrative Results

5.2.5.1 Scenario

In this section, we evaluate the performance, in terms of grade of service, of implementing frequency slices of different widths, namely, 50, 25, 12.5, and 6.25 GHz, with the objective of finding the best spectrum granularity. To this end, the dynamic RSA algorithm, defined in Section 5.2.2, was developed and integrated into an ad hoc event-driven simulator based on OMNeT++ [OMNet]. A dynamic network environment is simulated on three realistic national optical network topologies shown in Figure 5.1: the 21-node and 35-link Telefonica (TEL) topology, the 20-node and 32-link BT topology, and the 21-node and 31-link DT topology. In all the experiments, the width of the available spectrum is set to 800 GHz and guard bands are neglected.

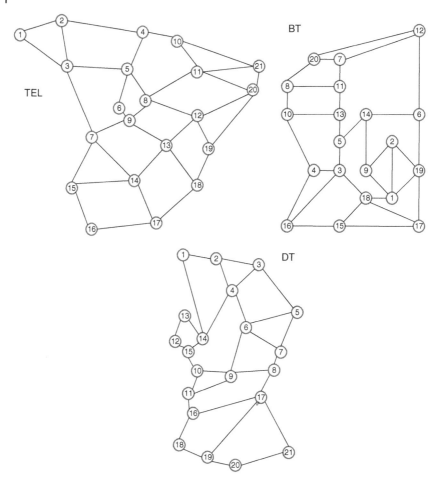

Figure 5.1 Sample optical network topologies used in this chapter: Spanish Telefonica (left), British Telecom (center), and Deutsche Telekom (right).

Since the studies performed in this section might be influenced by the mix of bitrates demanded by connection requests, different TPs are used (see details in Table 5.4). When TP-1 is selected, 400 Gb/s connections are not considered. Profiles TP-2 to TP-6 successively increase the requested on-average bitrate; this is performed at the expense of decreasing the amount of 10 Gb/s traffic while keeping constant the amount requested for each of the rest bitrates and so the amount of 400 Gb/s traffic increases. The weighted bitrate columns detail the contribution of each bitrate to the on-average bitrate of each profile.

Incoming connection requests arrive following a Poisson process and are sequentially served without prior knowledge of future incoming connection

Table 5.4 Traffic profiles.

Traffic profile	Average bitrate (Gb/s)	10 Gb/s		40 Gb/s		100 Gb/s		400 Gb/s	
		Connections (%)	Weighted bitrate	Connections (%)	Weighted bitrate	Connections (%)	Weighted bitrate	Connections (%)	Weighted bitrate
TP-1	22.2	74.1	7.4	18.5	7.4	7.4	7.4	0.0	0.0
TP-2	24.1	80.0	8.0	13.4	5.4	5.4	5.4	1.3	5.4
TP-3	38.4	60.0	6.0	27.0	10.8	10.8	10.8	2.7	10.8
TP-4	52.0	40.0	4.0	40.0	16.0	16.0	16.0	4.0	16.0
TP-5	66.1	20.0	2.0	53.4	21.4	21.4	21.4	5.3	21.4
TP-6	80.0	0.0	0.0	66.7	26.7	26.7	26.7	6.7	26.7

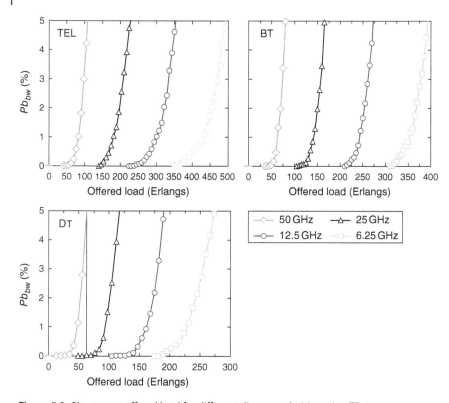

Figure 5.2 Pb_{bw} versus offered load for different slice granularities using TP-1.

requests. The holding time of connections is exponentially distributed with the mean value equal to 2 h. Source/destination pairs are randomly chosen with equal probability (uniform distribution) among all network nodes. Different values of offered network load are considered by changing the arrival rate while keeping the mean holding time constant. The bitrate demanded by each connection request is 10, 40, 100, or 400 Gb/s and it is chosen following one of the TPs previously described.

5.2.5.2 Optimal Slice Width Analysis

We conducted a number of simulations using the previously described configuration and TPs. To integrate blocking of different requested bitrates into a single figure of merit we use a bandwidth-weighted blocking probability (Pb_{bw}), which accounts for the total amount of blocked bitrate with respect to the total requested. For all the bandwidth-weighted blocking probability results, the simulation stopped when a confidence interval of at least 5% and a confidence level of 95% were achieved.

Figure 5.2 shows the influence of the slice width on the performance, in terms of Pb_{bw}, of each considered network topology as a function of the network load

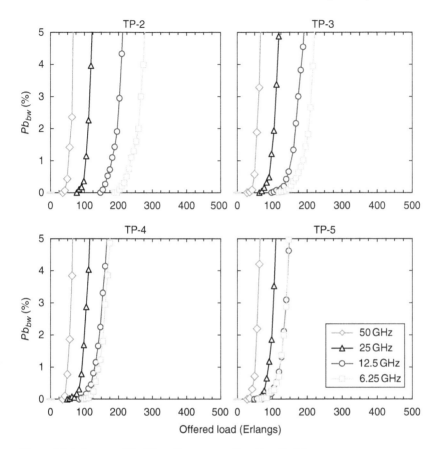

Figure 5.3 Influence of the TP on Pb_{bw} versus offered load in Erlangs.

(in Erlangs) using traffic profile TP-1. As illustrated, as soon as the frequency slice width is reduced, the amount of traffic served for a given Pb_{bw} notably increases. This is because more efficient spectrum utilization is achieved by reducing the slice width for this TP. Note that most of the connections request 10 Gb/s, which can be served using one single slice, even in the narrowest slice width scenario.

Graphs in Figures 5.3 and 5.4 study the influence of the TP on the performance of the TEL network. Four TPs are used (TP-2 to TP-5), each increasing the amount of 400 Gb/s traffic at the expense of that of 10 Gb/s, thus increasing the on-average bitrate requested. Plots in Figure 5.3 show the bandwidth-weighted blocking probability as a function of the offered load to the network. Pb_{bw} values remain similar for each of these TPs when the wider slice widths (50 and 25 GHz) are used. In contrast, they clearly decrease when the narrower slice widths (12.5 and 6.25 GHz) are used. The opposite can be observed in Figure 5.4, where the offered load to the network is expressed in terms of total

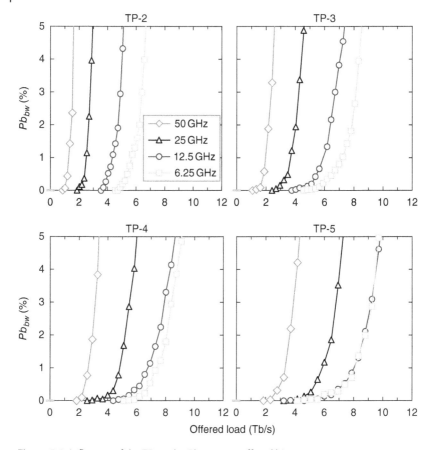

Figure 5.4 Influence of the TP on the Pb_{bw} versus offered bitrate.

bitrate offered to the network. In general, more bitrate is served (in any slice width option) when the traffic increases from TP-2 to TP-5.

The reasons are that, on the one hand, the effect of the more efficient spectrum utilization obtained for TP-1 is gradually reduced as the on-average requested bitrate is increased; on the other hand, spectrum fragmentation increases as a consequence of requesting connections for a larger number of frequency slices (up to 32 slices). In fact, as shown in Figure 5.4, in the case of requesting TP-5, the advantage of using the narrowest slice width (6.25 GHz) observed under other TPs is canceled and similar results as using 12.5 GHz are obtained.

Finally, from Figure 5.5 we can see that under TP-6 the performance when using 12.5 and 6.25 GHz slice widths is quite close for all three networks under study. In view of these results, spectrum granularity of 12.5 GHz is the clear candidate to be used in EONs.

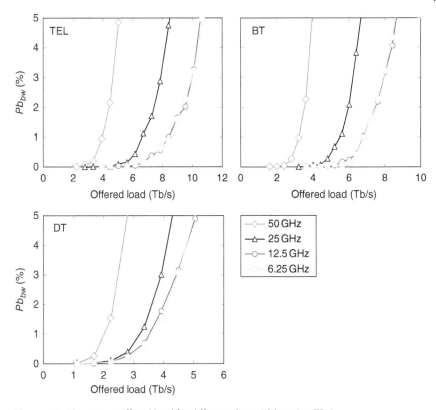

Figure 5.5 Pb_{bw} versus offered load for different slice widths using TP-6.

5.3 Provisioning in Translucent Networks

Several impairment-aware routing and wavelength assignment (IA-RWA) algorithms have been proposed in the literature for WDM networks, being roughly sorted according to the constraint used to estimate/evaluate the QoT value. The first category considers IA-RWA algorithms, which aim at not exceeding a maximum transparent distance [Sh02]. The second category includes those IA-RWA algorithms that compute QoT indirectly by estimating or analytically calculating one or several physical indicators, such as OSNR, estimated bit error rate (BER), or the polarization mode dispersion [Ya05.1], [Ya05.2].

With this in mind, in this section, we consider dynamic translucent networks, where a number of regenerators are available, and study the RSA problem taking into account physical impairments (IA-RSA). Similarly, as in [Ya05.1] for WDM, we assume a distributed regeneration approach, wherein hybrid nodes constitute the translucent EON, where each hybrid node is made up of an all-optical switching core and a pool of 3R regenerators at some optical ports. Consequently, a lightpath traversing a hybrid node can be optically switched to the outgoing port or electronically regenerated before going to the output port.

Before describing the IA-RSA problem, we need to introduce some additional notation. We are given a graph describing the network topology and the state of the resources (frequency slices at every link and regenerators availability at every node). Similarly, as in the previous sections, we represent that network topology by the graph $G(N, E)$. In addition, let S be the set of available frequency slices in each link $e \in E$. Let $N_R \subseteq N$ be the subset of nodes with regeneration capability, and conversely $N_T \subseteq N$ the subset of nodes without regeneration capability; thus $N = N_R \cup N_T$ and $N_R \cap N_T = \varnothing$. Additionally, we are given a pair of source and destination nodes $\{s, t\}$ for the connection being requested.

In our model, the OSNR represents the QoT parameter, which allows evaluating the feasibility of the transparent segments. Specifically, the computed $Total\_OSNR$ is compared to a threshold OSNR level ($OSNR\_thr$) to determine whether the physical impairment constraint is fulfilled at the receiver node:

$$Total\_OSNR_N \geq OSNR\_thr \qquad (5.2)$$

If equation (5.2) is not satisfied, available 3R regenerators are allocated at intermediate nodes to improve the signal quality.

The IA-RSA problem consists in finding a feasible set of lightpaths for the requested connection, so as to minimize the number of regenerators needed to guarantee the QoT for that connection. As a secondary objective, the length of the route, in terms of the number of hops, should be minimized.

To fulfill the regenerators minimization objective, we build an auxiliary digraph $\mathcal{D}_{st}(\mathcal{N}_{st}, \mathcal{A})$, where $\mathcal{N}_{st} = N_R \cup \{s, t\}$ and $\mathcal{A}$ is the set of directed arcs. Each arc $a = (u, v) \in \mathcal{A}$ connects two nodes $u, v \in \mathcal{N}_{st}$, where $u \in N_R \cup \{s\}$ and $v \in N_R \cup \{t\}$. An arc $a = (u, v)$ exists only if a lightpath can be found between u and v in G, where the route has an acceptable OSNR level.

For illustrative purposes, Figure 5.6 shows an example of auxiliary digraph construction. There, digraph $\mathcal{D}_{st}$ is built upon the reception of a connection request between nodes s and t. To this end, the set N_R with those nodes with regeneration capabilities currently available (grey circles) and nodes s and t belong to $\mathcal{N}_{st}$. Directed arcs are created connecting s and nodes in N_R to t and other nodes in N_R, provided that a feasible lightpath exists in G. In such scenario, routes $\{s, 2, t\}$ and $\{s, 5, t\}$ minimize the number of regenerators used, whereas other feasible routes such as $\{s, 2, 6, t\}$, need more regenerators than the previous ones. Note that every segment in the route represents a transparent lightpath that has to be computed on graph G. To expand every segment, we compute the RSA between their ends on G, where each node and link in G has been labeled with the inverse of its linear OSNR level to minimize OSNR.

Finally, to solve the IA-RSA problem, we use the Translucent IA-RSA algorithm described in Table 5.5. We first compute a set of kSPs between s and t in the auxiliary graph $\mathcal{D}_{st}$, so that the number of hops ($|\mathcal{A}(k)|$) equals that of

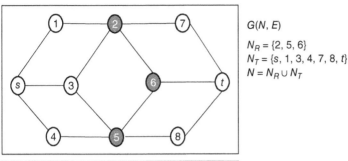

$G(N, E)$

$N_R = \{2, 5, 6\}$
$N_T = \{s, 1, 3, 4, 7, 8, t\}$
$N = N_R \cup N_T$

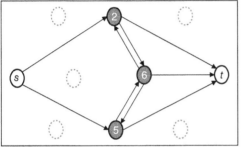

$\mathcal{D}_{st}(\mathcal{N}_{st}, \mathcal{A})$

$\mathcal{N}_{st} = \{s, 2, 5, 6, t\}$

Figure 5.6 Original (top) and auxiliary (bottom) graphs.

Table 5.5 Translucent IA-RSA algorithm.

INPUT: $<s, t, b>$, $G(N, E)$, $\mathcal{D}_{st}(\mathcal{N}_{st}, \mathcal{A})$
OUTPUT: *lightpath*

```
 1: SP ← Shortest Path (in terms of hops) in 𝒟st
 2: if SP not found then
 3:     return BLOCK_REQUEST
 4: K ← {k-shortest paths (hops) in 𝒟st: |𝒜(k)|=|𝒜(SP)|}
 5: Q ← ∅
 6: for each k ∈ K do
 7:     q ← ∅
 8:     for each a=(u,v) ∈ 𝒜(k) do
 9:         q ← q ∪ {getRSA(G, u, v, b)}
10:         if v ≠ t then
11:             use one regenerator in v
12:     Q ← Q ∪ {q}
13: return x∈Q: ∀q∈Q |E(x)| ≤ |E(q)|
```

---

the computed shortest path (lines 1–4). This ensures that the number of regenerators is minimized (first objective). Next, each route k is transformed to a translucent optical connection in graph G, expanding each transparent optical segment and using regenerators to stitch the segments (lines 5–12). The translucent connection minimizing the total number of hops $|E(x)|$ (second objective) is eventually returned (line 13).

A major concern regarding this algorithm is scalability. Note that an auxiliary digraph $\mathcal{D}_{st}$ needs to be built to solve the IA-RSA problem for each single connection request. To solve that issue, a set of auxiliary digraphs $\mathcal{D}_s(N, \mathcal{A})$ can be computed offline and stored to be used upon the reception of connection requests. Note that $|N|$ digraphs are built, one $\mathcal{D}_s$ for each node s in N. In addition, although each digraph $\mathcal{D}_s$ contains all the nodes, s being the source one, the rules to construct each of the directed arcs are the same as earlier; only s and nodes in N_R can be sources of arcs. At this stage, N_R contains the set of nodes where regeneration capabilities have been installed. When a connection request (s, t) arrives, the algorithm gets the precomputed digraph $\mathcal{D}_s$ and updates regeneration availability removing those arcs (u, v) in $\mathcal{A}$ where $u \neq s$ and u does not have regenerators currently available. Next, the algorithm in Table 5.5 is executed taking into account that solving the RSA problem for transparent segments may result in unfeasible routes (due to accumulated impairments or the lack of resources) as a consequence of the current state of the resources.

5.4 Dynamic Spectrum Allocation Adaption

Although optical networks can manage dynamic connection requests, the network ability to deal with elastic operations to support the dynamic elastic connectivity model needs to be studied. To that end, in this section, we study how connections requiring dynamic SA can be managed in EONs when their required bitrate varies once in operation.

In this context, adaptive SA with known a priori 24 h traffic patterns was addressed in [Ve12.2], [Kl13], [Sh11] and spectrum adaptation under dynamic traffic demands was studied in [Ch11], [Ch13]. Concurrently, different policies for elastic SA were proposed, including symmetric [Ve12.2], [Kl13], [Sh11] and asymmetric [Ve12.2], [Kl13], [Ch13] spectrum expansion/reduction around a reference frequency, as well as the entire spectrum reallocation policy [Ve12.2], [Kl13]. First, the authors in [Ve12.2] and later those in [Kl13] showed that in EON with time-varying traffic and under a priori known (deterministic) traffic demands, elastic SA policies performing lightpath adaptation (symmetric, asymmetric, or reallocation-based) allow to serve more traffic than when fixed SA is applied.

A different approach was described in [Ch11], where an EON based on Orthogonal Frequency-Division Multiplexing (OFDM) was compared against traditional WDM. The authors assumed that traffic between nodes was known based on a probabilistic traffic model and, accordingly, a set of subcarriers were prereserved between each pair of nodes. In addition, a set of not prereserved subcarriers that are shared among the different connections can be allocated/deallocated to/from connections to satisfy demands' requirements.

In this section, we assume that a TP is characterized by a guaranteed bitrate and a range of variation, given for a considered set of time periods. Therefore, lightpaths can be preplanned, based, for example, on the offline provisioning problems presented in Chapter 3, and then established on the network.

We allow that several parallel optical connections may exist between the same origin/destination pair to serve the requested bitrate. Once in operation, however, the established lightpaths must elastically adapt their capacity, so as to convey as much bitrate as possible from the demanded one. In particular, each change in traffic volume resulting in a change in the amount of optical spectrum of a lightpath (i.e., spectrum expansion/reduction) can require the network controller to find the appropriate SA for that lightpath. In response to this request, a dynamic SA algorithm implemented in the controller is in charge of adapting the SA in accordance with the SA policy that is used in the network.

5.4.1 Spectrum Allocation Policies

The considered SA policies for time-varying traffic demands in dynamic scenarios are shown in Figure 5.7. Each policy puts the following restrictions on the assigned central frequency (CF) and the allocated spectrum width, in particular:

- *Fixed* (Figure 5.7a): both the assigned CF and the spectrum width do not change in time. At each time period, demands may utilize either whole or only a fraction of the allocated spectrum to convey the bitrate requested for that period.
- *Semi-Elastic* (Figure 5.7b): the assigned CF is fixed, but the allocated spectrum may vary. Here, spectrum increments/decrements are achieved by allocating/releasing frequency slices symmetrically, that is, at each end of the already allocated spectrum while keeping invariant the CF. The frequency slices can be shared between neighboring demands, but used by, at the most, one demand in a time interval.
- *Elastic* (Figure 5.7c): asymmetric spectrum expansion/reduction (with respect to the already allocated spectrum) is allowed, and it can lead to short shifting of the CF. Still, the relative position of the lightpaths in the spectrum remains invariable, that is, no reallocation in the spectrum is performed.

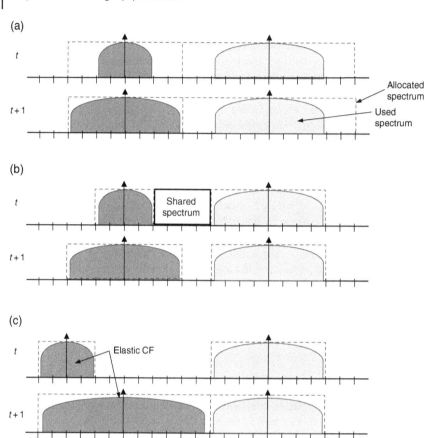

Figure 5.7 Three SA policies for time-varying traffic in a flexgrid network. Two time intervals are observed: t before and $t+1$ after spectrum adaptation has been performed. (a) Fixed, (b) semi-elastic, and (c) elastic.

5.4.2 Problem Statement

The problem of dynamic lightpath adaptation can be formally stated as:

Given:

- an EON topology represented by a graph $G(N, E)$;
- a set S of available slices of a given spectral width for every link in E;
- a set L of lightpaths already established on the network; each lightpath l is defined by the tuple $<R_l, f_l, n_l>$, where the ordered set $R_l \subseteq E$ represents its physical route, f_l its CF, and n_l the number of frequency slices;
- a lightpath $p \in L$ for which spectrum adaptation request arrives and the required number of frequency slices, $(n_p)^{req}$.

Output: the new values for the SA of the given lightpath p: $<R_p, f_p, (n_p)'>$ and $<R_p, (f_p)', (n_p)'>$, respectively, if the *Semi-Elastic* and *Elastic* policies are used.

Objective: maximize the amount of bitrate served.

5.4.3 Spectrum Adaption Algorithms

For the *Fixed* SA policy, the allocated spectrum does not change in time; therefore, any fraction of traffic that exceeds the capacity of the established lightpath is lost. Regarding the *Semi-Elastic* and *Elastic* policies, the corresponding lightpath adaptation algorithms are presented in Tables 5.6 and 5.7, respectively. In the following, we discuss the details of these algorithms:

- *Semi-Elastic* algorithm: the elastic operation is requested for a lightpath p and the required number of frequency slices to be allocated is given. Because f_p must be kept invariant, $(n_p)^{req}$ must be an even number. If elastic spectrum reduction is requested, the tuple for lightpath p is changed to $<R_p, f_p, (n_p)^{req}>$ (lines 1–2). In the opposite, when an elastic expansion is requested, the set of spectrum adjacent lightpaths at each of the spectrum sides is found by iterating on each of the links of the route of p (lines 4–7). The greatest value of available spectrum without CF shifting, n_{max}, is subsequently computed and the value of spectrum slices actually assigned to p, $(n_p)'$, is computed as the minimum between s_{max} and the requested one (lines 8–9). The tuple representing lightpath p is now $<R_p, f_p, (n_p)'>$.
- *Elastic* algorithm: the CF of p can be changed, so the difference with the *Semi-Elastic* algorithm explained earlier is related to that issue. Now, the value of

Table 5.6 Semi-elastic SA algorithm.

```
INPUT:  G(N, E), S, L, p, (n_p)^req
OUTPUT: (n_p)'

1:  if (n_p)^req ≤ n_p then
2:      (n_p)' ← (n_p)^req
3:  else
4:      L⁺ ← ∅, L⁻ ← ∅
5:      for each e ∈ R_p do
6:          L⁻ ← L⁻ ∪ {l ∈ L: e ∈ R_l, adjacents(l, p), f_l < f_p}
7:          L⁺ ← L⁺ ∪ {l ∈ L: e ∈ R_l, adjacents(l, p), f_l > f_p}
8:      n_max ← 2·min{min{availableSlicesLeft(f_p, f_l, n_l), l ∈ L⁻},
                      min{availableSlicesRight(f_p, f_l, n_l), l ∈ L⁺}}
9:      (n_p)' ← min{n_max, (n_p)^req}
10: return (n_p)'
```

Table 5.7 Elastic SA algorithm.

INPUT: $G(N, E)$, S, L, p, $(n_p)^{req}$
OUTPUT: $(f_p)'$, $(n_p)'$

1: **if** $(n_p)^{req} \leq n_p$ **then**
2:   $(n_p)' \leftarrow (n_p)^{req}$
3:   $(f_p)' \leftarrow f_p$
4: **else**
5:   $L^+ \leftarrow \emptyset$, $L^- \leftarrow \emptyset$
6:   **for each** $e \in R_p$ **do**
7:     $L^- \leftarrow L^- \cup \{l \in L: e \in R_l, \text{adjacents}(l, p), f_l < f_p\}$
8:     $L^+ \leftarrow L^+ \cup \{l \in L: e \in R_l, \text{adjacents}(l, p), f_l > f_p\}$
9:   $n_{max} \leftarrow \min\{\text{availableSlicesLeft}(f_p, f_l, n_l), l \in L^-\} +$
        $\min\{\text{availableSlicesRight}(f_p, f_l, n_l), l \in L^+\}$
10:   $(n_p)' \leftarrow \min\{n_{max}, (n_p)^{req}\}$
11:   $(f_p)' \leftarrow \text{findSA\_MinCFShifting}(p, (n_p)', L^+, L^-)$
12: **return** $(f_p)', (n_p)'$

---

n_{max} is only constrained by the number of slices available between the closest spectrum adjacent paths. Then, n_{max} is the sum of the minimum available slices along the links in the left side and the minimum available slices in the right side of the allocated spectrum (line 9). Finally, the returned value $(f_p)'$ is obtained by computing the new CF value so as to minimize CF shifting (line 11).

5.4.4 Illustrative Results

5.4.4.1 Scenario

In this study, we focus on the elastic capabilities of EONs supported by a network controller to facilitate the dynamic elastic connectivity model. We consider three scenarios representing different levels of traffic aggregation consisting in three different networks based on the TEL network presented in Figure 5.1: TEL21, TEL60, and TEL100, with 21, 60, and 100 nodes, respectively. The main characteristics of the topologies are presented in Table 5.8. For the ongoing experiments, the optical spectrum width was set to 1.2 THz.

For the sake of fairness in the following comparisons, each network topology was designed to provide approximately the same blocking performance at a given total traffic load. To this aim, the models and methods proposed in [Ve12.1] and [Ru13.1] to design core networks under dynamic traffic assumption were adapted and used to fit our preplanning needs. Note that, although the relation between traffic load and unserved bitrate is kept pretty similar in all defined scenarios, the number of core nodes strongly affects the level of

Table 5.8 Details of network topologies.

	Nodes	Links	Nodal degree
TEL21	21	35	3.33
TEL60	60	90	3.0
TEL100	100	125	2.5

traffic aggregation. Namely, since the analyzed core networks cover the same geographical area and the offered traffic is the same for each network, the lower the number of core nodes, the larger the aggregation networks, as well as the higher the flow grooming on lightpaths.

For the preplanning phase, the flows uniformly distributed among core node pairs were generated following the time-variant bandwidth distribution used in [Ve12.2]. In particular, three different TPs, defined by their h^{min} and h^{max} bitrate values ($TP_i = [h_i^{min}, h_i^{max}]$), were used to compute bitrate fluctuations of flows in time. The considered TPs with their bandwidth intervals are TP1 = [10, 100], TP2 = [40, 200], and TP3 = [100, 400].

Here, a granularity of 1 h was considered. Each flow belongs to one of these TPs (with the same probability), and bitrate fluctuations are randomly chosen following a uniform distribution within an interval defined by h^{min} and h^{max}. All traffic flows between a pair of core nodes are transported using the minimum number of lightpaths. Hence, more than one lightpath between each pair of nodes might be needed. Furthermore, the lightpaths' maximum capacity was limited to 100 Gb/s and to 400 Gb/s for the completeness of our study, and Quadrature Phase-Shift Keying was assumed as the unique modulation format.

The models proposed in [Kl13] were solved for each pair of generated traffic matrix and core network and for the proposed elastic policies in order to find the initial route and SA that minimize the amount of blocked bitrate. The solutions of the preplanning phase, that is, the routing and the initial SAs of the lightpaths, were used as input data for an OMNeT++-based EON simulator.

Similarly, as for the planning phase, a set of flows following the TPs defined earlier were configured in the simulator. For each flow, the requested bitrate varies randomly 10 times per hour, that is, a finer and randomly generated granularity both in time and in bitrate is considered in the simulator. Hence, bitrate variations appear randomly in the system where a grooming module ensures that all the bitrate can be served. If either more resources are needed for a lightpath or resources can be released, the grooming module of the aggregation network requests the network to perform an elastic operation on such lightpath. To this end, algorithms in Tables 5.6 and 5.7 are executed provided that either the *Semi-Elastic* or the *Elastic* policies are used; obviously, no lightpath adaptation algorithm is run when the *Fixed* policy is considered.

5.4.4.2 Efficiency of SA Policies

Firstly, the efficiency of SA policies is analyzed in the considered network scenarios assuming different lightpath capacity limits (100 and 400 Gb/s). Recall that the largest core network (i.e., TEL100) presents the lowest level of traffic aggregation and, as a consequence, the highest variability of traffic offered to the core.

Figure 5.8 illustrates the accumulated percentage of unserved bitrate as a function of the offered traffic load represented by the average load for each core network and lightpath capacity limit.

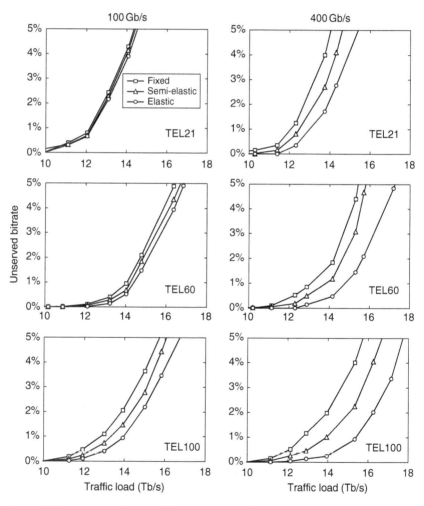

Figure 5.8 Percentage of unserved bitrate against traffic load for TEL networks when lightpaths' capacity is limited to 100 and 400 Gb/s.

Table 5.9 Load gain of adaptive SA policies versus fixed SA at 1% of unserved bitrate.

Maximum capacity	SA policy	TEL21	TEL60	TEL100
400 Gb/s	Elastic	7.73%	12.99%	20.99%
	Semi-elastic	3.25%	5.67%	9.34%
100 Gb/s	Elastic	0.92%	2.47%	8.83%
	Semi-elastic	0.60%	1.24%	3.85%

We observe that when the aggregation level is high (TEL21), and lightpath capacity is limited to 100 Gb/s, all the SA policies offer similar performance. The rationale behind that is the low traffic variability in this scenario and, therefore, the network does not benefit from adaptive SA. Furthermore, since the volume of aggregated traffic is large and the lightpath capacity is low, a large number of lightpath connections are established between each pair of nodes where a relatively large fraction of these lightpaths is always saturated; hence, elasticity is not required in that scenario. Nonetheless, as soon the aggregation level is gradually increased in TEL60 and TEL100 networks, the benefits of elasticity, although rather limited, also increase.

The advantages of elasticity clearly appear when the maximum capacity of the lightpaths is increased to 400 Gb/s. Starting from the limited benefits observed when the aggregation level is high (TEL21), unserved bitrate remarkably decreases when that level is reduced in TEL60 and especially in TEL100.

Regarding SA policies, in all the analyzed scenarios, the *Elastic* SA policy outperforms the *Semi-Elastic* SA policy which, on the other hand, performs better than *Fixed* SA.

Table 5.9 reports in detail the performance gain of elastic SA policies with respect to the fixed SA for the network load, which results in 1% of unserved bitrate. Interestingly, *Elastic* SA allows serving up to 21% of traffic more than *Fixed* SA in the TEL100 scenario with 400 Gb/s lightpath capacity limit. In the same scenario, the gain when using the *Semi-Elastic* SA policy is just over 9%.

5.4.4.3 Study of Lightpath Variability

Analysis of the relationship between traffic aggregation level and the variability of each lightpath along time is interesting. Toward this end, the set of operating lightpaths is divided into three subsets, named L1–L3, depending on the degree of bitrate variations in time:

- The L1 subset contains all those lightpaths whose bitrate variations are almost inappreciable, and thus the amount of spectrum resources is constant. This is mainly the case of fully loaded lightpaths grooming several demands, where statistical multiplexing keeps constant the total bitrate at maximum level.

Table 5.10 Distribution of lightpaths according to the level of variability.

Maximum capacity	Lightpath type	TEL21	TEL60	TEL100
100 Gb/s	L1	19.0%	15.2%	14.3%
	L2	34.9%	41.3%	42.9%
	L3	46.2%	43.5%	42.9%
400 Gb/s	L1	0.0%	0.0%	0.0%
	L2	94.8%	98.0%	100.0%
	L3	5.2%	2.0%	0.0%

- The L2 subset includes lightpaths where bitrate variations are dramatic within some range of minimum and maximum values. While aggregated traffic can reach the maximum bitrate at a certain time, a minimum amount of traffic is always present.
- The L3 subset contains those lightpaths where the highest relative bitrate fluctuations are observed, changing instantly from not being used to a peak where several spectrum slices are required.

Table 5.10 details the distribution of lightpaths belonging to each of the subsets defined earlier. These results are consistent with the discussion carried out in the analysis of the efficiency of SA policies, and similar arguments can be used.

Firstly, there is a higher percentage of lightpaths of type L1 in the scenario with low lightpath capacity limit (100 Gb/s) compared to the scenario where 400 Gb/s lightpaths are allowed. This fact considerably reduces the number of lightpaths needing elasticity and thus the performance of the fixed policy is close to the elastic SA policies. That percentage decreases when the aggregation level is decreased (see TEL100 versus TEL21).

Secondly, when 400 Gb/s lightpaths can be used, virtually all lightpaths belong to set L2, being just a small fraction of lightpath in set L3. Hence, lightpath capacity is always subject to changes, there are no saturated lightpaths, and only some lightpaths experience high capacity variations. In contrast, traffic fluctuations of lightpaths when the limit of 100 Gb/s is applied are much higher (see the percentage of lightpaths in set L3), especially in the TEL100 scenario in which traffic aggregation is low and traffic variability is higher.

5.5 Conclusions

In this chapter, several algorithms for dynamically provisioning p2p demands in EONs were presented. Specifically, both RSA and RMSA algorithms to be used in dynamic flexgrid network were reviewed. Their design included adaptive considerations involving the optical signal to be conveyed by the optical connection, such as the use of modulation formats, distance constraints, cascade filters, and guard bands.

The bulk RSA/RMSA heuristic algorithm that solves the RSA/RMSA problem for a set of demands was then presented and integrated into an event-driven simulator, where a dynamic network environment was simulated for three networks under study. The best frequency slice width was analyzed as a function of the TP to be served by each of the optical networks under consideration. Interestingly, the 12.5 GHz slice width proved to provide the best trade-off between performance and complexity of the hardware.

Next, an IA-RSA algorithm to compute lightpaths in translucent networks was presented. The algorithm made use of auxiliary digraphs to minimize regeneration usage and path length in terms of number of hops.

Finally, we reviewed elastic SA policies to dynamically adapt the spectrum allocated to a lightpath in response to changes in traffic demands. Two SA policies were analyzed, namely, a symmetric SA policy (referred to as *Semi-Elastic* SA) and an asymmetric SA policy (referred to as *Elastic* SA), and their performance against a *Fixed* SA policy, which does not allow for spectrum changes, was compared. For each elastic SA policy, a dedicated lightpath adaptation algorithm was presented. The evaluation was carried out by simulation on different network scenarios, which are characterized by different levels of traffic aggregation and lightpath capacity limits. Results showed that the effectiveness of lightpath adaptation in dealing with time-varying traffic highly depends on both the aggregation level and the maximum lightpath capacity. In a network with low traffic aggregation, the best performance gap of about 21% (versus the *Fixed* SA policy) was achieved for *Elastic* SA operating with high lightpath capacity limits.

6

Transfer-based Datacenter Interconnection

Adrián Asensio, Marc Ruiz and Luis Velasco

Universitat Politècnica de Catalunya, Barcelona, Spain

Although datacenter (DC) resource managers can request optical connections and control their capacity, a *notification-based* connectivity model brings noticeable costs savings. That connectivity model is useful when connectivity can be specified in terms of volume of data and completion time (we call this transfer-based connectivity).

To increase network resource availability, in this chapter, we propose an Application Service Orchestrator (ASO) as an intermediate layer between the cloud and the network. In this architecture, DC resource managers request data transferences using an application-oriented semantic. Then, those requests are transformed into connection requests and forwarded to the Application-based Network Operations (ABNO) in charge of the transport network. The proposed ASO can perform elastic operations on already established connections supporting transferences provided that the committed completion time is ensured. Polling- and notification-based models are compared in a scenario considering ASO.

Then, we propose not only using elastic operations incrementing transfer-based connections' capacity but also scheduling elastic operations, both incrementing and decrementing transfer-based connections' capacity. Transfer-based requests from different DC operators received by ASOs are routed and resource allocation scheduled to simultaneously fulfill their Service Level Agreement (SLA), which becomes the Routing and Scheduled Spectrum Allocation (RSSA) problem. An Integer Linear Programming (ILP) model is proposed, and an algorithm for solving the RSSA problem in realistic scenarios is designed. Results showing remarkable gains in the amount of data transported motivate its use for orchestrating DC interconnection.

Provisioning, Recovery, and In-operation Planning in Elastic Optical Networks,
First Edition. Luis Velasco and Marc Ruiz.
© 2017 John Wiley & Sons, Inc. Published 2017 by John Wiley & Sons, Inc.

Table of Contents and Tracks' Itineraries

6.1 Introduction

Datacenter (DC) federations allow cloud operators to reduce capital expenditures needed to build large infrastructures by accommodating dynamic cloud services workload in remote DCs, with the twofold objective of increasing their revenues from IT resources that would otherwise be underutilized and by reducing operational expenditures by reducing part of the huge energy consumption in DCs. Note that servers can be turned off when they are not used or turned on to satisfy increments in the demand [Ar10]. Therefore, cloud computing strongly depends on the simplicity to interconnect DCs, so that elastic cloud operations can be performed easily. Since the amount of data to be transported is huge, this is also an opportunity for network operators to increase their revenues; for example, Cisco global cloud index [CISCO12] forecasts inter-DC traffic to increase from $167\,EB$ ($1\,EB = 10^{18}$ bytes) in 2012 to $530\,EB$ in 2017. Part of inter-DC traffic is as a result of elastic cloud operations, where workloads, encapsulated in virtual machines (VMs), can be moved from one DC to another. Another source of inter-DC traffic is for synchronization among databases (DBs) placed in geographically distant DCs.

As a result of the huge volume of data being transferred, for example, several TB, during short periods of time, authors in [Ve13.2] proposed to use a flexgrid-based Elastic Optical Networks (EONs) to interconnect DCs; cloud resource managers can request connections of the desired bitrate, and tear them down as soon as they are not needed. Recall that the frequency slot width actually allocated to each flexgrid connection depends on the requested bandwidth and the used modulation format.

Owing to the fact that network resources might not be available at the requesting time, elastic connections were proposed in [Ve13.2] as a way to reduce the time needed to transfer the volume of data. As detailed in Chapter 5, elasticity can be implemented by dynamically modifying the spectrum allocated to an optical connection. The availability of programmable bandwidth-variable transponders (BVTs) capable of generating optical signals of the desired characteristics, including modulation format and bitrate, enables elastic connections.

Because dynamic operation frequently results in not optimal use of network resources, network reoptimization can be performed triggered by some event, for example, when a connection request cannot be served. Notwithstanding its feasibility, some other techniques can be applied to increase network efficiency to increase network operators' revenues.

As described in Chapter 4, scheduling algorithms inside DC resource managers are the entities taking decisions related to VM migration and DB synchronization. Once decided, connections to remote DCs need to be established; to this purpose, the Application-based Network Operations (ABNO) [ABNO] architecture includes a northbound interface for the service layer to request connections set up and tear them down. The allocated connection's capacity can be reduced with respect to the one requested in case not enough network resources are available at the requesting time, which increases time-to-transfer, which in turn frustrates cloud operators' expectations.

Given that scheduling algorithms usually run periodically to evaluate the performance of the services and the quality perceived by end users, connections can be scheduled in advance. In that regard, several works have proposed to use advance reservation, where requests specify a deadline to establish the connection; for example, the authors in [Lu13] propose algorithms for advance reservation requests in flexgrid-based networks. Nevertheless, the use of advance reservation involves both, delaying the connection's setup from the instant when data volume to be transferred is calculated and estimating the acceptable deadline, thus adding more complexity to the cloud.

A different approach is explored in this chapter; cloud resource managers request data transferences to be completed within a time interval (e.g., the period of scheduling algorithms). This application-oriented semantic liberates application developers from understanding and dealing with network specifics and complexity. Note that completion time relates to cloud operation, while setup deadline responds to network operation. In this architecture, DC resource managers request data transferences using an application-oriented semantic. An Application Service Orchestrator (ASO) on top of ABNO transforms transfer-based requests into connection requests that are forwarded to the ABNO in charge of the transport network.

In this chapter, we explore ASO functionalities by using scheduling techniques in addition to elastic operations, with the twofold objective of fulfilling Service Level Agreement (SLA) contracts and increasing the network operator's revenues by accepting more connection requests. To decide whether a transfer-based request is accepted or not, ASO solves the Routing and Scheduled Spectrum Allocation (RSSA) problem that augments the routing and spectrum allocation (RSA) problem introduced in Chapter 3. The RSSA problem decides the elastic operations to be done on already established connections so as to make enough room for the incoming request ensuring its requested completion time provided that the committed completion times of ongoing transferences are guaranteed.

6.2 Application Service Orchestrator

To provide an abstraction layer to the underlying network, a new stratum on top of the ABNO, the ASO, can be deployed (Figure 6.1). The ASO implements a northbound interface to request transfer operations. Those applications' operations are transformed into network connection requests. The northbound

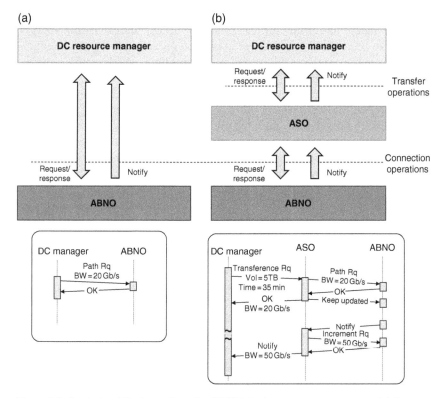

Figure 6.1 Control architecture where the ABNO interfaces resource managers (a). A proposed architecture where the ASO offers an application-oriented semantic interface (b).

interface uses application-oriented semantic, thus liberating application developers from understanding and dealing with network specifics and complexity.

As an example of that paradigm, we propose an operation where applications request transfers using cloud middleware native semantic, with the following:

- source DC,
- destination DC,
- the amount of data to be transferred,
- and completion time.

The ASO is in charge of managing inter-DC connectivity; if enough resources are not available at the requesting time, notifications (similar to interruptions in computers) are sent from the ABNO to the ASO each time specific resources are released. Upon receiving a notification, the ASO takes decisions on whether to increase the bitrate associated with a transfer. Therefore, we have effectively moved from a polling-based connectivity model where applications request connection operations to a notification-based connectivity model where transfer operations are requested to the ASO, which manages connectivity.

6.2.1 Models for Transfer-based Connections

We consider a scenario where a *follow-the-work* strategy for VM migration is implemented; VMs are moved to DCs closer to the users, aiming at reducing the user-to-service latency. Cloud scheduling algorithms run periodically taking VM migration decisions. Once the set of VMs to be migrated and DB synchronization needs are determined, cloud management performs those operations in collaboration with inter- and intra-DC networks.

Since scheduling algorithms run periodically, for example, every hour, VM migrations are required to be performed within each period. In fact, the shorter the transfer time, the better the offered network service is by having the VMs in their proper locations earlier.

ASO is deployed between the ABNO and DC resource managers (Figure 6.2) and implements *polling-based* and *notification-based* connectivity models.

In the polling-based model, each local DC resource manager manages connectivity to remote DCs, so as to perform VM migration in the shortest total time. The orchestrator in the source DC resource manager requests label-switched path (LSP) setup, teardown, as well as elastic operations to the ASO, which forwards them to the ABNO (Figure 6.3a). In this model, the ASO works as a proxy between applications and the network. After checking local policies, the ABNO performs LSP operations on the controlled network.

Although resource managers have full control over the connectivity process, physical network resources are shared with a number of clients, so LSP setup and elastic spectrum increments can be blocked as a result of lack of resources in the network. Hence, resource managers need to implement some periodical

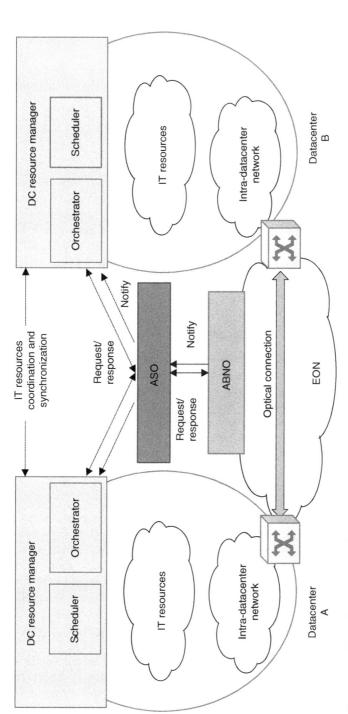

Figure 6.2 ASO implementing transfer-based operations.

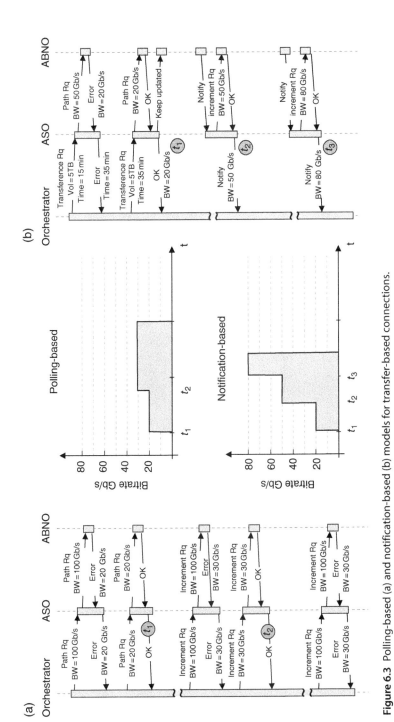

Figure 6.3 Polling-based (a) and notification-based (b) models for transfer-based connections.

retries to increase the allocated bandwidth until reaching the required level. These retries can negatively impact the performance of the inter-DC control plane and not ensure higher bandwidth.

In the notification-based model (Figure 6.3b), DC resource managers request data transferences instead of connection requests. The source DC resource manager sends a transfer request to the ASO specifying the destination DC, the amount of data that needs to be transferred, and the maximum completion time. Upon its reception, the ASO requests the ABNO to find the greatest spectrum width available, taking into account local policies and current SLA, and sends a response back to the cloud manager with the best completion time. The source cloud manager organizes the data transference and sends a new transfer request with the suggested completion time. A new connection is established, and its capacity is sent in the response message; in addition, the ASO requests the ABNO to keep it informed if more resources are left available on the route of that LSP. ABNO has access to both the traffic engineering database (TED) and the LSP-DB, so algorithms deployed in the ABNO monitor spectrum availability in those physical links. When resource availability allows increasing the allocated bitrate of some LSP, the ASO performs elastic spectrum operations, so as to ensure committed transfer completion times. Each time the ASO modifies bitrate by performing elastic spectrum operations, a notification is sent to the source cloud manager containing, among others, the value of the new throughput. The cloud manager then optimizes VM migration as a function of the actual throughput while delegating, ensuring completion transfer time to the ASO controller.

Table 6.1 presents the algorithm implemented in the ASO for transfer requests. It translates requested data volume and completion time into the required bitrate (line 1). Next, an optical connection request is sent toward the

Table 6.1 Algorithm for transfer-based connection requests.

INPUT: *source, destination, dataVol, rqTime*
OUTPUT: *Response*

1: *minBitrate* ← translateAppRequest(*dataVol, rqTime*)
2: *netResp* ← requestConnection(*source, destination, minBitrate*)
3: **if** *netResp* = *KO* **then**
4:    *maxBitrate* ← getMaxBitrate(*source, destination*)
5:    *minTime* ← translateNetResponse(*maxBitrate, dataVol*)
6:    **return** {*KO, minTime*}
7: requestSubscription(*netResp.connId.route*)
8: *time* ← translateNetResponse(*netResp.connId.bitrate, dataVol*)
9: **return** <*OK, netResp.connId, netResp.connId.bitrate, time*>

ABNO specifying source and destination of the connection and the bitrate (line 2). In the case of lack of network resources (lines 3–6), the maximum available bitrate between source and destination DCs is requested to the ABNO, and its result is translated into the minimum completion time, which is used to inform the requesting DC resource manager. If the connection can be established, the ASO requests a subscription to the links in the route of the connection, aiming at being aware of available resources as soon as they are released in the network (line 7). Finally, the actual completion time is recomputed taking into consideration the connection's bitrate and both bitrate and time are communicated back to the requesting DC resource manager.

6.2.2 Illustrative Results

For evaluation purposes, we developed scheduling algorithms in an OpenNebula-based cloud middleware emulator [OPENNEBULA]. Federated DCs are connected to an ad hoc event-driven simulator developed in OMNeT++ [OMNet]. The simulator implements the ASO and the flexgrid-based EON with the ABNO and includes the algorithms described in Chapter 5 for elastic connection operations.

In the simulations, we assume the global 11-node topology depicted in Figure 6.4. Each node corresponds to locations that are used as sources for traffic between users and DCs, referred to as user-to-datacenter (U2DC) traffic. In addition, four DCs are strategically placed in Illinois, Spain, India, and Taiwan. Traffic between DCs (DC2DC) and U2DC traffic compete for resources in the physical network. We fixed the optical spectrum width to 4 THz, the capacity for the ports connecting DCs to 1 Tb/s, the number of VMs to 35,000 with an image size of 5 GB each, and we considered 300,000 DBs with a differential image size of 450 MB and a total size of 5 GB each, half the size of

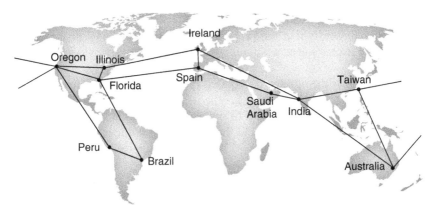

Figure 6.4 Global inter-DC topology.

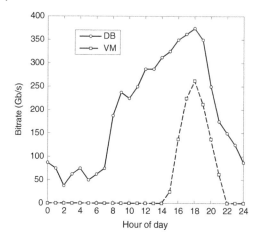

Figure 6.5 Required bitrate in a day.

Wikipedia [WIKISIZE] at the time when simulations were carried out. Additionally, Transmission Control Protocol (TCP), IPv4, Multi-Protocol Label Switching (MPLS), and Ethernet headers have been considered.

Figure 6.5 shows the required bitrate to migrate VMs and to synchronize DBs in 30 min. VM migration is performed as a follow-the-work strategy, and thus connectivity is used during just part of the day. In contrast, DB synchronization is performed along the day, although the required bitrate depends on the amount of data to be transferred that is related to users' activity.

Figure 6.6a depicts the assigned bitrate for DB synchronization and VM migration, respectively, between two DCs during a 24-h period when the polling-based model is used. Figure 6.6b shows the assigned bitrate when the notification-based model is used. Note that, in general, intervals tend to be narrower using the notification-based model, while polling-based model tends to provide longer transfer times as a result of not obtaining additional bitrate in retries. The reason is that the proposed transfer-based model assigns additional bitrate to ongoing connections as soon as resources are released by other connections.

Table 6.2 shows the number of required requests messages per hour needed to increase bitrate of connections for the whole scenario. As illustrated, only 50% of those requests succeeded to increase connections' bitrate under the polling-based model, in contrast to the 100% reached under the notification-based model. Additionally, Table 6.2 shows that when using the notification-based model both the maximum and average required time-to-transfer are significantly lower than when the polling-based model is used. The longest transfers could be done in only 28 min when the notification-based model was used compared to just under 60 min using the polling-based model. Note that in scenarios with enough resources where all connection requests are accepted,

(a) Polling-based

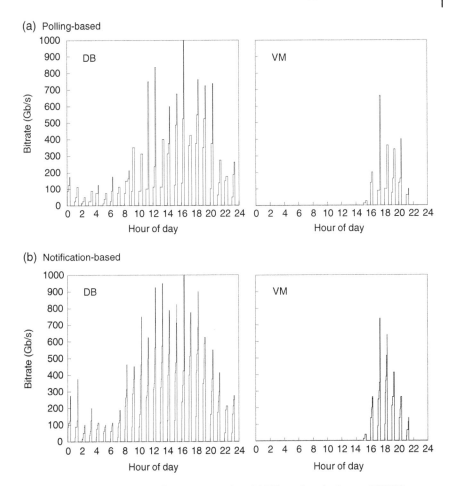

(b) Notification-based

Figure 6.6 DC2DC connection bitrate versus time: (a) DB synchronization and (b) VM migration.

Table 6.2 Performance results.

| | Requests (#/h, %success) | Maximum/avgerage time-to-transfer (min) | |
		DBs	VMs
Polling-based	43.1, 53.5%	58.0/29.0	54.0/28.2
Notification-based	65.3, 100%	28.0/22.4	27.0/22.2

the amount of requested bitrate is the same for both models; otherwise, transference overlapping might occur, increasing the amount of data to transfer at given scheduling periods and affecting cloud performance.

6.3 Routing and Scheduled Spectrum Allocation

In this section, we explore ASO functionalities by using scheduling techniques in addition to elastic operations, with the twofold objective of fulfilling SLA contracts and increasing network operator revenues by accepting more connection requests. To decide whether a transfer-based connection request is accepted or not, the ASO controller solves the RSSA problem. The RSSA problem decides the elastic operations to be performed on already established connections, so as to make enough room for the incoming request ensuring its requested completion time provided that the committed completion times of ongoing transferences are guaranteed.

To generalize the scenario, we consider that other connections not supporting DC transfer-based connections can also be established on the network. For those connections, we assume that its duration (*holding time*) is known and its bitrate is fixed, that is, no elastic operations are allowed. Moreover, it is worth mentioning that ASO requires some additional elements to solve the RSSA problem, such as a specific service database containing information about the services deployed and a scheduled TED (sTED) to manage scheduling schemes.

6.3.1 Managing Transfer-based Connections

For illustrative purposes, let us assume two DCs connected through a flexgrid-based EON. At a given time, a scheduling algorithm inside the resource manager of one of the DCs (DC-1) decides to migrate a set of VMs toward the other DC (DC-2). Imagine that the total data volume is 54 TB and, because the scheduler runs periodically, 30 min is required to perform the whole migration.

As shown in Section 6.2.2, transfer-based connections reduce time-to-transfer remarkably; cloud resource managers request transfers using their native semantic, that is, specifying source and destination DCs, the amount of data to be transferred, and the completion time. The ASO in charge of the inter-DC network transforms the incoming transfer-based connection request into a connection request toward ABNO. A scheme of the control architecture is shown in Figure 6.7a.

To transfer 54 TB from DC-1 to DC-2 in 30 min, a 300 Gb/s connection needs to be established. Let us assume that no resources are available at requesting time (t^0) to guarantee that completion time; in such case the ASO might try to reduce the resources allocated to other transfers currently in progress without

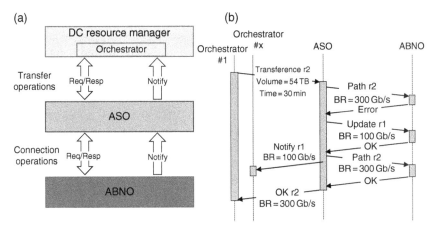

Figure 6.7 Control architecture scheme (a). An example of messages exchanged to request connection operations (b).

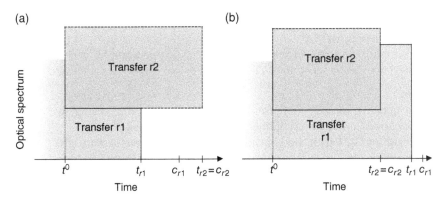

Figure 6.8 Scheduling schemes for SA. (a) Bitrate squeezing only and (b) with scheduled resource reservation.

exceeding their committed completion time, so as to make enough room for the incoming transfer. Figure 6.8 shows examples of the described operation. The bitrate of optical connection supporting ongoing transfer $r1$ in Figure 6.8a can be squeezed, thus increasing its scheduled completion time t_{r1}, but without violating the committed completion time, c_{r1}. Released resources can be used to set up an optical connection to support the requested transfer $r2$.

Figure 6.7b represents the sequence of messages exchanged to set up the optical connection for the requested transfer $r2$ including updating the bitrate of the optical connection supporting transfer $r1$. Note that the originating service client for transfers $r1$ and $r2$ are different, that is, to serve a request from one client, we are reducing the resources allocated to another. Notwithstanding, this is a *best practice* since in EONs elastic operations on established connections can be done without traffic disruption and the committed completion time is not exceeded. In addition, the originating service client is notified of any update in the connection bitrate to help the involved scheduler to organize the underlying data transference.

Another example is depicted in Figure 6.8b. Here, to ensure the committed completion time for transfer $r1$, some resources need to be again allocated to it as soon as transfer $r2$ is completed. In view of the example, it is obvious that spectral resources need to be scheduled, so as to be allocated to some transfer in the future. It is clear now that elastic operations include increasing and decreasing connections' bitrate.

6.3.2 The RSSA Problem

In this section, we focus on the RSSA problem to be solved each time a transfer-based request arrives at ASO. Note that the RSA problem, as defined in Chapter 3, can be solved to find a route and SA for fixed-bitrate connections requests provided that the scheduling data is supplied to the PCE within the ABNO architecture.

To simplify the RSSA problem, only one elastic operation for each ongoing transference can be scheduled. The RSSA problem can be formally stated as:

Given:

- a network topology represented by a graph $G(L, E)$, L being the set of locations and E the set of fiber links connecting two locations;
- a subset $D \subseteq L$ with those locations' source of transfer-based traffic (DCs);
- a subset $U \subseteq L$ with those locations' source of fixed-bitrate traffic;
- the characteristics of the spectrum for each link in E: a set S of available spectrum slices of a given spectrum width;
- the capacity and number of flows of the optical transponders equipped in each location;
- a set R with the transferences currently in progress in the network. For each transference $r \in R$, the tuple $<o_r, d_r, v_r, c_r, r_r, s_r^0, s_r^1, t_r^1, t_r>$ specifies the origin (o_r) and destination (d_r) DCs, the remaining amount of data (v_r) to be transferred, the requested completion time (c_r), the route (r_r) and frequency slot currently allocated (s_r^0), the scheduled frequency slot allocation (s_r^1) to be performed at time t_r^1, and the scheduled completion time (t_r);
- a set C with fixed-bitrate connections currently established in the network. For each connection $c \in C$, the tuple $<o_c, d_c, c_c, r_c, s_c>$ specifies the origin (o_c) and destination (d_c) locations, the requested completion time (c_c), the route (r_c), and the allocated slot (s_c);
- the new transfer-based request described by the tuple $<o_r, d_r, v_r, c_r>$. Let t^0 denote the current time.

Output:

- the route (r_r), the frequency slot allocation (s_r^0), and the scheduled completion time (t_r) for the new transference request. The bitrate of the connection is given by $b(s_r^0, l(r_r))$, where $l(r_r)$ is the total length of the route r_r.
- the new Spectrum Allocation, SA, (w_r^0), scheduled reallocation (w_r^1, t_r^1), and scheduled completion time (t_r) for each transference request to be rescheduled.

Objective: Minimize the number of connections to be rescheduled to make room for the incoming request.

6.3.3 ILP Formulation

The problem consists of rescheduling some of the transferences in progress. A rescheduling for any transference r includes two SAs: s_r^0 from t^0 to t_r^1 and s_r^1 from t_r^1 to t_r, so that the combined area of the two rectangles defined by those segments allows conveying the remaining amount of data v_r, while ensuring that $t_r \leq c_r$. It is worth noting that the capacity of the connections is also limited by the available capacity in the optical transponders.

It is clear that scheduled SAs entail deciding both dimensions, spectrum and time, which might lead to nonlinear constraints. To avoid nonlinearities, we propose precomputing any feasible combination of rectangles. Owing to that fact, and without loss of generality, we need to discretize time into time intervals.

Figure 6.9 shows three feasible rescheduling solutions for a given transference r. Note that the total area of the three pairs of rectangles is identical and $t_r \leq c_r$ in the three solutions.

Table 6.3 presents an algorithm to compute the sets of feasible rectangles for each ongoing transference. Sets of rectangles A_0 and A_1 are defined: set A_0 contains rectangles specifying reallocation w_r^0 from time t^0 to t_r^1, whereas rectangles in set A_1 specify rescheduled allocation w_r^1 from t_r^1 to t_r. In addition, matrix β_{01} indicates whether rectangle pair $a_0 \in A_0$ and $a_1 \in A_1$ is a feasible rescheduling.

The *getRectangles* function (line 1 in Table 6.3) finds rectangles for set A_0. To prevent any transference from stopping, the area of rectangles in this set must be

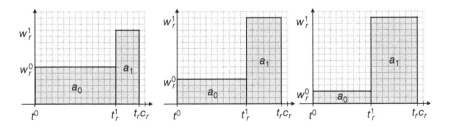

Figure 6.9 Three feasible rescheduling plans for a given transference.

Table 6.3 Algorithm for computing rectangle sets.

INPUT: r

OUTPUT: A_0, A_1, β_{01}

```
1: A₀ ← getRectangles(r)
2: A₁ ← Ø
3: for each a₀ ∈ A₀ do
4:    A₁ ← A₁ U getRectangles(r, a₀)
5: for each a₀ ∈ A₀ do
6:   for each a₁ ∈ A₁ do
7:     β₀₁ ← computeFeasibility(r, a₀, a₁)
8: return A₀, A₁, β₀₁
```

---

strictly greater than zero. In addition, SAs increasing current allocation contain current allocation and some contiguous spectrum, whereas those decreasing the allocation are created by including part of the current allocation. Additional constraints are added to ensure that SA does not exceed the capacity of source and destination transponders and to guarantee that the time defined by the rectangles does not exceed the committed completion time for the given request.

Next, for each found rectangle a_0, the set of rectangles A_1 to complete transference r is computed (lines 3–4). Note that one rectangle with area equal to 0 can be added, provided that the transference can be completed with rectangle a_0. Finally, the *computeFeasibility* function (line 7) checks rectangle compatibility to create matrix β_{01}. Following a similar procedure, a set of precomputed rectangles A for the incoming transfer request is also found, ensuring that both the requested completion time and the amount of data to transfer are satisfied.

As for the Integer Linear Programming (ILP) model, the following sets and parameters are defined:

Sets

E Set of fiber links in the network, index e.

S Set of frequency slices, index s.

T Set of time intervals, index t.

R Set of ongoing transferences, index r.

P Set of routes between origin and destination for the new request, index p.

A Set of candidate areas for the requested transference, index a.

$A_0(r)$ Set of candidate areas for transference r. Each area starts in t^0 and is defined by t_r^1 and an SA.

$A_1(r)$ Set of candidate areas for transference r. Each area is defined by a starting interval t_r^1, a scheduled completion time t_r, and an SA.

Parameters

ω_{ar} 1 if transference r was assigned to area a; 0 otherwise.

δ_{as} 1 if area a includes slice s; 0 otherwise.

γ_{at} 1 if area a includes time interval t; 0 otherwise.

ρ_{pe} 1 if route p uses link e; 0 otherwise.

ρ_{re} 1 if transference r uses link e; 0 otherwise.

$\beta_{raa'}$ 1 if pair of areas $a \in A_0$ and $a' \in A_1$ for transference r are feasible; 0 otherwise.

δ_{kest} 1 if fixed-bitrate connection k uses slice s in link e at time interval t; 0 otherwise.

α_{est} 0 if slice s in edge e is used by any fixed-bitrate connection at time interval t, that is, $\alpha_{est} = 1 - \sum_{k \in K} \delta_{kest}$; 1 otherwise.

Additionally, the following variables are defined:

x_{ap} Binary; 1 if the new transference is assigned to area a through route p; 0 otherwise.

y_r Binary, 1 if transference r is rescheduled; 0 otherwise.

x_{ar}^0 Binary; 1 if transfer r is assigned to area $a \in A_0$; 0 otherwise.

x_{ar}^1 Binary; 1 if transfer r is assigned to area $a \in A_1$; 0 otherwise.

The ILP model for the RSSA problem is as follows:

$$\text{(RSSA) minimize} \quad \sum_{r \in R} y_r \tag{6.1}$$

subject to:

$$\sum_{a \in A} \sum_{p \in P} x_{ap} = 1 \tag{6.2}$$

$$\sum_{a \in A_0(r)} x_{ar}^0 = 1 \quad \forall r \in R \tag{6.3}$$

$$\sum_{a \in A_1(r)} x_{ar}^1 = 1 \quad \forall r \in R \tag{6.4}$$

$$\sum_{a' \in A_1(r)} \beta_{raa'} \cdot x_{a'r}^1 \geq x_{ar}^0 \quad \forall r \in R, a \in A_0(r) \tag{6.5}$$

$$x_{ar}^0 - \omega_{ar} \leq y_r \quad \forall r \in R, a \in A_0(r) \tag{6.6}$$

$$\begin{aligned}
\sum_{r \in R} \sum_{a \in A_0(r)} \delta_{as} \cdot \gamma_{at} \cdot \rho_{re} \cdot x_{ar}^0 + \\
\sum_{r \in R} \sum_{a \in A_1(r)} \delta_{as} \cdot \gamma_{at} \cdot \rho_{re} \cdot x_{ar}^1 + \quad \forall e \in E, s \in S, t \in T \\
\sum_{a \in A} \sum_{p \in P} \delta_{as} \cdot \gamma_{at} \cdot \rho_{pe} \cdot x_{ap} \leq \alpha_{est}
\end{aligned} \tag{6.7}$$

The objective function (6.1) minimizes the number of transferences whose bitrate needs to be reduced so that the new transfer request can be served.

Constraint (6.2) guarantees that the transfer request is served assigning it to a route p and a rectangle a, which defines the SA and the scheduled completion time.

Constraints (6.3)–(6.5) select a feasible pair of rectangles for each ongoing transference r. Constraints (6.3) and (6.4) ensure that one rectangle in set A_0 and another in set A_1 are selected, whereas (6.5) ensures that the selected pair of rectangles creates a feasible solution for transference r, that is, committed completion time and the amount of data to be transferred are satisfied. Constraint (6.6) stores whether transference r needs to be reallocated comparing current and assigned rectangle a_0.

Finally, constraint (6.7) guarantees that each frequency slice in each link is not used by more than one connection during each time interval.

The RSSA optimization problem is *NP-hard* since it includes the RSA problem. Regarding its size, the number of variables is $O(|P|\cdot|A| + |R|\cdot(|A_0| + |A_1|))$ and the number of constraints is $O(|R|\cdot|A_0| + |E|\cdot|S|\cdot|T|)$. Considering a realistic scenario on a network with 30 links, 640 frequency slices, and 3600 one-second time periods, the number of variables and the number of constraints are in the order of 10^8, which makes this ILP model untractable even using state-of-the-art computer hardware and the latest commercially available solvers, for example, CPLEX [CPLEX].

As a result of the size of the RSSA problem, we propose an algorithm to solve the RSSA problem in realistic scenarios next.

6.3.4 Algorithms to Manage Transfer-based Requests

In this section, we present a heuristic algorithm to solve the RSSA problem to be executed upon a transference request arrives at ASO. In contrast, the RSA problem is solved for fixed-bitrate connection requests. In addition, a policy to assign resources to ongoing transferences as soon as they are released is presented.

6.3.4.1 Algorithm for the RSSA Problem

Table 6.4 shows the algorithm proposed for solving the RSSA problem. Firstly, the minimum admissible bitrate for the transference is found (line 1) considering both data volume and completion time. In lines 2–5, the algorithm solves an RSA problem (we use a shortest path for route selection). In case a feasible solution is found, the slot with maximum width is assigned in an attempt to reduce its scheduled completion time.

The remaining of the algorithm in Table 6.4 is specific for RSSA, and it is executed in case no feasible solution for the RSA problem is found. All shortest routes with the same length in hops than the shortest one are explored in the

Table 6.4 Algorithm for the RSSA problem.

INPUT: *o, d, v, c*
OUTPUT: *route, slot*

```
 1: br ← getMinBitrate(v, c)
 2: {route, slot} ← computeRSA(o, d, br, sTED)
 3: if route ≠ ∅ then
 4:    slot ← getMaxSA(route, sTED)
 5:    return {route, slot}
 6: P ← computeKSP(o, d)
 7: for each p ∈ P do
 8:   sw ← getSlotWidth(p.length, br)
 9:   S ← getFreeSlots(p, sTED)
10:   sort(S, slot width, DESC)
11:   for each s ∈ S do
12:      Q ← getAdjacentPaths(p, s, sTED)
13:      R ← findSlot(Q, s, sw, sTED)
14:      if R ≠ ∅ then
15:         doReschedule(R)
16:         slot ← getSA(p, br, sTED)
17:         return {p, slot}
18: for each p ∈ P do
19:   sw ← getSlotWidth(p.length, br)
20:   S ← getFeasibleBorderPathSlots(p, sTED)
21:   sort(S, slot width, DESC)
22:   repeat from line 11 to line 17
23: return {∅, 0}
```

---

hope of finding a feasible slot by increasing the set of available slots (whose spectrum width is smaller than the required) that are continuous along a given route. For each available slot, adjacent connections supporting ongoing transferences are considered as a candidate to squeeze its current bitrate provided that the committed completion time is satisfied (lines 6–17). Specifically, the *findSlot* function tries to increase the slot width in an attempt to make room for the incoming transference. The needed reschedulings are afterward stored in the *sTED* database, and connection update requests are sent to the ABNO controller in charge of the EON. To compute the maximum bitrate squeezing that can be applied, scheduling schemes introduced in Section 6.3.1 are considered.

As a final attempt to find a feasible solution, the computed routes are again explored (lines 18–22). This time, those slices currently allocated to connections supporting ongoing transferences that are at the border of their allocated

Table 6.5 Algorithm for resource assignment.

```
INPUT: R, t⁰
```

```
1: for each  r ∈ R do
2:    if t¹ᵣ = t⁰ then
3:       doReSchedule (r)
4: sort(R, tᵣ - t⁰, ASC)
5: for each  r ∈ R do
6:    expandMax(r)
```

slot are considered. Then, a slot is generated by reducing the SA of those connections (considering the committed completion time for the transference) (line 20). Next, a procedure similar to that described for available slots is followed.

6.3.4.2 Resource Assignment Policy

After a connection has been torn down, released resources need to be allocated to those connections supporting ongoing transferences, so as to satisfy the planned scheduling. For example, in Figure 6.8b, some resources were scheduled to be allocated to transference r_1. Then, after connection supporting transference r_2 is torn down, those resources need to be allocated. Lines 1–3 in Table 6.5 describe the proposed algorithm, where t^0 specifies the current time.

However, some resources might remain unallocated even after scheduled allocations are done. Aiming at maximizing resource utilization, connections supporting ongoing transferences are offered the opportunity to increase allocated resources, thus reducing its scheduled completion time (lines 4–6). Functions in lines 3 and 6 send the appropriate requests to the ABNO controller. Note that the benefit from the latter is reduction in completion time, which also facilitates that those connections can be torn down sooner. That effect might be amplified by assigning released resources to those connections whose scheduled completion time is closer to the current time.

6.3.5 Illustrative Results

Aiming at evaluating the performance of the proposed transfer-based connections, we developed an ad hoc event-driven simulator in OMNeT++. In this section, we present the results obtained, where several network topologies have been considered. A dynamic network environment was simulated, where incoming connection requests arrive at the system following a Poisson process and are sequentially served without prior knowledge of future incoming connection requests.

6.3.5.1 Heuristic versus ILP Model

Before evaluating the performance of the proposed transfer-based connections, let us compare the ILP model and the heuristic algorithm proposed to solve the RSSA underlying problem.

The ILP model was implemented in Matlab and CPLEX was used to solve the selected problem instances. Regarding the heuristic algorithm, it was implemented in C++ and integrated into OMNeT++. Both methods were run in a 2.4 GHz Quad-Core computer with 16 GB RAM memory under the Linux operating system. To generate problem instances, we used the simulation environment described earlier and implemented a small network topology with six nodes and eight edges, while setting the spectrum width to 1 THz, the BVT capacity to 800 Gb/s, and the maximum completion time to 30 s.

Table 6.6 summarizes some representative parameters for a set of instances and the obtained solution and time-to-solve when the ILP and the heuristic algorithm were applied. For each problem instance, the number of ongoing transferences and the maximum size of sets A_0 and A_1 for every transference are detailed. Regarding solving times, the values in the ILP column are for CPLEX solving time; that is, the problem generation time (in the order of tens of seconds) and rectangle sets computation (hundreds of milliseconds) are not included.

Results show that several seconds are needed to solve really small problem instances using the ILP model. In contrast, time-to-solve is in the order of tens of milliseconds when the proposed RSSA heuristic algorithm is used, thus illustrating its applicability to real scenarios. In view of this, the algorithm is used to solve the RSSA problem in the simulation experiments presented next. Notwithstanding the computation time, the algorithm provides virtually the same solution than the ILP model; in only one instance did the algorithm reduce the SA of two established connections, while the ILP limited that number to only one.

Table 6.6 ILP model versus heuristic comparison.

$\|R\|$	Maximum ($\|A_0\|$)	Maximum ($\|A_1\|$)	Time-to-solve		Solution	
			ILP	Algorithm	ILP	Algorithm
7	691	3409	10.3 s	5 ms	1	1
8	2173	7203	15.4 s	13 ms	1	1
11	1442	3693	19.8 s	9 ms	1	1
14	1592	5845	25.9 s	31 ms	1	1
15	1447	4571	26.8 s	6 ms	1	1
17	1463	5381	19.8 s	27 ms	1	2

6.3.5.2 RSA versus RSSA Comparison

Three realistic national network topologies have been considered (Figure 6.10): the 21-node Telefonica (TEL), the 20-node British Telecom (BT), and the 21-node Deutsche Telekom (DT) networks. For these experiments, the optical spectrum width is fixed to 4 THz in each link and the modulation format is restricted to Quadrature Phase-Shift Keying. We assume that an optical cross-connect with a 1 Tb/s BVT is installed in each location, being the source of both fixed-bitrate and transfer-based connections.

Fixed-bitrate connections request 100 Gb/s during a fixed holding time of 500 s; an exponential interarrival time is considered. Source/destination pairs are randomly chosen with equal probability (uniform distribution) among all nodes. Different values of the offered network load are created by changing the interarrival rate. We assume that no retrial is performed; that is, if a fixed-bitrate connection request cannot be served, it is immediately blocked.

In contrast, transfer-based offered load is related to VM migration among DCs. The number of VMs to migrate between two DCs is proportional to the number of considered VMs and randomized using a uniform probability distribution function. Different values of the offered network load are created by changing the total amount of VMs and the size of every VM. The number of VMs ranges from 15,000 to 45,000, whereas VMs size ranges from 2.5 to 7.5 GB. Data volume is translated into required network throughput considering TCP, IP, Ethernet, and MPLS headers. In this case, we assume that transfer-based requests are generated asking to complete the transference in 30 min. If the request cannot be served, one retrial is performed requesting resources so to keep completion time limited to 1 h. If the request cannot be served, it is finally blocked.

The ability to deal with transfer-based connection requests of the RSSA algorithm in Table 6.4 was compared against that of the RSA running lines 1–5 in the algorithm in Table 6.4. To that end, we used the simulation environment described earlier, where we implemented the ASO on top of the ABNO

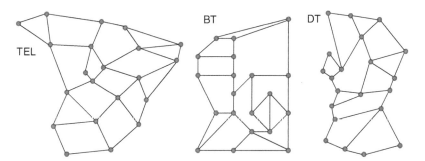

Figure 6.10 The Telefonica (TEL), British Telecom (BT), and Deutsche Telekom (DT) network topologies used in this chapter.

architecture controlling the EON. Finally, all the results were obtained after requesting 15,000 connections and each point in the results is the average of 10 runs.

To compare both the RSA and the RSSA algorithms, we first run simulations considering that only transfer-based traffic is requested. Figure 6.11a shows the percentage of unserved bitrate for several offered loads and for each of the considered network topologies. VMs' size was fixed to 5 GB. For the sake of comparability between topologies, offered loads have been normalized to the value of the highest load.

The proposed RSSA algorithm is able to serve more traffic in all three network topologies compared to using just RSA. Interestingly, the gain from using the RSSA is small under low loads since enough resources are available to serve all the requests. As soon as the load increases, the ability of RSSA to perform elastic operations on the established optical connections, together with the scheduling to ensure that the committed completion time is not exceeded, brings clear benefits. Obviously, for high loads, those benefits would cancel as few opportunities to reduce resources would exist. If we focus on 1% of unserved bitrate, we observe that noticeable gains in the offered load are obtained. Table 6.7 presents those values, which range from 12% to more than 30% traffic gain when the RSSA algorithm is used.

Additionally, for the sake of providing a thorough study, we fixed the number of VMs while varying their size from 2.5 to 7.5 GB. To that end, we selected the load, unleashing 1% of unserved bitrate when VMs' size was fixed to 5 GB. Figure 6.12a shows the percentage of unserved bitrate, while Figure 6.12b plots transference completion times. As expected, the results obtained are in line with those in Figure 6.11 obtained by varying the number of VMs; the amount of data to be transported is the underlying parameter being increased.

Because the number of elastic operations performed on each connection is an important factor, Table 6.8 summarizes the number of elastic operations (spectrum reduction and expansion) performed over each established connection supporting a transference. The percentage of connections experiencing elastic operations is also summarized.

The average number of spectrum reductions and spectrum expansions that each connection experiences is just over one and arrives at a maximum of four reductions and six expansions. However, if we analyze the percentage of connections experiencing elastic operations, we realize that the percentage of expansions doubles that of spectrum reductions. Recall that spectrum reductions are performed when a connection request arrives, whereas spectrum expansions are performed when connections are torn down to implement scheduled SAs and to reduce transference completion time. It is clear that scheduled allocations are as a result of previous spectrum reduction and thus the number of reductions should be greater or equal to the number of expansions. Therefore, in view of Table 6.8, the algorithm for resource assignment

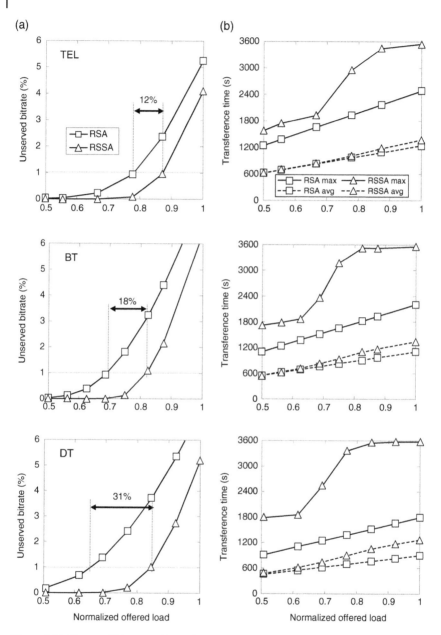

Figure 6.11 Unserved bitrate and transference completion time versus load (in terms of number of VMs). Only transfer-based requests are considered.

Table 6.7 Offered load gain and transference time comparison.

		TEL	BT	DT
Gain at 1% unserved bitrate		12.2%	17.8%	30.9%
Transference time in seconds (RSSA)	Maximum	3444	3523	3543
	Average	1180	1092	1048
Transference time in seconds (RSA)	Maximum	2167	1822	1518
	Average	1085	912	761

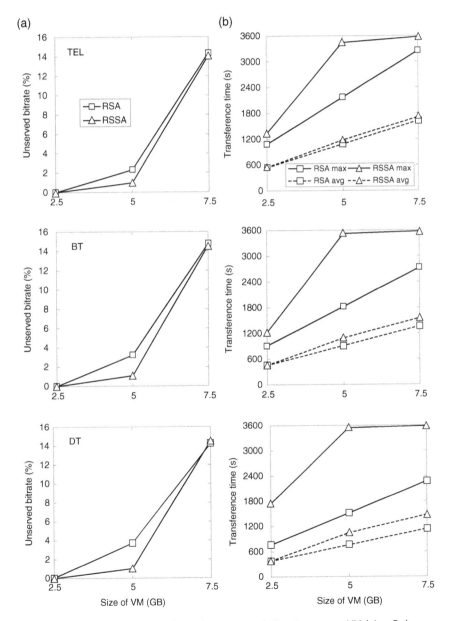

Figure 6.12 Unserved bitrate and transference completion time versus VMs' size. Only transfer-based requests are considered. Unserved bitrate (a) and transference time (b).

Table 6.8 Elastic operations at 1% of unserved bitrate.

		TEL	BT	DT
Number of spectrum reductions	Maximum	3.5	3.6	3.9
	Average	1.09	1.13	1.18
Number of Spectrum expansions	Maximum	4.1	4.8	5.6
	Average	1.18	1.27	1.37
Percentage of connections experiencing elastic operations	Reductions	9.3	18.8	24.4
	Expansions	23.6	43.5	53.4

proposed in Table 6.5 is responsible for performing extra spectrum expansions to reduce transference completion time and to increase resource utilization.

Finally, we study the effect of mixing fixed-bitrate and transfer-based traffics. Note that spectral resources allocated to fixed-bitrate connections must remain invariant along connections' life. We set the transfer-based load to one of the lower values in Figure 6.11 and increased fixed-bitrate traffic load. Figure 6.13a plots the percentage of unserved transfer-based bitrate, whilst Figure 6.13b shows transference completion times as a function of fixed-bitrate traffic load. As in the previous study, offered loads have been normalized to the value of the highest load. Results show that the proposed RSSA algorithm leads to remarkable gains, ranging from 140 to 500% with respect to the RSA, in the amount of fixed-bitrate offered traffic. As for completion times, average values are similar for both algorithms and in line with those in Figure 6.11b. Plots for maximum times follow a similar trend for both RSA and RSSA, although slightly higher for the RSSA (Figure 6.13b). This is due to the fact that fixed-bitrate traffic reduces the possibility of making room for incoming requests by reducing SAs on transfer-based connections. To improve that, defragmentation techniques, such as those described in Chapter 8, can be applied.

6.4 Conclusions

In this chapter, ASO implementing a northbound interface with application-oriented semantic has been proposed as a new abstraction layer between DC resource managers and the ABNO in charge of EON-based DC interconnection network. Each resource manager can request transfer operations specifying source and destination DCs, the amount of data to be transferred, and the desired completion time.

The polling-based and the notification-based connectivity models have been compared in a DC federation scenario considering ASO. In the polling-based model, resource managers request connection operations to the ASO, which

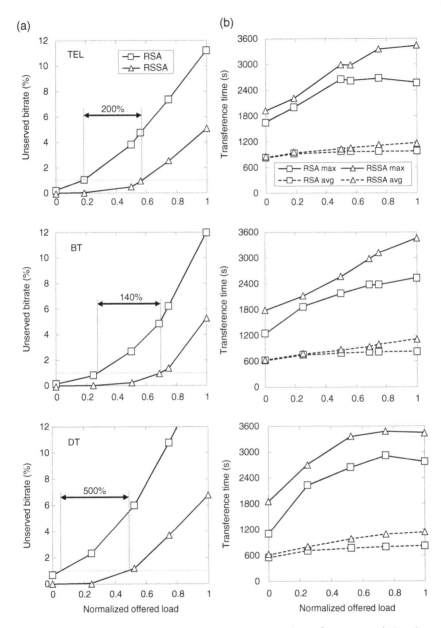

Figure 6.13 Unserved bitrate of transfer-based requests and transference completion time versus offered load of fixed-bitrate traffic. Unserved bitrate (a) and transference time (b).

forwards them to the ABNO in charge of the interconnection network. Moreover, it needs periodical retries requesting to increase the connection's bitrate, which does not translate into immediate bitrate increments and could have a negative impact on the performance of the inter-DC control plane. In contrast, in the notification-based model, DC resource managers request transfer-based operations to ASO and it takes advantage of the use of notify messages. ASO is then aware of DC resource managers' necessities, as well as network resources availability, and is thus able to reduce time-to-transfer by approximately 20% on average and up to 50% for maximum values.

After the benefits of time-to-transfer when ASO is responsible for managing DC interconnection were demonstrated, we considered the scenario where a set of customers request transfer operations to ASO. We defined the RSSA problem to decide the set of already established connections for which their current SA can be reduced without violating the committed completion time, so as to make enough room for the incoming transfer-based request.

We formally stated the RSSA problem and modeled it using an ILP formulation. In view of the complexity of the problem and owing to the fact that it needs to be solved in real time, a heuristic algorithm was proposed. Aiming at taking advantage of the resources released when connections have been torn down, a policy to assign them to ongoing transferences was eventually proposed. The policy gives more priority to those connections with the closest scheduled completion time.

Illustrative results were obtained by simulation on three telecom network topologies. The performance of the proposed RSSA algorithm when dealing with transfer-based requests was compared against that of the RSA. Results showing remarkable gains, as high as 30 and 500% in the amount of transfer-based and fixed-bitrate traffic, respectively, were obtained for the considered scenarios.

Since RSSA entails reducing and scheduling increments in spectrum allocated to established connections, maximum completion times were increased to just under committed completion times. Notwithstanding, average completion times obtained when RSSA was applied showed values similar to that of the obtained with RSA.

It is worth highlighting that connections supporting transferences experienced few elastic operations, just above one reduction and one expansion.

To conclude, in this chapter, we have shown the benefits of using ASO on top of ABNO to manage inter-DC connectivity considering transfer-based connections and using scheduling techniques and elastic operations to fulfill SLAs from a set of customers simultaneously.

7

Provisioning Multicast and Anycast Demands

Lluís Gifre[1], Marc Ruiz[1], Victor López[2] and Luis Velasco[1]

[1] Universitat Politècnica de Catalunya, Barcelona, Spain
[2] Telefonica Investigación y Desarrollo, Madrid, Spain

New applications require conveying huge bitrate from a source node to one or multiple destinations, and, as a result, multicast and anycast connections need to be established on the optical layer. In this chapter, the point-to-multipoint (P2MP) routing and spectrum allocation (RSA) problem is proposed to find feasible light-trees to serve multicast demands. Given that creating transparent light-trees noticeably improves the performance obtained by serving the demand using a set of lightpaths, that approach is experimentally validated. Finally, the anycast RSA problem is studied and experimentally assessed.

Table of Contents and Tracks' Itineraries

Provisioning, Recovery, and In-operation Planning in Elastic Optical Networks,
First Edition. Luis Velasco and Marc Ruiz.
© 2017 John Wiley & Sons, Inc. Published 2017 by John Wiley & Sons, Inc.

7.1 Introduction

New generation applications, such as cloud applications, emerging content distribution for multimedia applications, and datacenter (DC) interconnection services, are changing the patterns of the traffic that optical networks need to transport. For instance, ultra-high-definition video transmission and data synchronization among distributed databases running inside geographically distant DCs require that the same data, sourced from a single ingress node, arrives at a set of different egress nodes; we call this type of demands *multicast*. As a result of the huge amount of data being conveyed (e.g., uncompressed real-time 8k transmission needs 72 Gb/s connections [Na11]), these demands need to be transported directly over the optical layer.

A multicast demand (or connection request) d can be defined by the tuple <o, T, b>, where o represents the source node, T the set of destination nodes, and b the requested bitrate. Multicast demands can be served by establishing $|T|$ directed point-to-point (P2P) optical connections (*lightpaths*) from the source node to every destination node. Nonetheless, as demonstrated in Ref. [Li13] for optical networks based on Wavelength Division Multiplexing (WDM) technology, establishing *light-trees*, that is, point-to-multipoint (P2MP) connections in the optical layer, improves the performance of establishing a set of *lightpaths*. Note that a path is a particular case of a tree, where the set of destination nodes contains just one node.

As an evolution of the WDM technology, Elastic Optical Networks (EONs) are changing optical transport networks; the combination of a finer spectrum granularity and advanced modulation formats together with the use of Sliceable Bandwidth-Variable Transponders (SBVTs) able to manage traffic flows from/ to multiple destinations provides the possibility to leverage the characteristics of light-trees.

To compare the performance of P2MP connections in EONs with that in WDM networks, the authors in [Wa12] proposed two heuristic algorithms; they assumed *broadcast- and select*-based optical nodes. The feasibility of implementing and controlling light-trees in EONs was demonstrated in [Sa13], where the authors presented two node architectures. The first node architecture uses passive light splitters and requires that the P2MP connection satisfies the spectrum continuity constraint, whereas the second one takes advantage of frequency conversion.

The authors in [Go13] present an Integer Linear Program (ILP) formulation to compute the routing and spectrum allocation (RSA) problem for a set of multicast demands. Their formulation is based on precomputing a set of paths and ensuring the contiguity of the allocated frequency slices. As a result of its huge complexity, a heuristic is proposed.

In this chapter, we compare the performance of lightpaths and light-trees to deal with multicast traffic in EONs. To this end, we state the P2MP-RSA problem that aims at finding the optimal RSA for a multicast demand using different path and tree routing schemes. We adapt the RSA formulation in Chapter 3 that relies on sets of precomputed slots of contiguous slices to serve multicast demands, thus reducing the problem's complexity. The performance of different schemes is afterward compared in terms of grade of service and cost of transponders. Note that multicast demands are asymmetrical in contrast to unicast demands and so using light-trees to serve multicast demands should bring some cost savings. A similar case was studied in [Wo13], where the authors showed that because IP traffic is mainly directed, savings could be obtained by establishing unidirectional lightpaths.

Regarding anycast provisioning, as for multicasting, a demand d can be defined by the tuple $<o, T, b>$. However, in contrast to multicast demands, only one destination needs to be selected from the set T and therefore *anycast* demands can be served by establishing one single *lightpath* from the source node to one of the destination nodes.

The authors in [Gh12] consider both available network resources and IT resource requirements to compute an optimal anycast route for dynamic connections. RSA algorithms were proposed in [Wa14] to serve anycast connections in EONs; to select the destination of the anycast connection, a set of candidate paths was computed. Since the approaches available might entail long computation times, in this chapter, we review a novel mechanism for provisioning anycast connections that avoids the computation of candidate paths by using an auxiliary graph.

7.2 Multicast Provisioning

Figure 7.1 illustrates the routing schemes considered in this chapter to serve multicast demands in the optical layer. Client routers/switches, for example, emulating DCs, are connected to optical cross-connects (OXCs) by means of SBVTs. Let us assume that 400 Gb/s SBVTs consist of four modulators, so they can be shared among four connections. In the case of unidirectional paths, SBVTs can be shared by up to four incoming connections and up to four outgoing connections. In this scenario, the multicast demand <A, {B, C, D}, 100 Gb/s> can be served by three lightpaths or, alternatively, by one single light-tree.

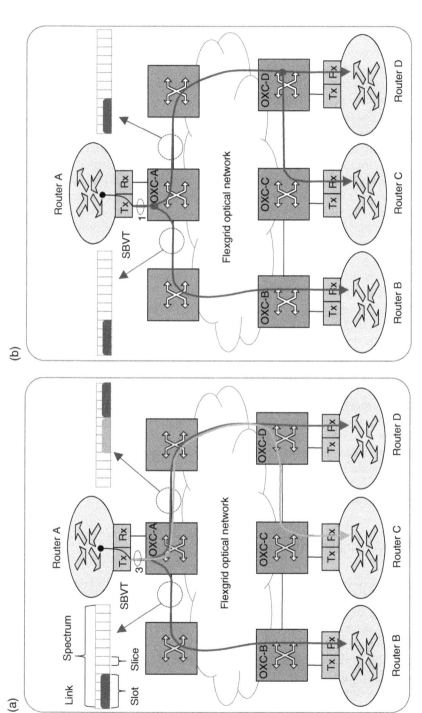

Figure 7.1 Routing schemes for multicast traffic. (a) Path scheme and (b) tree scheme.

Figure 7.1a shows the *path scheme*, where three 100 Gb/s unidirectional lightpaths (also known as *subpaths*), are set up. Each subpath consumes one SBVT Tx modulator from those available in the source router and one SBVT Rx modulator in the destination router. Furthermore, one spectrum slot is allocated for each subpath along the route. Although the path scheme requires many resources (SBVT modulators and spectrum), every subpath can be routed independently from the other ones serving the same demand, which provides high flexibility in the case where spectral resources are scarce or fragmented.

To reduce resource utilization, the *tree scheme* can be implemented (Figure 7.1b) setting up a single 100 Gb/s signal, which remarkably reduces the required number of SBVT modulators to just one in the source node. Besides, spectral resources can be shared on some segments of the tree (e.g., from OXC-A to OXC-D), which also reduces spectral resource utilization. Notwithstanding, a single spectrum allocation is required for the whole tree, which might increase the demands blocking ratio in case no continuous spectrum allocation is available along the tree, for example, as a result of spectrum fragmentation.

7.2.1 P2MP-RSA Problem Statement

In this section, we formally state the P2MP-RSA problem.

Given:

- A network topology represented by a graph $G(N, E)$, where N is the set of nodes including routers (N^R) and OXCs (N^O) and E the set of links connecting two nodes;
- a set S of available frequency slices of a given spectral width in each link;
- a set of available SBVTs in every router;
- a multicast demand to be served identified by the tuple $<o, T, b>$.

Output: the RSA for the connection.

Objective: minimize the amount of resources needed to serve the request, including the number of used SBVT modulators.

7.2.2 ILP Formulation

In this section, we present an ILP formulation to solve the P2MP-RSA problem. The ILP formulation is based on the RSA modeling for unicast demands described in Chapter 3. The solution entails computing Steiner trees, that is, minimum weight trees that contain the source and the destinations and may include additional nodes [Ch99]. Routing constraints build directed Steiner trees, while spectrum allocation is performed by assigning precomputed slots to each computed Steiner tree. To obtain directed Steiner trees, one path and one slot is found for each subconnection and those paths sharing links and slots are merged into a light-tree.

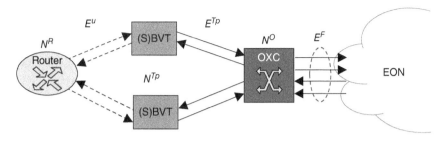

Figure 7.2 Augmented topology.

To guarantee the feasibility in the use of SBVTs, the graph is augmented (Figure 7.2) considering SBVT modulators as nodes. Those transponders with free capacity at the source and the destinations of the request are added as intermediate nodes (N^{Tp}) between routers and OXCs. Links connecting transponders and OXCs (E^{Tp}) are configured with the same spectrum structure as fiber links connecting OXCs (E^F), whereas those links connecting transponders and routers (E^u) are uncapacitated. SBVTs are included in the augmented topology provided that they have one or more modulators available and remaining capacity to support the requested bitrate.

The following sets and parameters are defined:

Topology

N	Set of network nodes.
$N^O \subseteq N$	Subset of OXCs.
$N^{Tp} \subseteq N$	Subset of transponders.
$N^R \subseteq N$	Subset of routers.
E	Set of directed network links.
$E^F \subseteq L$	Subset of fiber links connecting two OXCs.
$E^{Tp} \subseteq L$	Subset of uncapacitated links connecting a transponder with an OXC.
$E^u \subseteq L$	Subset of links connecting a router with a transponder.
km(e)	Length (in km) of link e.
$\delta_{ne}^{\rightarrow}$	1 if link e leaves from node n.
$\delta_{ne}^{\leftarrow}$	1 if link e arrives at node n.

Request

$o \in N^R$	Request's source node.
$T \subseteq N^R$	Request's target nodes.
b	Request's bitrate (in Gb/s).

Spectrum and Reach

S Set of spectrum slices.

C Set of spectrum slots for the request.

f_{es} Equal to 1 if slice s in link e is free.

m Maximum optical reach (in km).

The decision variables are:

x_{tec} Binary, equal to 1 if destination t is reached through link e and slot c; 0 otherwise.

w_{ec} Binary, equal to 1 if slot c in link e is used; 0 otherwise.

y_c Binary, equal to 1 if slot c is allocated; 0 otherwise.

The P2MP-RSA ILP formulation is as follows:

$$\left(P2MP - RSA\right) \quad \min \sum_{e \in E^F} \sum_{c \in C} w_{ec} \tag{7.1}$$

subject to:

$$\sum_{e \in E} \sum_{c \in C} \left(\delta_{ne}^{\rightarrow} - \delta_{ne}^{\leftarrow} \right) \cdot x_{tec} = \begin{cases} 1, & \forall t \in T, n = o \\ 0, & \forall t \in T, n \in N \neq \{o,t\} \\ -1, & \forall t \in T, n = t \end{cases} \tag{7.2}$$

$$\sum_{e \in E} \sum_{c \in C} km(l) \cdot x_{tec} \leq m, \quad \forall t \in T \tag{7.3}$$

$$\sum_{t \in T} x_{tec} \leq |T| \cdot w_{ec}, \quad \forall e \in E, c \in C \tag{7.4}$$

$$\sum_{e \in E} \sum_{c \in C} \delta_{ne}^{\leftarrow} \cdot w_{ec} \leq 1, \quad \forall n \in N^O \tag{7.5}$$

$$\sum_{e \in E} \left(\delta_{ne}^{\rightarrow} - \delta_{ne}^{\leftarrow} \right) \cdot x_{tec} = 0, \quad \forall t \in T, n \in N \neq \{o,t\}, c \in C \tag{7.6}$$

$$\sum_{e \in E} w_{ec} \leq |L| \cdot y_c, \quad \forall c \in C \tag{7.7}$$

$$\sum_{c \in C} y_c = 1 \tag{7.8}$$

$$\sum_{c \in C} q_{cs} \cdot w_{ec} \leq f_{es}, \quad \forall e \in E^F \cup E^{Tp}, s \in S \tag{7.9}$$

The objective function (7.1) minimizes the total amount of spectrum used by the light-tree. Note that only the spectrum in links connecting OXCs is considered.

Constraint (7.2) deals with subconnection routing by finding a directed path from source to destination nodes using multicommodity flow conservation equalities. It is worth mentioning that if a subconnection cannot be served, the problem becomes infeasible and therefore the request will be rejected. Additionally, since the total distance covered by an optical connection is limited, constraint (7.3) ensures that the maximum reach is not exceeded.

Constraint (7.4) is in charge of building a light-tree by including every resource used by any subconnection. Note also that a slot in a link can be shared among several subconnections. Besides, constraint (7.5) does not allow intermediate OXCs to receive several tree branches. The opposite is allowed, that is, an incoming tree branch can be split into multiple outgoing tree branches.

Constraints (7.6)–(7.8) ensure the spectrum continuity constraint along the light-tree. Constraint (7.6) guarantees that each subconnection meets this condition by ensuring the same slot at each intermediate node of the path. Constraint (7.7) stores the slots used in the light-tree, whereas constraint (7.8) forces the number of used slots to be equal to one.

Finally, constraint (7.9) makes sure that the selected frequency slot was available at every link used by the light-tree.

The size of the problem is $O(|T| \cdot |E| \cdot |C|)$ variables and $O(|C| \cdot |T| \cdot |N| + |E| \cdot |S|)$ constraints, which cannot be practically solved in real time for realistic network scenarios. For this very reason, an efficient heuristic to solve the P2MP-RSA problem is introduced next.

7.2.3 Heuristic Algorithm

Table 7.1 shows the main algorithm that consists of a routing phase that builds a tree followed by a spectrum allocation phase. Although routing and spectrum allocation are independent, the routing phase tries to guarantee the availability of at least one end-to-end frequency slot for the request.

Before starting the routing phase, the set of all possible frequency slots with enough bandwidth to allow serving the requested bitrate is precomputed (line 2 in Table 7.1). Next, an iterative procedure that finds and merges routes from source to destinations into a tree under construction is executed (lines 3–14). The procedure runs until all the destination nodes are added to the tree (line 3). At each iteration, all remaining destinations are evaluated as candidates to be inserted into the tree. Once some parameters are initialized (lines 4–5), the route with the minimum cost for each remaining destination is found (lines 6–7). Among all destinations, the one with the highest minimum cost is selected (lines 8–10). The solution is infeasible if no route is available for any

Table 7.1 P2MP-RSA algorithm

INPUT: $<o, T, b>$, S, R, $<ch, ct, cs>$
OUTPUT: $solution$

1: $Solution \leftarrow \emptyset$
2: $slots \leftarrow computeSlots(S, b)$
3: **while** $T \neq \emptyset$ **do**
4:    $maxCost \leftarrow -\infty$
5:    $r_{sel} \leftarrow \emptyset$
6:    **for each** $t \in T$ **do**
7:      $<route, cost> \leftarrow$ P2MP-R $(R(o, t), S, slots, T, solution,$
$<ch, ct, cs>)$
8:      **if** $cost > maxCost$ **then**
9:        $maxCost \leftarrow cost$
10:       $r_{sel} \leftarrow route$
11:    **if** $r_{sel} = \emptyset$ **return** $infeasible$
12:    $solution \leftarrow updateTree(solution, r_{sel})$
13:    $slots \leftarrow updateSlots(slots, solution)$
14:    $T \leftarrow updateT(T, solution)$
15: $solution \leftarrow SA(slots)$
16: **return** $solution$

destination (line 11). On the contrary, the tree is updated adding the new links in the selected route (line 12). The set of frequency slots is updated eliminating those frequency slots that cannot be selected after adding the new route (line 13). Finally, since a route can cover more than one remaining destination, the set T must also be updated at the end of each iteration (line 14). Once the tree has been built, an available frequency slot is chosen from the set of end-to-end available slots along every tree edge (line 15) and the light-tree is eventually returned.

The success of the algorithm in Table 7.1 strongly depends on the P2MP-R algorithm that finds the candidate routes to be added to the tree. Table 7.2 shows the details of such algorithm, which receives, in addition to other data and parameters previously described, a set R of precomputed routes from the source to the destination being evaluated over the optical layer. After initializing the output variables (line 1 in Table 7.2), the cost of each route is evaluated with the aim of finding the route with minimum cost, taking into account the already built tree and frequency slot availability.

Before computing the cost of a candidate route, the number of frequency slots in common with the current tree is computed (lines 3–4). If some frequency slot is found in common, the number of hops in the route not used in

Table 7.2 P2MP-R algorithm.

INPUT: `R, S, slots, T, solution, <ch, ct, cs>`
OUTPUT: `route, cost`

```
 1: route ← ∅; cost ← ∞
 2: for each r in R do
 3:    nslots ← computeAvailableSlots(r, slots, S, solution)
 4:    if nslots = 0 then continue
 5:    nhops ← computeHopsNotInTree(r, solution)
 6:    ndests ← computeIntermediateDestinations(r, T)
 7:    costIte ← ch*nhops - ct*ndests - cs*nslots
 8:    if cost < costIte then
 9:       route ← r
10:       cost ← costIte
11: return <route, cost>
```

---

the current tree is computed (line 5). Next, the number of intermediate destinations not included in the current tree that are intermediate nodes in the route to the end destination is obtained (line 6). These variables are used to compute the cost of the route (line 7). If cost coefficients are nonnegative, the cost of the route is minimized when:

- the route adds few new links to the existing tree;
- the number of intermediate destinations increases; and
- the number of available frequency slots increases.

It is worth mentioning that the weight of each cost coefficient will affect the selected route and, consequently, the quality of the obtained light-tree. In the next section, we find the configuration producing the minimum request blocking.

Since the P2MP-RSA algorithm basically consists in computing costs of already precomputed routes, its complexity is low. Assuming the worst case, where only one destination is added at each iteration, the complexity of the algorithm in Table 7.1 is $O(|T|^2)$.

The next section evaluates the performance of path and tree schemes.

7.2.4 Illustrative Numerical Results

The performance of the two approaches is compared on three realistic national network topologies: the 12-node Deutsche Telekom (DT), the 22-node British Telecom (BT), and the 30-node Telefonica (TEL). For these experiments, the optical spectrum width is fixed to 2 THz in each link and all connections use the Quadrature Phase-Shift Keying modulation format. We assumed that

SBVTs with four modulators providing a total capacity of 400 Gb/s are equipped in the DC routers. Over such optical topologies, six DCs have been placed in areas with high population density; we consider that connections are requested among those DCs.

A dynamic network environment was simulated where incoming multicast requests arrive at the system following a Poisson process and are sequentially served without prior knowledge of future incoming requests. The number of nodes in each multicast request was selected in the range [3, 6], and the DCs were chosen using a uniform distribution among all DCs. The bitrate of every request was fixed to 100 Gb/s. The time between two consecutive request arrivals and the holding time of an accepted request were randomly generated following an exponential distribution. Different values of offered network load, computed as the total bitrate of the connections being served, were considered by changing the arrival rate.

The P2MP-RSA algorithm was used to solve the multicast provisioning problem for the tree scheme. For comparison purposes, the path scheme was also considered in the experiments. A multicast request involving n destinations is divided into n unicast requests from the source to each single destination. A dynamic RSA algorithm, based on the bulk path computation algorithm proposed in Chapter 5, was implemented. Note that, for the sake of a fair comparison, a multicast request is rejected if one or more unicast requests in the bulk cannot be served.

Figure 7.3 plots the blocking probability (Pb) as a function of the offered load for the path and tree approaches on the three national topologies considered (DCs are collocated with colored nodes). For this set of experiments, we assumed that enough transponders are equipped at each node to ensure that blocking is as a result of the lack of spectrum resources.

The tree scheme provides the best performance on every topology as a result of its efficient resource usage. However, the tree scheme does not achieve an excellent performance in every topology; the rationale behind this is related to the spectrum continuity constraint and the size of the network. In fact, finding a light-tree under the spectrum continuity constraint poses a challenge in large networks, where the number of hops increases.

Let us now study the cost of the schemes related to transponders. Aiming at comparing transponder cost for a given multicast traffic load under the different schemes, we accounted for the number of transponders to ensure $Pb \leq 1\%$. Figure 7.4 shows total transponders' cost for several SBVT/BVT cost ratios for the highest normalized loads unleashing $Pb \leq 1\%$ for each considered network topology. It is clear, in view of Figure 7.4, that the tree scheme brings cost savings compared to the path one since the number of transponders that need to be installed in the DC routers is lower; cost savings of the tree scheme range between 5 and 15%.

Once the performance of the light-trees has been proved, the next section proposes a workflow and studies the support of Path Computation Element

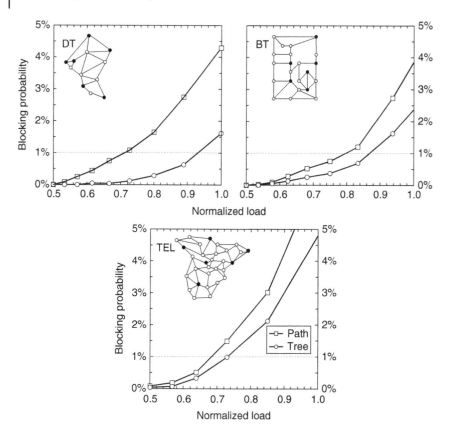

Figure 7.3 Blocking probability versus normalized load.

(PCE) protocol (PCEP) for light-tree provisioning. We assume the front-end/ back-end PCE (fPCE/bPCE) architecture described in Chapter 4.

7.2.5 Proposed Workflows and Protocol Issues

We assume that an Applications Service Orchestrator (ASO) module (see Chapter 6) is responsible for managing multiple DC networks as a single entity. Thus, it generates the connection requests and forwards them to the Application-Based Network Operations (ABNO) [ABNO] controller for unicast and multicast services between Ethernet switches. Figure 7.5 presents the proposed workflow to serve multicast connection requests under the light-tree approach.

Once a multicast connection request arrives at the ABNO controller, it requests a P2MP path computation to fPCE (PCReq message 1 in Figure 7.5).

When the light-tree approach is followed, upon reception of message (1) and because P2MP path computation might take a long time, fPCE delegates it to

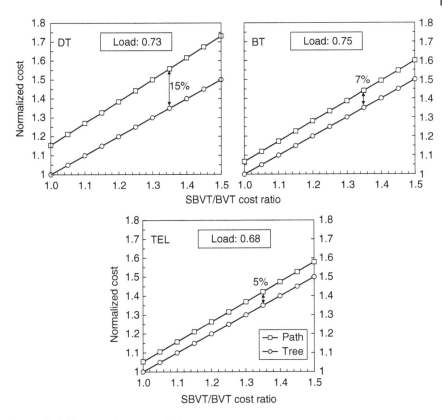

Figure 7.4 Transponder cost analysis.

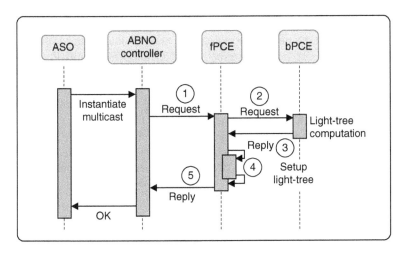

Figure 7.5 Workflow for the light-tree approach.

the specialized bPCE (2). Once the computation ends, bPCE sends back the solution to fPCE (3), which sends the appropriate commands to the underlying data plane (4). When fPCE receives the confirmation from the data plane, a PCRep message (5) is sent back to the ABNO controller.

Let us now focus on analyzing the PCReq and PCRep messages (2) and (3) in the light-tree approach. These messages follow PCEP standards for P2MP LSPs [RFC6006]. PCReq message (2) requests a single P2MP Label-Switched Path (LSP); the end points can be specified using an END-POINT class 3 (P2MP IPv4) object, which includes the IP addresses of the source and the destination (*leaves* in RFC6006 terminology) nodes.

PCRep message (3) specifies the computed RSA for the P2MP LSP using Explicit Route Object (ERO) and secondary ERO (SERO) objects; one single ERO object is used to define the route and spectrum allocation from the source node to one of the leaves, while additional SERO objects define the route and spectrum allocation for each of the rest of leaves; the starting node in each SERO object can be whatever node has been previously defined in ERO or SERO objects.

7.2.6 Experimental Assessment

The experimental assessment was carried out on the test bed depicted in Figure 7.6. The DC orchestrator is OpenStack [OpenStack], and Neutron plugin is in charge of providing the local overlay networks; it interacts with the local Software Defined Networking (SDN) controller and ASO. ASO and each of the OpenStack instances exchange information via the Neutron plugin, so ASO is aware of which virtual machines in each DC belong to the same network. The L2 data plane consists of HP 5406zl Ethernet switches; a Floodlight [Floodlight] controller configures the switches using OpenFlow. The optical data plane includes four programmable wavelength-selective switches; to complete the network topology, six node emulators are additionally deployed. Nodes are handled by colocated Generalized Multi-Protocol Label Switching (GMPLS) controllers running Resource Reservation Protocol-Traffic Engineering (RSVP-TE) with flexgrid extensions [Cu13]. GMPLS controllers communicate with the L0 PCE by means of PCEP. Finally, RSA algorithms are implemented in C++ and deployed in bPCE.

The details of messages (2) and (3) in the proposed workflow are given in Figure 7.7. PCReq message (2) requests a single P2MP LSP, where the end points are specified using an END-POINT object that includes the IP addresses of the source node (X1) and the leaves (X2, X3, and X4).

PCRep message (3) contains the computed RSA for the LSP. In line with [RFC6006], aiming at describing P2MP routes in an efficient way, we use one single ERO object and additional SERO objects to define the route and spectrum allocation. To illustrate that, the route defined by the ERO in Figure 7.7 is

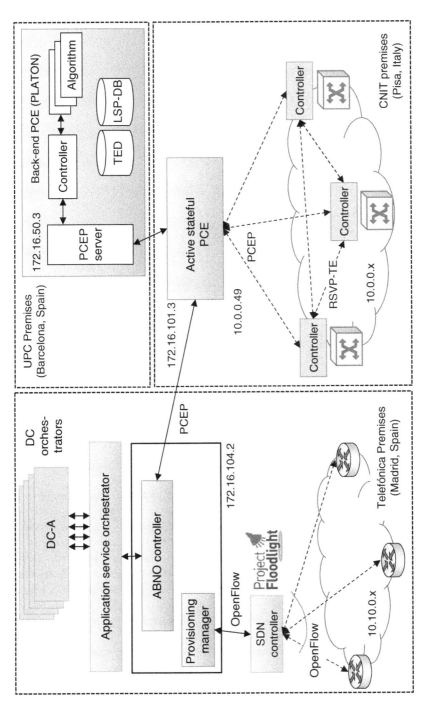

Figure 7.6 Test bed setup. IP addresses are specified.

```
②END-POINT object
    Object Class: END-POINT OBJECT (4)
    Object Type: P2MP IPv4 (3)
    Leaf Type: New leaves to add (1)
    Source IPv4 address: 10.0.0.1
    Destination IPv4 address 1: 10.0.0.2
    Destination IPv4 address 2: 10.0.0.3
    Destination IPv4 address 3: 10.0.0.4
```

```
③EXPLICIT ROUTE object (ERO)
   ▸SUBOBJECT: Unnumbered Interface ID: 10.0.0.1:6
   ▾SUBOBJECT: Label n = -4    m = 4
      Label: 6a 00 ff fc 00 04 00 00
   ▸SUBOBJECT: Unnumbered Interface ID: 10.0.0.6:2
   ▸SUBOBJECT: Label Control
   ▸SUBOBJECT: IPv4 Prefix: 10.0.0.2/32
 ▾SECONDARY EXPLICIT ROUTE object (SERO)
   ▸SUBOBJECT: Unnumbered Interface ID: 10.0.0.6:9
   ▸SUBOBJECT: Label Control
   ▸SUBOBJECT: Unnumbered Interface ID: 10.0.0.9:3
   ▸SUBOBJECT: Label Control
   ▸SUBOBJECT: IPv4 Prefix: 10.0.0.3/32
 ▸SECONDARY EXPLICIT ROUTE object (SERO)
```

Figure 7.7 Details of P2MP objects in PCEP messages (2) and (3).

for leaf X2 and includes X6 as an intermediate node. The first SERO object is for leaf X3 and starts in X6. Since we computed transparent light-trees, the spectrum allocation must be continuous alongside the whole light-tree.

7.3   Anycast Provisioning

In this section, let us focus on a content distribution scenario where a set of core DCs store copies of the same content, for example, an ultra-high-definition uncompressed film. When the content is requested from a remote location (assume large size files), for example, a metro DC that downloads the film, transcodes it into the appropriate encoding formats, and serves it to end users, an *anycast* connection can be established between that metro DC and any core DC. Recall that we defined an anycast connection by the tuple $<o, T, b>$, where o represents the source node, T the set of destination nodes, and b the requested bitrate. Only one destination needs to be selected from the set T, and therefore anycast demands can be served by establishing one single *lightpath* from the source node to one of the destination nodes. Figure 7.8 illustrates an example of anycast provisioning where content is replicated in three different core DCs, that is, $o = \text{MetroDC5}$ and $T = \{\text{CoreDC1, CoreDC2, CoreDC4}\}$.

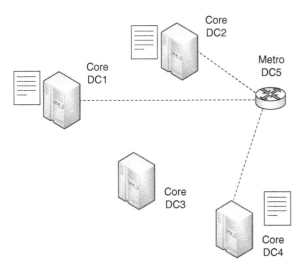

Figure 7.8 Example of anycast provisioning.

7.3.1 Optical Anycast (AC_RSA) Problem Statement

The AC_RSA problem aims at finding an anycast optical connection between a selected source and a set of destinations. In our content distribution scenario, the source node is a metro DC that requests the establishment of a bidirectional LSP, whereas the destinations are a set of core DCs that store the desired content. The problem can be stated as follows:

Given:
- the network topology represented by a graph $G(N, E)$,
- the number of available slices in each link $e \in E$,
- the source metro DC $o \in N$,
- the set T of core DCs (destinations),
- the number of slices required for the anycast connection.

Output: the RSA for the bidirectional anycast connection between o and one core DC in T.

Objective: minimize the number of optical resources required.

7.3.2 Exact Algorithm for the AC_RSA Problem

The approach that can be devised to solve this problem consists in building an auxiliary graph with links connecting the metro DC (source) to all the possible core DCs (destinations). Those links in the auxiliary graph will be supported by optical connections, so the metric of each link is a function of the estimated number of optical hops for the lightpath supporting it. The graph is augmented adding one *dummy* node and links connecting each destination to the dummy

(a)

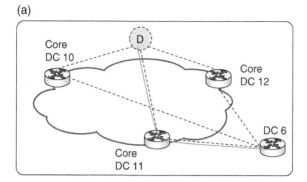

(b)

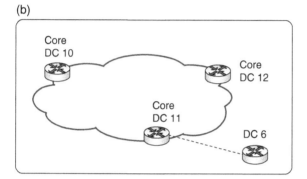

Figure 7.9 Anycast provisioning. (a) Path computed on the augmented graph and (b) final anycast connection.

node; see Figure 7.9a, where the dummy node is labeled as D. Then, the shortest path is computed between the source metro DC and the dummy node, and the resulting path contains the optimal route for the anycast connection to be established (Figure 7.9b).

7.3.3 Illustrative Numerical Results

The algorithm presented in the previous section was evaluated on an ad hoc event-driven simulator based on OMNeT++ [OMNet]. To that end, the 24-node 43-link Large Carrier US topology [Hu10], reproduced in Figure 7.10, is used; metro DCs were placed in every node location, while six core DCs were strategically placed in New York, Florida, Illinois, Texas, Washington, and California. Each core DC is connected to the optical transport network through a switch equipped with 100 Gb/s transponders.

A dynamic network environment was simulated, where 100 Gb/s anycast connection requests arrive following a Poisson process. The holding time of the connection requests is exponentially distributed with a mean value equal to

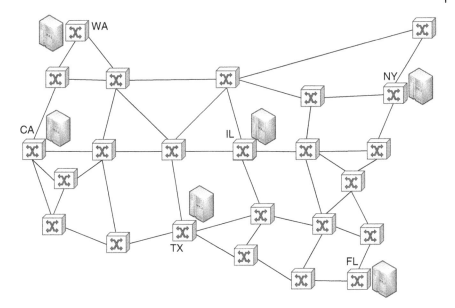

Figure 7.10 Large carrier US network topology.

1 h. For each connection, a metro DC is randomly chosen with equal probability (uniform distribution). Different loads are created by changing the arrival rate while keeping the mean holding time constant.

Figure 7.11a shows the blocking probability as a function of the offered load. Each plot in the graph is for a fixed number of core DCs (destinations), ranging from 1 to 5. As observed, the effects of increasing the number of possible destination DCs has a direct effect on the blocking probability; assuming a target 1% blocking probability, 66% more traffic can be served when two core DCs are placed compared to one single core DC. The rationale behind this is related to the length of the anycast connections (Figure 7.11b); the number of hops to reach the core DCs decreases when more core DCs are considered, thus reducing resource utilization, especially when the selected core DCs are geographically distributed.

7.3.4 Proposed Workflow

When a metro DC needs to synchronize contents with the core DCs, the local cloud resource manager issues an anycast connection request to the ASO (Figure 7.12).

The ASO issues a message to the ABNO controller with the request (message 1 in Figure 7.12). The controller creates a PCReq message and forwards it to the fPCE (1), which computes the route for the connection using the AC_RSA algorithm and issues a PCInit request for the LSP to the Provisioning Manager. When

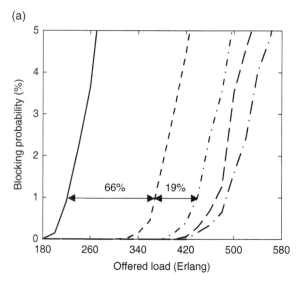

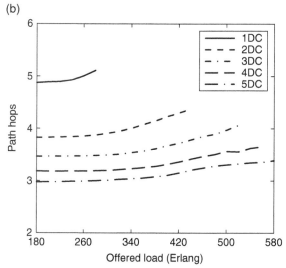

Figure 7.11 Anycast connection provisioning: (a) blocking probability and (b) number of hops versus offered load.

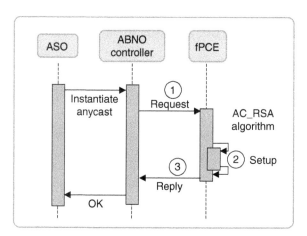

Figure 7.12 Workflow for anycast provisioning.

①
```
▼Path Computation Element communication Protocol
  ▶Path Computation Request Header
  ▶RP object
  ▼END-POINT object
    Object Class: END-POINT OBJECT (4)
    0011 .... = Object Type: Point-to-MultiPoint IPv4 (3)
    ▶Flags
    Object Length: 24
    Leaf Type: New anycast leave to add (5)
    Source IPv4 Address: 10.0.1.1 (10.0.1.1)
    Destination IPv4 Address: 10.0.1.10 (10.0.1.10)
    Destination IPv4 Address: 10.0.1.11 (10.0.1.11)
    Destination IPv4 Address: 10.0.1.12 (10.0.1.12)
  ▶BANDWIDTH object
```

③
```
▼Path Computation Element communication Protocol
  ▶Path Computation Reply Header
  ▶RP object
  ▼EXPLICIT ROUTE object (ERO)
    Object Class: EXPLICIT ROUTE OBJECT (ERO) (7)
    0001 .... = Object Type: 1
    ▶Flags
    Object Length: 120
    ▶SUBOBJECT: Unnumbered Interface ID: 10.0.1.1:1
    ▶SUBOBJECT: Label Control
    ▶SUBOBJECT: Label Control
    ▶SUBOBJECT: Unnumbered Interface ID: 10.0.0.1:3
    ▶SUBOBJECT: Label Control
    ▶SUBOBJECT: Label Control
    ▶SUBOBJECT: Unnumbered Interface ID: 10.0.0.3:110
    ▶SUBOBJECT: Label Control
    ▶SUBOBJECT: Label Control
    ▶SUBOBJECT: IPv4 Prefix: 10.0.1.10/32
  ▶BANDWIDTH object
```

Figure 7.13 Anycast provisioning PCReq and PCRep messages.

the LSP has been set up, the fPCE replies to the controller (3), which in turn, replies to the ASO. Finally, the ASO updates its database and replies to the cloud resource manager confirming the availability of the requested connection.

Note that for the anycast path computation, the set of candidate core DCs has to be specified. To this end, we propose extending the P2MP END-POINT object defined in [RFC6006] with a new leaf type to select one-of-many destinations.

7.3.5 Experimental Assessment

Figure 7.13 shows the P2MP END-POINT object included in PCReq message (1); we used the proposed leaf type 5 labeled as *New anycast leave to add* to request an anycast connection from metro DC 1 (10.0.1.1) to any of the core

DCs: 10, 11, and 12. In the PCRep message (3), an ERO object specifies the route from metro DC 1 to core DC 10 through the optical layer. Note that since the request was for an anycast connection, the route to one single destination is provided. Moreover, since a bidirectional LSP was established, two Label Control subobjects are included in the PCRep message.

The control contribution to the anycast provisioning workflow processing time was less than 3 ms, including for the messages exchange and the AC_RSA algorithm in the fPCE.

7.4 Conclusions

This chapter focused on multicast and anycast optical connections. For the multicast optical connections, two approaches have been proposed to serve large capacity (e.g., 100 Gb/s) multicast connectivity services on the optical layer, where a set of DCs is connected through an EON. In the first approach, each multicast request is served by establishing a set of lightpaths. In the second approach, each multicast request is served by establishing a single light-tree. Results obtained by simulation validated the tree scheme as the best option to serve multicast traffic. Next, the feasibility of implementing the tree approach using standardized control plane protocols was studied, and a workflow was proposed. Multicast connections (including light-trees) are supported in PCEP using P2MP extensions.

Next, the anycast provisioning problem was studied on a content distribution architecture consisting of metro and core DCs. An exact algorithm that finds anycast shortest paths on an auxiliary graph was proposed. Exhaustive simulations were carried out to evaluate the performance of the proposed algorithm when the number of possible core DCs ranged from 1 to 5. The results showed that high efficiency could be achieved when increasing the number of core DCs; more than 66% load increment was observed.

Aiming at experimentally assessing the proposed anycast provisioning problem, an ABNO-based architecture was considered. The ASO module was responsible for service management and issues requests to the ABNO. A workflow was proposed, and the PCEP feasibility was studied; PCEP extensions for anycast connection provisioning were proposed.

Part III

Recovery and In-operation Planning
in Single Layer Networks

8

Spectrum Defragmentation

Alberto Castro[1], Lluís Gifre[2], Marc Ruiz[2] and Luis Velasco[2]

[1] University California Davis, Davis, CA, USA
[2] Universitat Politècnica de Catalunya, Barcelona, Spain

Dynamic operation of Elastic Optical Networks (EONs) might cause the optical spectrum to be divided into fragments, which makes it difficult to find the contiguous spectrum of the required width along the paths for incoming connection requests, thus leading to an increased blocking probability. To alleviate to some extent that spectrum fragmentation, we propose applying spectrum defragmentation, triggered before an incoming connection is blocked, to change the spectrum allocation of already established optical connections so as to make room for the new connection. Two different defragmentation schemes are presented: (i) *spectrum reallocation* and (ii) *spectrum shifting*. Spectrum reallocation changes the central frequency of already established connections without rerouting them. Because spectrum reallocation might cause traffic disruption, spectrum shifting can be applied as a hitless spectrum defragmentation procedure, so as to shift the central frequency of already established connections to create wider contiguous spectrum fragments for incoming connections. After studying the two spectrum defragmentation schemes (Integer Linear Programming (ILP) formulations and heuristic algorithms are proposed), we explore how the Application-based Network Operations (ABNO) architecture can deal with spectrum defragmentation while the network is in operation. A workflow involving several elements in the ABNO architecture is proposed and experimentally assessed.

Provisioning, Recovery, and In-operation Planning in Elastic Optical Networks,
First Edition. Luis Velasco and Marc Ruiz.
© 2017 John Wiley & Sons, Inc. Published 2017 by John Wiley & Sons, Inc.

Table of Contents and Tracks' Itineraries

8.1 Introduction

In dynamic scenarios, spectrum fragmentation appears in Wavelength Division Multiplexing (WDM)-based networks, subject to the wavelength continuity constraint as a consequence of the unavailability of spectrum converters. Fragmentation increases the blocking probability of connection requests, making the network grade of service worse.

The first work addressing the fragmentation problem in WDM-based networks was presented in [Le96], where a strategy named *move-to-vacant wavelength retuning* was proposed. It is based on reallocating the already established optical connections to other wavelengths in order to make enough room for new connections that would otherwise be blocked. To reduce the disruption period, a parallel connection in the same route is first created in a free wavelength for each connection to be rerouted. This strategy provides the advantage that old and new routes share the same optical nodes, so their computation and setup are facilitated. Authors in [Mo99] presented an algorithm to improve that in [Le96], introducing rerouting in addition to simple wavelength retuning. Several rerouting strategies are investigated in [Ch08], where the authors also proposed intentional rerouting, which runs periodically. The fragmentation problem also appears in grooming-capable optical networks, and some works have already addressed this. The work in [Ya08] presented two rerouting algorithms to alleviate resource inefficiency and improve the network throughput. Authors in [Ag09] presented a centralized flow reallocation module to minimize the overall network cost. The standardized *make-before-break* rerouting technique (similar to the move-to-vacant one) is also experimentally tested establishing the new path for rerouting and transferring traffic from the current path to the new one before the old path is finally torn down. In this way, reallocation procedures do not cause any working traffic disruption.

Two main strategies have been proposed to reallocate (including rerouting and wavelength reassignment) already established paths in the context of WDM-based networks: periodic defragmentation and path-triggered defragmentation. The former strategy focuses on minimizing fragmentation itself all over the network at a given period of time, whereas the latter focuses on making enough room for a given connection request if it cannot be established with the current resources allocation. Periodic defragmentation, requiring long computation times as a result of the amount of data to be processed, is essentially performed during low activity periods, for example, during nights [Ag09]. Conversely, path-triggered defragmentation, involving only a limited set of already established connections, might provide solutions in shorter times and can be run in real time. It is worth noting that incoming connection requests or tear-downs arriving while network reoptimization or defragmentation operations are running must be kept apart until the latter operation terminates.

The problem of fragmentation is worse in Elastic Optical Networks (EONs) since, in addition to spectrum continuity, spectrum contiguity needs to be guaranteed. Some works recently addressed this issue [We11], [Am11], [Ta11]. Authors in [We11] proposed a spectrum defragmentation algorithm that takes advantage of Wavelength Selective Switches (WSSs) that support wavelength conversion. Blocking probability can be reduced by 50% at the expense of notably increasing the cost of the photonic layer. Experimental wavelength conversion and its application to reduce spectrum fragmentation were presented in [Am11]. In [Ta11], the authors proposed an algorithm that performs spectrum defragmentation using rerouting instead of wavelength conversion. Since rerouting usually causes service disruption, they used the make-before-break technique to minimize disruption.

In this chapter, we apply reoptimization to EONs to solve the problem of spectrum fragmentation. Specifically, spectrum defragmentation is applied after a request for a new connection cannot be served.

8.2   Spectrum Reallocation and Spectrum Shifting

Aiming at improving the grade of service of EONs, that is, reducing blocking probability, in this chapter we review routing and defragmentation algorithms that reallocate already established connections in the spectrum without rerouting them, so as to make enough room for incoming connection requests.

The scheme follows a path-triggered scheme, where the defragmentation algorithm is triggered whenever enough resources for an incoming connection request have not been found. Every link in the route from a set of *k*-shortest routes connecting source and destination nodes is checked to know whether the number of available frequency slices is equal to or higher than the required number for the incoming connection request. If enough frequency slices are available in one of the shortest routes, a defragmentation algorithm is run to find a set of already established path reallocations that create enough room for the connection request in the selected route (*newP*); otherwise, the connection request is blocked. Figure 8.1 summarizes the scheme for routing and spectrum defragmentation.

Figure 8.2 illustrates the spectrum fragmentation on a small network consisting of 9 nodes and 11 links (Figure 8.2a), where the entire spectrum width corresponds to 16 frequency slices. Figure 8.2b represents the utilization of

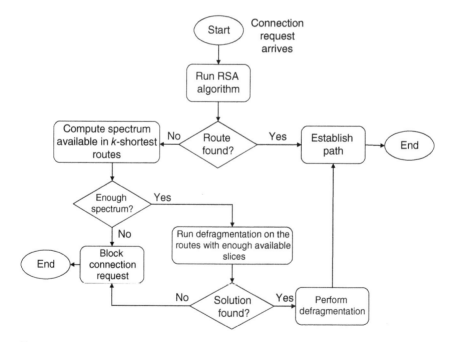

Figure 8.1 Routing and spectrum defragmentation.

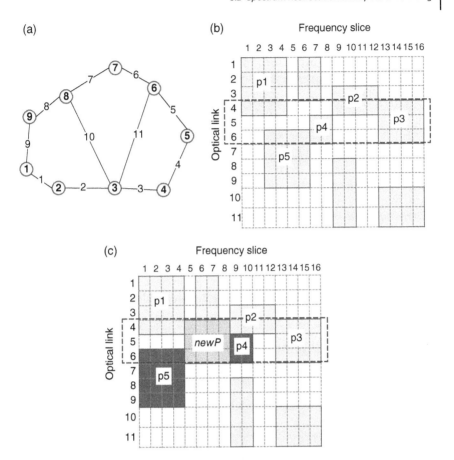

Figure 8.2 Example of spectrum defragmentation. (a) 9-node 11-link network where the entire spectrum width corresponds to 16 frequency slices. (b) Utilization of each frequency slice in the network. (c) Reallocation to make room for *newP*.

each frequency slice in the network, where a number of paths are already established. In this scenario, the connection request between nodes 4 and 7 requesting a four-slice slot cannot be served. Notwithstanding, each link in the shortest route 4–5–6–7 (links 4–5–6) had at least four free slices, and then the request could be established reallocating some of the established paths. In Figure 8.2c, paths p_4 and p_5 are reallocated making enough room for the new optical connection *newP*.

Aiming at reducing path setup delays, we impose a number of constraints to this approach:

1) We use $k = 1$, that is, only the links in the shortest path of the connection request are checked for spectrum availability.

2) Already established paths can be reallocated in the spectrum, but their routes are not modified (rerouting is not used).

3) Only those paths using any link in the route of *newP* are considered as a candidate to be reallocated, thus preventing any other paths to be reallocated.

4) The number of paths that are actually reallocated is restricted by a threshold.

Notice that in the example in Figure 8.2, spectrum defragmentation entailed reallocating paths p_4 and p_5 to a nearby spectrum slot, where the spectrum in between the initial and the final slices was not allocated to any other path. In general, spectrum reallocation might entail changing the path to an arbitrary new frequency slot. Assuming that, in the next section, we propose the *SPectrum REaLLOcation* (SPRE (LLO) → (SSO)) problem that reallocates already established connections in the spectrum. However, that reallocation might entail using the *make-before-break* rerouting technique to avoid traffic disruption. A different approach to avoid traffic disruption is using spectrum shifting for defragmentation. To that end, we propose the *SPectRum shiftING* (SPRING) problem to solve the defragmentation problem.

8.3 Spectrum Reallocation: The SPRESSO Problem

8.3.1 Problem Statement

The SPRESSO problem can be formally stated as follows:

Given:
- an optical network, represented by a graph $G(N, E)$, N being the set of nodes and E the set of optical links connecting two nodes,
- a set S of frequency slices available in each link $e \in E$,
- a set P of already established paths,
- a new path (*newP*) to be established in the network. A route for the path has been already selected, but there is no feasible spectrum allocation,
- the threshold number of paths to be reallocated.

Output:
- new spectrum allocation for each path to be reallocated,
- the spectrum allocation for *newP*.

Objective: Minimize the number of paths to be reallocated so as to fit *newP* in.

8.3.2 ILP Formulation

An Integer Linear Programming (ILP) formulation for the SPRESSO problem is presented next that takes as input the set of already established paths that share common links with *newP*.

The following sets and parameters are defined:

E	Set of optical links, index e.
P	Set of already established paths, index p.
$newP$	New path to be established in the network.
$E(p)$	Subset of E with those links in the route of path p.
$P(e)$	Subset of P with those paths using optical link e.
$P(p)$	Subset of P with those paths sharing at least one link with path p. $P(p) = \bigcup_{e \in E(p)} P(e)$.
Pm	Subset of P with the candidate paths. $Pm = P(newP)$.
S	Set of frequency slices, index s.
C	Set of slots, index c. Each slot c contains a subset of contiguous slices.
$C(p)$	Subset of C with those slots that can be assigned to path p.
$S(p)$	Subset of S with those frequency slices allocated to path p.
δ_{cs}	1 if slot c uses slice s, 0 otherwise.
ω_{pc}	1 if path p was using slot c, 0 otherwise.
η_{es}	1 if slice s in optical link e is free, 0 otherwise. Note that to compute η_{es} only paths in $P \backslash Pm$ are considered.
$\eta_{es}(p, c)$	1 if reallocation of path p to slot c is feasible, 0 otherwise.
$maxR$	The threshold number of paths to be reallocated.

Additionally, the decision variables are:

x_{pc}	Binary, 1 if slot c is assigned to path p, 0 otherwise.
y_p	Binary, 1 if path p is reallocated, 0 otherwise.
r	Integer with the number of already established paths to reallocate.

Then, the ILP model for the SPRESSO problem is as follows:

$$(\text{SPRESSO}) \quad \text{minimize} \quad r \tag{8.1}$$

subject to:

$$r = \sum_{p \in Pm} y_p \tag{8.2}$$

$$\sum_{c \in C(p)} x_{pc} = 1 \quad \forall p \in Pm \cup \{newP\} \tag{8.3}$$

$$x_{pc} - \omega_{pc} \le y_p \quad \forall p \in Pm \cup \{newP\}, c \in C(p) \tag{8.4}$$

$$\sum_{p \in P(e)} \sum_{c \in C(p)} \delta_{cs} \cdot x_{pc} \le \eta_{es} \quad \forall e \in E, s \in S \tag{8.5}$$

$$r \le maxR \tag{8.6}$$

The objective function (8.1) minimizes the number of paths that need to be reallocated in the spectrum so *newP* can be allocated. Constraint (8.2) counts the number of paths that need to be reallocated. Note that only *newP* and those paths sharing at least one common optical link with *newP* are in the set of candidate paths. Constraint (8.3) assigns one slot to every candidate path and *newP*. Constraint (8.4) stores whether path *p* is reallocated comparing current and assigned slots. Constraint (8.5) guarantees that each frequency slice in a given optical link is assigned to one path at the most. Finally, constraint (8.6) limits the number of paths to be reallocated.

The number of variables and constraints is $O(|Pm|\cdot|C|)$ and $O(|Pm|\cdot|C| + |E|\cdot|S|)$, respectively. Although the size of the problem is limited, it must be solved in real time (in the order of milliseconds) to minimize the optical connections setup delay. In our experiments over the networks described in Section 8.5, the problem takes several seconds on average to be solved; more than 1 min in the worst case. Therefore, in the next section, we propose a heuristic algorithm that provides near-optimal solutions in practical computation times, short enough to be used in the management plane of real EONs.

8.3.3 Heuristic Algorithm

In this section, we propose a Greedy Randomize Search Procedure (GRASP)-based heuristic (see Chapter 2). In our implementation, a multi-greedy heuristic consisting of three different constructive algorithms is used to provide diversification. A multi-start local search procedure [Fe95] is used to find locally optimal solutions.

The GRASP procedure is hybridized with a Path Relinking (PR) heuristic, as a strategy to integrate both intensification and diversification. PR generates new solutions by exploring the trajectories connecting high-quality solutions. The evaluated path starts at a so-called *initiating* solution and moves toward a so-called *guiding* solution, which is usually taken from a stored set of good-quality solutions called the *elite set* (*ES*). *ES* is initially populated with three GRASP solutions resulting from applying the three greedy constructive algorithms plus the corresponding local search. Each subsequent iteration of the GRASP algorithm produces a new feasible solution using one randomly selected greedy constructive algorithm. The decision to add a new feasible solution to the ES is taken based on both, the cost function value and the diversity it adds to *ES*.

Table 8.1 reproduces the GRASP+PR algorithm for the SPRESSO problem, where a set of values in the interval [0, 1], named *GreedyProportions*, is used to run the required proportion of each constructive algorithm.

To cope with the stringent computation time requirement, it is essential to produce as diverse GRASP solutions as possible with the purpose of reducing the number of iterations and thus the computation time needed to find

Table 8.1 GRASP+PR heuristic algorithm for the SPRESSO problem.

INPUT: *newP*, α, β, *maxR*, *GreedyProportions*, *numIter*
OUTPUT: *Sol*

```
 1: Incumbent ← ∅
 2: ES ← ∅
 3: for iter = 1…numIter do
 4:     r ← getRandom()
 5:     if r < GreedyProportions.alg1 then
 6:         Sol ← Algorithm-1(newP, α, maxR)
 7:     else if r < GreedyProportions.alg2 then
 8:         Sol ← Algorithm-2(newP, α, maxR)
 9:     else Sol ← Algorithm-3(newP, α, β, maxR)
10:     Sol ← LocalSearch(Sol)
11:     {Sol, ES} ← PathRelinking(Sol, ES)
12:     if |Sol| < |Incumbent| then Incumbent ← Sol
13: return Incumbent
```

high-quality near-optimal solutions. In view of that, we have developed three different greedy randomized constructive algorithms, each building really different feasible solutions.

Table 8.2 describes the greedy randomized constructive Algorithm-1. It involves reallocating those already established paths whose routes intersect that of *newP*, so as to generate enough room for the latter. Equation (8.7) is used to quantify the quality of reallocating a given path *i*; it explores every feasible spectrum reallocation for path *i* and accounts the greatest amount of contiguous space generated in the specific sets of optical links and slots for *newP*. A Restricted Candidate List (RCL) containing those paths with the best quality is used. Parameter α in the real interval $[0, 1]$ determines the paths in the RCL.

$$q(i) = \max_{c1 \in C(i)} \max_{c2 \in C(newP)} \sum_{e \in E(newP)} \sum_{s \in S(c2)} \eta_{es}(i, c_1) \tag{8.7}$$

As stated before, Algorithm-1 builds solutions by successively reallocating already established paths. Note that this is in contrast to removing all paths intersecting *newP* and reallocating them next to minimize the number of paths effectively reallocated, which is the approach adopted to solve the SPRESSO ILP model previously introduced.

To avoid, up to some extent, blocking among reallocations that the sequentiality of the algorithm might cause, we allow paths to be reallocated multiple times. Note, however, that aiming at minimizing disruption we could take advantage of sequentiality to implement the make-before-break mechanism.

Table 8.2 Greedy randomized constructive Algorithm-1.

INPUT: $newP$, α, $maxR$

OUTPUT: Sol

```
 1: Sol ← ∅
 2: Initialize the candidate set: Q ← P(newP)
 3: spaceNeeded ← |E(newP)| · |S(newP)|
 4: while Q ≠ ∅ AND |Sol| ≤ maxR do
 5:     Evaluate the quality q(i) for all i∈Q using equation (8.7)
 6:     qmin ← min{q(i) | i∈Q, i feasible}
 7:     qmax ← max{q(i) | i∈Q, i feasible}
 8:     RCL ← {i∈Q | q(i) ≥ qmax - α(qmax - qmin)}
 9:     if RCL = ∅ then
10:         Sol ← ∅
11:         break
12:     if qmax ≥ spaceNeeded then
13:         i ← arg max (RCL)
14:         Sol ← Sol ∪ {i}
15:         Perform the spectrum reallocation defined by element i
16:         break
17:     Select an element i from RCL at random
18:     Q ← Q \ {i}
19:     Sol ← Sol ∪ {i}
20:     Perform the spectrum reallocation defined by element i
21: if |Sol| ≤ maxR then return Sol
22: return ∅
```

However, special attention must be paid when paths are allowed to be reallocated multiple times to avoid cycling, that is, repeating the same sequence of reallocations periodically. In this regard, in addition to limit the length of the solutions as previously stated, two measures have been adopted in Algorithm-2: firstly, a single path cannot be reallocated in two consecutive iterations. This is implemented by storing the path that has been reallocated in the current iteration so that it remains in the candidate list Q, but it can be added to RCL in the next iteration; secondly, any path is prevented from being reallocated back to its previous slot, and then the last slot allocated to each path needs to be stored.

Moreover, quality quantification has been slightly modified as shown in equation (8.8) to avoid a given slot c to be considered as a feasible reallocation option for path i. Table 8.3 shows the constructive algorithm updated thus.

$$q(i,c) = \max_{\substack{c1\in C(i) \\ c1\neq c}} \max_{c2\in C(newP)} \sum_{e\in E(newP)} \sum_{s\in S(c2)} \eta_{es}(i,c_1) \tag{8.8}$$

Table 8.3 Greedy randomized constructive Algorithm-2.

INPUT: *newP*, *α*, *maxR*
OUTPUT: *Sol*

```
 1: Sol ← Ø
 2: Initialize the candidate set: Q ← P(newP)
 3: for each p∈Q do p.lastSlot ← p.slot
 4: spaceNeeded ← |E(newP)| · |S(newP)|
 5: lastRealloc ← Ø
 6: while |Sol| ≤ maxR do
 7:     Evaluate the quality q(i) for all i∈Q using equation (8.8)
 8:     qᵐⁱⁿ ← min{q(i, i.lastSlot) | i∈Q, i feasible ∩ i ≠
        lastRealloc}
 9:     qᵐᵃˣ ← max{q(i, i.lastSlot) | i∈Q, i feasible ∩ i ≠
        lastRealloc}
10:     RCL ← {i∈Q | q(i) ≥ qᵐᵃˣ - α(qᵐᵃˣ - qᵐⁱⁿ)}
11:     if RCL = Ø then
12:         Sol ← Ø
13:         break
14:     if qᵐᵃˣ ≥ spaceNeeded then
15:         i ← arg max (RCL)
16:         Sol ← Sol U {i}
17:         break
18:     Select an element i from RCL at random
19:     Sol ← Sol U {i}
20:     lastRealloc ← i
21:     i.lastSlot ← i.slot
22:     Perform the spectrum reallocation defined by element i
23: if |Sol| ≤ maxR then return Sol
24: return Ø
```

Finally, the more complex constructive Algorithm-3, illustrated in Table 8.4, exploits a totally different method; path reallocation is confined to incremental sizes of the spectrum, starting from the number of slices required by *newP*. For a given number of slices *n*, equation (8.9) is used to initialize the candidate set $Q_1(n)$ with the sets of already established paths whose route intersects that of *newP* and are using, at least in part, the area of the spectrum considered.

$$Q_1(n) = \left\{ \{ p \in P(newP) : S(c) \cap S(p) \neq \varnothing \}, \quad \forall c \in C : |S(c)| = n \right\} \qquad (8.9)$$

Table 8.4 Greedy randomized constructive Algorithm-3.

INPUT: $newP$, α, β, $maxR$

OUTPUT: Sol

```
 1: Sol ← ∅
 2: n ← |S(newP)|
 3: while n ≤ |S| do
 4:     Initialize the candidate set Q₁(n) as defined in
        equation (8.9)
 5:     Evaluate the quality q(i) = |i| for all i ∈ Q₁(n)
 6:     while Q₁(n) ≠ ∅ do
 7:         q₁ᵐⁱⁿ ← min{q(i) | i ∈ Q₁(n)}
 8:         q₁ᵐᵃˣ ← max{q(i) | i ∈ Q₁(n)}
 9:         RCL1 ← {i ∈ Q₁(n) | q(i) ≤ q₁ᵐⁱⁿ + α(q₁ᵐᵃˣ−q₁ᵐⁱⁿ)}
10:         Select an element i from RCL1 at random
11:         Q₁(n) ← Q₁(n) \ {i}
12:         for each p ∈ i do
13:           Release network resources used by p
14:           Initialize the candidate set Q₂ ← i ∪ {newP}
15:           while Q₂ ≠ ∅ AND |Sol| ≤ maxR do
16:             Evaluate the quality q(p) = |E(p)|·|S(p)| for all
                p ∈ Q₂
17:             if ∃ p ∈ Q₂ | p not feasible then
18:               Sol ← ∅
19:               break
20:             q₂ᵐⁱⁿ ← min{q(p) | p∈Q₂}
21:             q₂ᵐᵃˣ ← max{q(p) | p∈Q₂}
22:             RCL2 ← {p∈Q₂ | q(p) ≥ q₂ᵐᵃˣ − β(q₂ᵐᵃˣ−q₂ᵐⁱⁿ)}
23:             Select an element p from RCL2 at random
24:             Q₂ ← Q₂ \ {p}
25:             if ∃ c∈C(p) | ηₑₛ=1 ∀e∈E(p), s∈S(c) then
26:               Sol ← Sol ∪ {p}
27:               Perform the spectrum reallocation
28:             else
29:               Sol ← ∅
30:               break
31:           if |Sol| >maxR then return ∅
32:           if Sol ≠ ∅ then return Sol
33:           Restore network resources
34:     n++
35: return ∅
```

Algorithm-3 uses two nested Rcls. An outer $RCL1$ is built from $Q_1(n)$ using the size of each element to quantify its quality. After selecting at random one element i from $RCL1$, every path in that element is removed from the network and its allocated resources are released. Next, a second candidate set is initialized with those paths in i plus *newP*. All these paths must be reallocated, and the inner $RCL2$ is used to find a reallocation order; the required spectrum space is used as a criterion in this case. If some paths cannot be reallocated, another element i is selected from $RCL1$ until $Q_1(n)$ is empty; when a new iteration of the algorithm starts and the size of the spectrum area is increased.

Due to the fact that a feasible solution *Sol* resulting from the constructive phase has no guarantee of being locally optimal, GRASP heuristics apply a local search procedure starting at *Sol* in the hope of finding a better solution in its neighborhood.

In our problem, a feasible solution is defined as the set of paths to be reallocated and their reallocation positions to make room for *newP*. To avoid the sequentiality problem inherent to the constructive algorithms, all paths with multiple reallocations in *Sol* are removed. Next, every path is checked to be moved back to its original allocation, thus reducing the size of the solution and resulting in a solution x.

As mentioned earlier, a multi-start local search procedure to find the optimal solution in the neighborhood of x ($N(x)$) is started. $N(x)$ contains those feasible solutions that can be reached from x by one move, defined as an exchange between a path p_2 not in x and another path p_1 in x, such that after doing it some other path p in x can be moved back to its original allocation. The local search procedure ends when no feasible movement can be done in x.

After a solution is returned from the GRASP phase, PR is applied between the current solution and a selected solution from *ES*. Then, the best solution found in this iteration is a candidate for inclusion in *ES* and it is only added if certain quality and diversity criteria are met. To include a new solution in *ES* a trade-off between quality and diversity is usually evaluated [Re10]. Solution x enters *ES* if it is better than the best solution $x' \in ES$. Alternatively, if x is better than the worst $x' \in ES$, the distance $d(x, ES) = \min\{x \ominus x': \ x' \in ES\}$ between the solution x and *ES* is computed to ensure enough diversity in *ES*. Only in the case where $d(x, ES) \geq d(ES) = \min\{x' \ominus x'': x', x'' \in ES\}$ the solution x is added to *ES*. It is worth noticing that to compute symmetrical differences, only reallocated paths are considered omitting target slots differences. To maintain the size of *ES* constant, before adding a solution x to *ES*, another solution $x' \in ES$ must be removed from that set. To keep quality and diversity in *ES*, the closest solution to x among those worse than x is selected.

8.4   Spectrum Shifting: The SPRING Problem

8.4.1   Problem Statement

The SPRING problem can be formally stated as follows:

Given:
- an optical network, represented by a graph $G(N, E)$, N being the set of nodes and E the set of optical links connecting two nodes,
- a set S of frequency slices available in each link $e \in E$,
- a set P of already established paths,
- a new request ($newP$) to be established in the network. A route for the path has been already selected but there is no feasible spectrum allocation.

Output:
- for each path to be shifted, its new spectrum allocation,
- the spectrum allocation for $newP$.

Objective: Minimize the number of paths to be shifted to fit $newP$ in.

An ILP model for the SPRING problem is presented next.

8.4.2   ILP Formulation

The ILP model takes as input the set of established paths that share common links with the shortest route for $newP$.

In addition to the sets and parameters introduced in Section 8.3.2, the following parameter is defined:

$\beta_{pp'}$
- 0 if originally the index of the first slice allocated to path p was lower than that of the first slice allocated to path p', provided that p' was sharing at least one link with path p, that is, $c < c'$, $c \in C(p)$, $p' \in P(p)$, $c' \in C(p')$.
- 1 if $c > c'$.

Note that $\beta_{pp'}$ is not defined for those paths not sharing any link.

Then, the formulation for the SPRING problem is as follows:

$$(\text{SPRING}) \qquad \min \sum_{p \in Pm} y_p \qquad\qquad (8.10)$$

subject to:

Constraints (8.3)–(8.5) in the SPRESSO ILP formulation

$$\sum_{c' \in C(p')} c' \cdot x_{p'c'} - \sum_{c \in C(p)} c \cdot x_{pc} < \beta_{pp'} \cdot |S| \quad \forall p \in Pm, p' \in P(p) \qquad (8.11)$$

$$\sum_{c' \in C(p')} c' \cdot x_{p'c'} - \sum_{c \in C(p)} c \cdot x_{pc} > (\beta_{pp'} - 1) \cdot |S| \quad \forall p \in Pm, p' \in P(p) \qquad (8.12)$$

The objective function (8.10) minimizes the number of paths that need to be reallocated in the spectrum so *newP* can be served.

Constraints (8.11) and (8.12) ensure feasible shifting by taking care of relative spectral position between path pairs sharing any link; that is, if two paths p and p' sharing a common link were originally allocated to slots with indexes $c < c'$ ($\beta_{pp'} = 0$), then that relative spectral position must be kept in the solution. The same must be ensured when $c > c'$ ($\beta_{pp'} = 1$). Note that when paths p and p' do not share any link, these constraints do not apply.

The number of variables is $O(|Pm| \cdot |C|)$, and the number of constraints is $O(|Pm|^2 + |Pm| \cdot |C| + |E| \cdot |S|)$. Similarly as for SPRESSO, although the size of the SPRING problem is limited, it must be solved in real time, for example, milliseconds, to minimize setup delay of *newP*. For this very reason, we propose a heuristic algorithm in next section.

8.4.3 Heuristic Algorithm

In this section, we propose to use the heuristic algorithm described in Table 8.5 to solve the SPRING problem.

Table 8.5 Algorithm for the SPRING problem.

INPUT: E, n

OUTPUT: *Solution*

```
 1: Solution ← ∅
 2: numPaths ← INFINITE
 3: for s = 1...|S| do
 4:   P⁺ ← ∅; P⁻ ← ∅; Pˢ ← ∅
 5:   for each e ∈ E(newP) do
 6:     if p(e, s-1) ≠ p(e, s) then
 7:       P⁻ ← P⁻ ∪ {p(e, s⁻)}
 8:       P⁺ ← P⁺ ∪ {p(e, s⁺)}
 9:     else
10:       Pˢ ← Pˢ ∪ {p(e, s)}
11:   {shift, conn} ← getMaxShift(P⁻, Pˢ, P⁺, s)
12:   if shift ≥ n AND conn < numPaths then
13:     Solution.P⁻ ← P⁻
14:     Solution.Pˢ ← Pˢ
15:     Solution.P⁺ ← P⁺
16:     Solution.s ← s
17:     numPaths ← conn
18: return Solution
```

The algorithm iterates on every frequency slice s to find the set of paths in the route of *newP* allocated using the closest slice with index lower than s (s^-), the set of paths allocated using the closest slice with index greater than s (s^+), and the set of paths allocated using s (lines 3–10 in Table 8.5).

Procedure *getMaxShift* finds the largest continuous slot that can be generated by shifting paths (line 11). Paths in set P^- are left shifted, paths in P^+ are right shifted, and paths in P^s are shifted left and right and the option generating the widest slot is chosen. If a slot with at least n contiguous slices by shifting paths is found and the set of paths involved is lower than that of the best solution found so far, the set of paths is stored as the best solution, and the number of paths involved is updated (lines 12–17). The best solution found is eventually returned (line 18).

8.5 Performance Evaluation

In this section, we concentrate on evaluating the performance of the spectrum reallocation approach.

Aiming at studying the performance of the proposed algorithms in realistic network scenarios, three national optical network topologies are considered: the 21-node and 35-link Spanish Telefónica (TEL) topology, the 20-node and 32-link British Telecom (BT) topology, and the 21-node and 31-link Deutsche Telecom (DT) topology. Figure 8.3 shows such topologies. In the studies that follow, the size of the entire spectrum is set to 800 GHz, and guard bands are neglected. In addition, since the studies performed in this section might be influenced by the mix of bitrates demanded by connection requests, the six traffic profiles (TPs) defined in Chapter 5 are used.

The GRASP+PR heuristic algorithm described in Section 8.3.3 was developed and integrated into an ad hoc event-driven simulator based on OMNeT++ [OMNet]. A dynamic network environment is simulated for the networks under study, where incoming connection requests arrive following a Poisson process and are sequentially served without prior knowledge of future incoming connection requests. The holding time of connections is exponentially distributed with the mean value equal to 2 h. Source/destination pairs are randomly chosen with equal probability (uniform distribution) among all network nodes. Different values of offered network loads are considered by changing the arrival rate while keeping the mean holding time constant. The bitrate demanded by each connection request is 10, 40, 100, or 400 Gb/s and it is chosen following one of the TPs considered.

8.5.1 SPRESSO Heuristics Tuning

Aiming at analyzing the performance of the GRASP+PR heuristic proposed to solve SPRESSO compared to the proposed ILP model, 1000 problem instances were saved from the OMNeT simulator (instances for the TEL, BT, and the DT

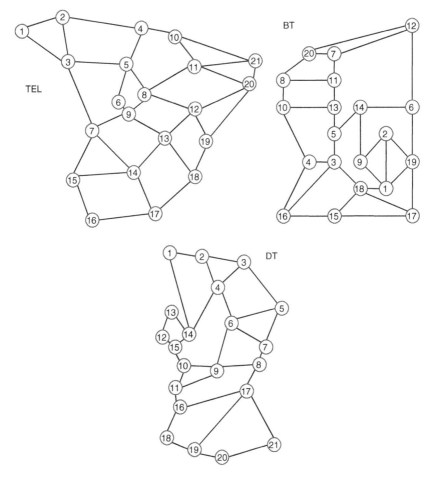

Figure 8.3 Sample optical network topologies: Spanish Telefónica (left), British Telecom (center), and Deutsche Telecom (right).

topologies, for that network load unleashing weighted blocking probability (Pb_{bw}) equal to 1% and 5% under TP-6).

To find the optimal value of parameter α, the three greedy randomized constructive algorithms described in Section 8.3.3 were compared. Each of the heuristics was followed by the local search phase; PR was disabled in these tests. The number of iterations was limited to 30, and each heuristic was run 10 times. Table 8.6 shows the average number of reallocations in the solution. As shown, the optimal value of α ranges from 0.5 for greedy Algorithm-1 to 0.1 for Algorithm-2 and Algorithm-3. We assumed $\beta = \alpha$ in Algorithm-3.

Next, the proposed multi-greedy heuristic with PR was run for the previous set of problem instances using the same value of α for all the greedy algorithms.

Table 8.6 On-average number of reallocations as a function of α.

α	Algorithm-1	Algorithm-2	Algorithm-3	Multi-greedy+PR
0.1	7.43	6.78	6.83	6.28
0.3	7.88	6.99	7.04	6.01
0.5	6.99	7.36	7.04	6.12
0.7	7.48	7.64	7.04	6.76

Table 8.7 ILP versus heuristic performance comparison.

Pb_{bw}	Number of reallocations	Heuristic versus ILP performance gap	Average ILP time (s)	Maximum ILP time (s)	Average heuristic time (s)	Maximum heuristic time (s)
1%	5.9	3.7%	28	72	0.098	0.145
5%	6.1	4.6%	33	87	0.137	0.178

As shown in Table 8.6, the best solutions were obtained when $α = 0.3$ was used. It is worth noticing that when PR is enabled, the proposed multi-greedy algorithm provides better solutions than each of the greedy algorithms alone; this is as a consequence of increasing the diversity of the solutions.

8.5.2 Heuristics versus the ILP Model

The problem instances discussed earlier were solved using the ILP model implemented in MatLab and using CPLEX v.12 [CPLEX] on a 2.4 GHz Quad-Core machine with 4 GB RAM running Linux. Table 8.7 shows the on-average results.

As shown, the proposed heuristic provides solutions less than 5% over the optimal values on average in times in the order of 100 ms and thus it can be used for real-time applications. To reach these shorter solving times, the number of iterations of the heuristic algorithm was limited to 10, which illustrates the effectiveness of the proposed constructive algorithms in providing diversification to the GRASP+PR-based heuristic.

8.5.3 Performance of the SPRESSO Algorithm

After evaluating the proposed heuristic algorithm, we integrated it into the general algorithm for routing and spectrum (re)allocation depicted in Figure 8.1.

Figure 8.4 shows the performance of the algorithm on the three network topologies under traffic profiles TP-2, TP-4, and TP-6 compared to a normal provisioning procedure, that is, without SPRESSO; we observe that remarkable gains are achieved when spectrum reallocation is applied.

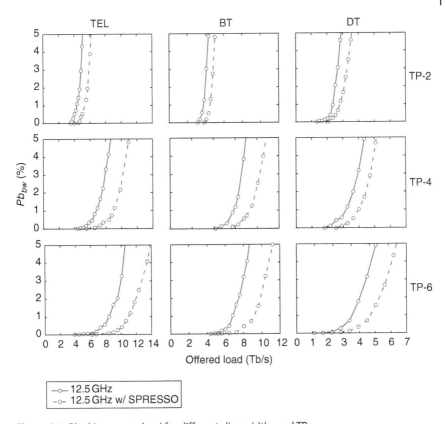

Figure 8.4 Blocking versus load for different slice widths and TPs.

Table 8.8 Load gain of using SPRESSO ($Pb_{bw} = 1\%$).

	TEL	BT	DT
TP-1	28.2%	20.8%	15.9%
TP-2	23.4%	19.1%	19.8%
TP-3	29.5%	19.2%	29.0%
TP-4	32.7%	25.3%	25.6%
TP-5	25.0%	35.3%	24.1%
TP-6	30.3%	34.1%	33.0%

Table 8.8 summarizes the load gain when using SPRESSO for each network topology and TP; presented values are for $Pb_{bw} = 1\%$. Interestingly, spectrum reallocation provides noticeable gains, especially in TPs 2–6; where the number of requests for 400 Gb/s is important there SPRESSO provides improvements

in the range between 20 and 31%. These results confirm the advantages of the proposed mechanism.

8.6 Experimental Assessment

In this section, we first work on adapting the proposed defragmentation algorithms to the specifics of the Application-based Network Operations (ABNO) architecture introduced in Chapter 4 and define the workflow that needs to be followed. In line with such architecture, let us assume that a back-end PCE (bPCE) is in charge of solving optimization problems, so the defragmentation algorithm is implemented in that bPCE and can be invoked from the front-end PCE (fPCE) using a specific code in OF objects. Therefore, we define the specific contents and semantic of PCReq and PCRep messages to deal with defragmentation. Next, we experimentally validate our proposal for in-operation spectrum defragmentation.

8.6.1 Proposed Workflow and Algorithm

From a control and management perspective, the defragmentation process maps into a set of state changes of active connections. Such state changes are reflected in the change of connections' attributes, in our case spectrum allocation, by changing the nominal central frequency of the slot allocated to a connection.

The optimization process can be triggered manually by a network operator through the Network Management System (NMS), by an automated process triggered by some threshold, or in a periodical fashion. In our case, let us assume that the defragmentation procedure is triggered after the fPCE fails to find a suitable route for a provisioning request, as defined in Section 8.2.

The request for a new connection is originally issued by the NMS and received by the ABNO controller through the north-bound interface. In such case, the ABNO controller is responsible for the coordinating connection setup, which composes and sends a specific request toward the fPCE in charge of computing and finally coordinates connection establishment.

The workflow that represents the provisioning-triggered defragmentation use case is detailed in Figure 8.5. As already mentioned, it starts with a network operator requesting a new connection provisioning through the NMS. Establishing a flexible optical connection includes computing and provisioning a continuous slot between two nodes in the data plane. The request is received by the ABNO controller via its north-bound interface (step 1 in Figure 8.5). When the ABNO controller receives the request, it asks the policy agent to check about rights of the received request (2). If access is granted, the ABNO controller requests the fPCE to compute the route and eventually set up the optical connection (3).

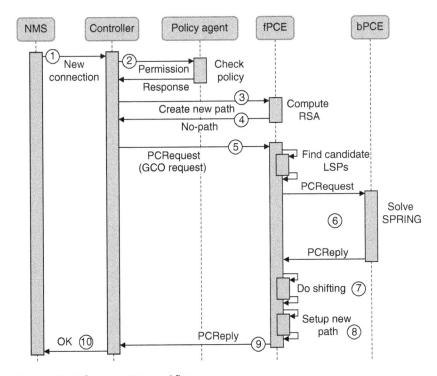

Figure 8.5 Defragmentation workflow.

Let us assume that as a result of spectrum fragmentation no end-to-end continuous slot is found (4). In that case, the ABNO controller may autonomously decide to perform a defragmentation process and sends a message to fPCE (5), which checks its feasibility and gathers information to create a Global Concurrent Optimization (GCO) request [RFC5557] that is sent toward bPCE to solve the optimization problem (6).

The GCO contains the set of already established Label-Switched Paths (LSPs) candidate to be part of the defragmentation process. The procedure, presented in Table 8.9, computes both the shortest path between the source and destination nodes of the requested connection and the needed slot width (lines 2–3 in Table 8.9). Next, it verifies that every link in the found route has enough available spectrum resources and creates the set of candidate LSPs (P) as the LSPs using any of those links (lines 4–7). Both the route and the set of LSPs are eventually returned (line 8).

After the bPCE sends the solution back to the fPCE, it proceeds to execute the defragmentation, that is, to shift some of the candidate LSPs (7).

To that end, a PCUpd message is sent toward each ingress Path Computation Client (PCC) of an LSP that needs to be shifted (step 7); PCCs report updating

Table 8.9 Procedure Find_Candidate_LSPs.

INPUT: *TED, LSP-DB, source, destination, bitrate*
OUTPUT: *P, newP*

1: *P* ← ∅
2: *newP* ← shortestPathByLenght(*TED, source, destination*)
3: *m* ← computeSlotSize(*newP, bitrate*)
4: **for each** *e* ∈ *E(newP)* **do**
5:   **if** freeSlices(*e*) < *m* **then**
6:     **return** {∅, ∅}
7:   *P* ← *P* ∪ getLSPs(*LSP-DB, e*)
8: **return** {*P, newP*}

Table 8.10 PCReq message contents.

```
<PCReq Message>::= <Common Header>
                   <SVEC>
                   <OF>
                   <re_opt request-list>
                   <request>
where:
   <re_opt request-list>::= <re_opt request>[<re_opt request-list>]

   <re_opt request>::= <RP>
                       <END-POINTS>
                       <RRO>
   <request>::= <RP>
                <END-POINTS>
                <BANDWIDTH>
                <IRO>
```

---

completion back using PCRpt messages. Finally, when every LSP has been updated, the PCE proceeds to establish the requested connection by sending a PCInit message to the source PCC (step 8). Upon its completion, the fPCE notifies the ABNO controller (9), which in turn notifies the NMS (10).

8.6.2 PCEP Issues

Owing to the fact that to solve the defragmentation problem the route and the set of candidate LSPs need to be specified, such input data has to be encoded and sent in a PCReq message. Table 8.10 specifies the contents to convey the needed parameters.

Table 8.11 PCRep message contents.

```
<PCRep Message> ::= <Common Header>
                    <response-list>
where:
  <response-list>::= <response>[<response-list>]

  <response>::= <RP>
                <NO-PATH> | <ERO>
```

Table 8.12 Procedure Do_Shifting.

```
INPUT: LSP-DB, P

for each p ∈ P do
if p.ERO ≠ getERO(LSP_DB, p) then
   update(p)
return
```

A list named `re_opt request-list` contains the candidate LSPs. The information includes source and destination nodes for the LSP (END-POINTS object) and its current route and spectrum allocation in the `Record Route Object` (RRO). With that information, the bPCE can find the associated LSP in its own LSP-DB. Information regarding the new connection requested follows, including source and destination nodes, the requested bitrate (BANDWIDTH object), and the computed route (`Include Route Object (IRO)`).

Since all the requests have to be jointly considered, they are grouped by a Synchronization Vector (SVEC) object. Finally, the PCReq message uses the `Objective Function` (OF) object to specify the objective function requested, spectrum defragmentation in our case.

Once the defragmentation problem is solved, the solution is coded in a PCRep message. Table 8.11 specifies the contents to convey the solution.

Since all the received requests in a PCReq message have to be replied, the `response-list` contains an `Explicit Route Object` (ERO) of each of the reoptimization requests. Although the route in the ERO will be invariant, since no rerouting is performed, the spectrum allocation can be either the same or a different one. Hence, an algorithm needs to be used in the fPCE to find the LSPs to be updated; such an algorithm is detailed in Table 8.12.

Regarding the request for the new connection, an ERO object can be received provided that a feasible solution for the defragmentation problem has been found. Note that the ERO object must follow the route specified in the incoming IRO. A NO-PATH object is included if no feasible solution is found.

The proposed algorithm for defragmentation and the described workflow have been experimentally validated in a test bed. The next section describes the scenario and the tests performed.

8.6.3 Experiments

The setup deployed, depicted in Figure 8.6, consists of an ABNO controller, an fPCE, and a bPCE that communicate through Path Computation Element Communication Protocol (PCEP) interfaces. The fPCE has been extended to enable back-end computation, functionally separated from PCEP sessions

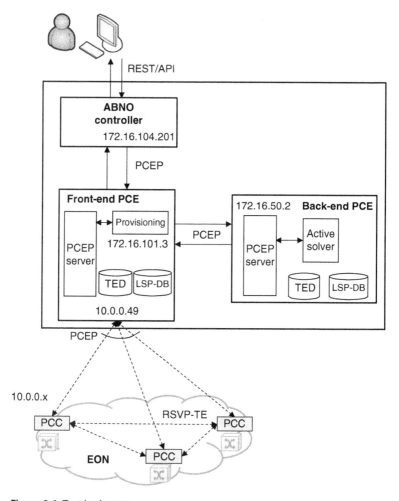

Figure 8.6 Test-bed setup.

No.	Time	Source	Destination	Prot	Le	Info
3	56 181.0738	172.16.104.201	172.16.101.3	PCEP	110	PATH COMPUTATION REQUEST MESSAGE
4	58 181.0739	172.16.101.3	172.16.104.201	PCEP	98	PATH COMPUTATION REPLY MESSAGE
5	60 181.1461	172.16.104.201	172.16.101.3	PCEP	110	PATH COMPUTATION REQUEST MESSAGE
6	61 181.1471	172.16.101.3	172.16.50.2	PCEP	730	PATH COMPUTATION REQUEST MESSAGE
	66 181.2823	172.16.50.2	172.16.101.3	PCEP	574	PATH COMPUTATION REPLY MESSAGE
7	68 181.3027	10.0.0.49	10.0.0.5	PCEP	78	UPDATE MESSAGE ⎤ LSP5
	69 181.3057	10.0.0.5	10.0.0.49	PCEP	78	REPORT MESSAGE ⎦
	72 183.5327	10.0.0.49	10.0.0.6	PCEP	78	UPDATE MESSAGE ⎤ LSP6
	73 183.5356	10.0.0.6	10.0.0.49	PCEP	78	REPORT MESSAGE ⎦
8	82 192.4523	10.0.0.49	10.0.0.4	PCEP	194	INITIATE MESSAGE ⎤ newP
	83 192.4555	10.0.0.4	10.0.0.49	PCEP	78	REPORT MESSAGE ⎦
9	85 195.4268	172.16.101.3	172.16.104.201	PCEP	190	PATH COMPUTATION REPLY MESSAGE

Figure 8.7 PCEP messages exchanged.

established locally with the data plane nodes. The Path Solver has been extended with the *Find_Candidate_LSPs* procedure in Table 8.9.

The data plane includes four programmable WSSs and four node emulators to complete the network topology. Nodes are handled by colocated Generalized Multi-Protocol Label Switching (GMPLS) controllers running Resource Reservation Protocol-Traffic Engineering (RSVP-TE) with flexgrid and push–pull operation extensions [Cu13]. GMPLS controllers communicate with the fPCE by means of PCEP. The GMPLS controllers connected to a WSS (by means of a Universal Serial Bus (USB) interface) run a dedicated programmable configuration tool for automatic filter reshaping with a resolution of 1 GHz.

Aiming at illustrating the spectrum defragmentation problem, the example in Figure 8.2 was reproduced. The capture in Figure 8.7 presents the resulting PCEP message exchange after a request for a new connection between nodes 4 and 7 requiring four slices is issued from the NMS to the ABNO controller. In this scenario, the lack of consecutive optical spectrum resources prevents serving the connection request. Notwithstanding, each link in the shortest route of the new optical connection *newP* (through links 4–5–6) had at least four free slices and then the request could be established, shifting some of the established connections. The capture includes PCEP messages exchanged inside the ABNO architecture (IP subnetwork 172.16.x.x) as well as those exchanges between the fPCE and the PCCs (IP subnetwork 10.10.x.x). For the sake of clarity, messages are correlated to the steps described in the workflow in Figure 8.5.

Figure 8.8 presents the contents of the PCEP messages exchanged between the ABNO controller and the fPCE. PCReq (message 3) includes the IP address of the end nodes and the requested bitrate (50 Gb/s). When no solution is found in the fPCE for the request, the PCRep message (4) including a NO-PATH object is sent back to the ABNO controller.

The controller then sends a PCReq message (5) to the fPCE that includes an OF object with the objective function code "defragmentation" (see the details in Figure 8.9). The fPCE computes the set of candidate LSPs that will

```
③ PATH COMPUTATION REQUEST MESSAGE Header
  ▶ RP object
  ▼ END-POINT object
      Object Class: END-POINT OBJECT (4)
      Object Type: 1
    ▶ Flags
      Object Length: 12
     ┌─────────────────────────────────────┐
     │ Source IPv4 Address: 10.10.0.4       │
     │ Destination IPv4 Address: 10.10.0.7  │
     └─────────────────────────────────────┘
  ▼ BANDWIDTH object
      Object Class: BANDWIDTH OBJECT (5)
      Object Type: 1
    ▶ Flags
      Object Length: 8
     ┌─────────────────────┐
     │ Bandwidth: 50.000000 │
     └─────────────────────┘
```

```
④ PATH COMPUTATION REPLY MESSAGE Header
  ▶ RP object
  ▶ NO-PATH object
```

Figure 8.8 PCEP messages 3 and 4.

```
⑤ PATH COMPUTATION REQUEST MESSAGE Header
  ▶ RP object
  ▼ END-POINT object
      Object Class: END-POINT OBJECT (4)
      Object Type: 1
    ▶ Flags
      Object Length: 12
      Source IPv4 Address: 10.10.0.4
      Destination IPv4 Address: 10.10.0.7
  ▼ BANDWIDTH object
      Object Class: BANDWIDTH OBJECT (5)
      Object Type: 1
    ▶ Flags
      Object Length: 8
      Bandwidth: 50.000000
  ▼ Unknown object (21)
      Object Class: OBJECTIVE FUNCTION OBJECT (OF) (21)
      Object Type: 1
    ▶ Flags
      Object Length: 8
     ┌──────────────────────┐
     │ OF-Code: Unknown (39) │ Defragmentation
     └──────────────────────┘
```

```
⑨ PATH COMPUTATION REPLY MESSAGE Header
  ▶ RP object
  ▶ BANDWIDTH object
  ▼ EXPLICIT ROUTE object (ERO)
      Object Class: EXPLICIT ROUTE OBJECT (ERO) (7)
      Object Type: 1
    ▶ Flags
      Object Length: 100
    ▶ SUBOBJECT: Unnumbered Interface ID
    ▼ SUBOBJECT: Label Control
      L=0 Strict Hop
      Type: SUBOBJECT LABEL (3)
      Length: 12
      U=0 Downstream Label
      Reserved: 0
      C-Type: 2       n=-2  m=2
      Label: 6a 00 ff fe 02 00 00 00
    ▶ SUBOBJECT: Unnumbered Interface ID: 10.10.0.5:2
    ▶ SUBOBJECT: Label Control
    ▶ SUBOBJECT: Unnumbered Interface ID: 10.10.0.6:3
    ▶ SUBOBJECT: Label Control
    ▶ SUBOBJECT: IPv4 Prefix: 10.10.0.7/32
```

Figure 8.9 PCEP messages 5 and 9.

participate in the optimization process and includes them in a PCReq message (6) indicating the current route and spectrum allocation. For instance, the RRO object for LSP 5 is detailed in Figure 8.10 (left). Together with the candidate LSPs, the request for the new connection is included, constrained to the route specified in the IRO object.

After the optimization problem has been solved, the bPCE sends back the PCRep (6) shown in Figure 8.10 (right) to the fPCE, where an ERO object is included in response to every already established LSP. To illustrate the defragmentation (in our experiments, we defined as reference frequency that between slices 8 and 9), we can observe that the original spectrum allocation for LSP5 was $n = -1$ and $m = 1$ (slices 7 and 8) and the computed one was $n = +1$ and $m = 1$ (slices 9 and 10), that is, LSP5 has been shifted two slices to make room for the new connection.

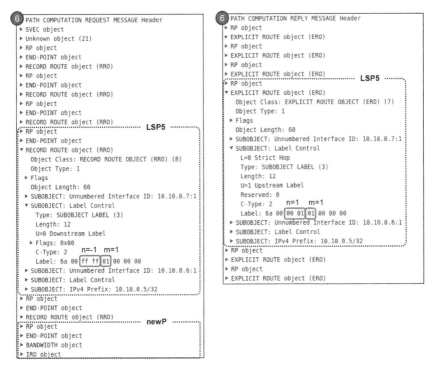

Figure 8.10 PCEP message 6. PCReq (left) and PCRep (right).

Once the fPCE receives the response with the solution, it finds the LSPs that need to be updated and starts updating them by sending PCUpd messages (7) to the source PCCs. Note, in view of messages in Figure 8.7, that defragmentation is performed sequentially following the order specified in RP objects. After the LSP updating process ends, the new connection is set up using a PCInit message (8).

When the defragmentation process finishes, the fPCE replies to the controller with a PCRep message (9) that includes the route and spectrum allocated for the new connection specifying that the new connection has been established using slices 5–8 ($n = -2$, $m = 2$).

8.7 Conclusions

Looking at making enough room for an incoming connection request, in this chapter the SPRESSO reoptimization mechanism was proposed to reallocate already established paths in the spectrum. The SPRESSO mechanism was modeled as an ILP problem. As a consequence of the real-time computation requirements, a GRASP+PR-based heuristic algorithm was proposed to obtain near-optimal solutions for SPRESSO in practical computation times.

In view of the traffic disruption that spectrum reallocation might cause, the spectrum shifting (SPRING) problem, an adaptation of SPRESSO for hitless spectrum defragmentation, has been studied. The hitless condition was ensured by constraining to shift connections between adjacent free frequency slices to make room to serve the new connection requested. The problem was formally stated and the SPRESSO ILP formulation was extended by adding constraints ensuring that connection reallocations are done in a hitless manner.

The SPRESSO approach was implemented, and the obtained results demonstrated that the performance of the networks in terms of blocking probability could be improved in a range from 20 to 31%.

Defragmentation was then experimentally assessed. An algorithm to select the candidate LSPs to be reoptimized was proposed to be run in the fPCE, whereas the algorithm proposed to solve the SPRING problem was run within the bPCE to solve the spectrum shifting problem for the set of candidate LSPs. A workflow defining the interactions between the ABNO components was devised and implemented to demonstrate the spectrum shifting use case experimentally.

9

Restoration in the Optical Layer

Alberto Castro[1], Lluís Gifre[2], Marc Ruiz[2] and Luis Velasco[2]

[1] University California Davis, Davis, CA, USA
[2] Universitat Politècnica de Catalunya, Barcelona, Spain

In this chapter, we explore point-to-point connection recovery schemes specifically designed for Elastic Optical Networks (Eons), in particular taking advantage of sliceable bandwidth variable transponders (SBVTs) that provide additional flexibility allowing reconfiguration of optical connections to be performed. The bitrate squeezing and multipath restoration (BATIDO) problem is proposed to improve restorability of large bitrate connections. BATIDO aims at maximizing the amount of restored bitrate by exploiting the available spectrum resources along multiple routes. BATIDO is formally stated and next modeled using an Integer Linear Programming (ILP) formulation. As a result of the stringent time to compute a solution, a heuristic algorithm providing a better trade-off between optimality and complexity is proposed to solve the problem. Extensive simulation results show the relevant restoration capabilities achieved by the proposed multipath recovery and bitrate squeezing scheme. Next, the BATIDO problem is extended by considering modulation formats and transponder availability. An algorithm to solve the modulation format-aware restoration (MF-Restoration) problem is presented and a workflow to implement MF-Restoration in the Application-based Network Operations (ABNO)-based architecture is proposed and experimentally assessed. Finally, in the context of a telecom Cloud infrastructure, anycast optical connections need to be recovered in case of failure. The AC_RECOVER problem is proposed, modeled using an ILP formulation, and solved using a heuristic algorithm. A workflow to implement AC_RECOVER in the ABNO-based architecture is proposed and experimentally assessed.

Provisioning, Recovery, and In-operation Planning in Elastic Optical Networks,
First Edition. Luis Velasco and Marc Ruiz.
© 2017 John Wiley & Sons, Inc. Published 2017 by John Wiley & Sons, Inc.

Table of Contents and Tracks' Itineraries

9.1 Introduction

In Elastic Optical Networks (EONs) implementing the flexgrid optical technology, sliceable bandwidth variable transponders (SBVTs) [Sa15] have been proposed to enable adaptable bitrate and spectrum allocation according to actual traffic needs (see Chapter 2). Each subcarrier module in the SBVT can be independently configured with a different spectrally efficient modulation format, including Dual Polarization (DP) Quadrature Phase Shift Keying (QPSK), DP 8-symbol Quadrature Amplitude Modulation (QAM8), and DP 16-symbol QAM (QAM16). The flexibility provided by using SBVTs has been considered in the design of future EONs [Ve13.1], [Da14].

Regarding the specific case of restoration upon failure occurrence, the use of SBVTs has already been considered in [Ca13] for multilayer IP/MPLS-over-EON scenarios. The possibility of reducing the allocated spectral resources to partially restore those connections affected by a failure was proposed in [So11], where the spectral resources reduction was called as *bitrate squeezing*. The use of multiple parallel lightpaths (named as *subconnections*) to restore each single connection was proposed in [Ca12.1], where the authors illustrate its benefits in static network scenarios.

From a control plane perspective, when a failure occurs, the Operations, Administration, and Maintenance (OAM) handler in the ABNO architecture is responsible for taking the appropriate action [ABNO], in particular requesting to the Path Computation Element (PCE) the restoration of the affected Label Switched Paths (LSPs). A restoration algorithm running in the PCE can compute a global solution for all the affected LSPs. As introduced in Chapter 4, specifically for in-operation planning problems that might require high computational effort, the PCE can be split into a front-end PCE (fPCE) and an in-operation planning tool, running as a backend PCE (bPCE), able to solve complex optimization problems, such as restoration algorithms.

9.2 Bitrate Squeezing and Multipath Restoration

To illustrate the different restoration schemes that can be applied to the set of connections affected by a failure in the context of EONs, Figure 9.1 shows a simple network topology where a lightpath is set up to serve a 200 Gb/s request between nodes *s* and *t*. In normal conditions (Figure 9.1a), let us assume that the lightpath uses a three-slice frequency slot with DP-QAM8 (spectral efficiency 6 bits/symbol).

Let us imagine that a link failure occurs, and the lightpath is affected. If a restoration lightpath, including route and spectrum allocation, can be found in the network for the required slot width, the disrupted lightpath is obviously restored using such lightpath. This is the normal restoration scheme that has been traditionally used in optical networking; we call this scheme as *single path restoration* (Figure 9.1b).

However, in contrast to protection schemes, the availability of contiguous frequency slices at failure time is not generally guaranteed in restoration schemes. In such case, the disrupted lightpath can be restored in part by allocating a narrower frequency slot. This case, named as *single path restoration with bitrate squeezing*, is illustrated in Figure 9.1c. Note that the restoration lightpath uses just two frequency slices and thus, the conveyed bitrate has been squeezed to 100 Gb/s.

Another possibility is to use several parallel subconnections, each conveying part of the total bitrate, so the original bitrate of the disrupted lightpath is fully restored (Figure 9.1d). In the example, a second lightpath with two frequency slices conveying 100 Gb/s is established. Note that, although restoration lightpaths recover the original bitrate entirely, one additional frequency slice is required. This illustrates the fact that spectral efficiency decreases when subconnections are used. Although for that very reason network operators prefer not using multipath for provisioning, it can be exploited to improve restorability, provided that the number of parallel lightpaths is kept limited.

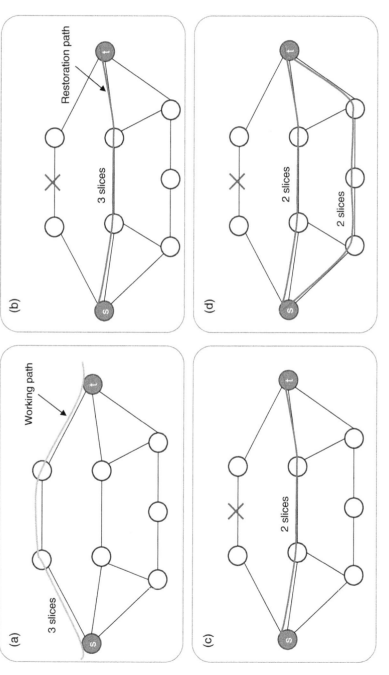

Figure 9.1 Bitrate squeezing and multipath restoration. (a) Normal conditions, (b) single path restoration, (c) single path restoration with bitrate squeezing, and (d) multipath restoration.

9.2.1 The BATIDO Problem

The problem can be formally stated as follows:

Given:
- a network topology represented by a graph $G(N, E)$, where N is the set of optical nodes and E is the set of fiber links connecting two optical nodes,
- a set S of available frequency slices in each fiber link in E,
- a set D of failed demands to be restored, each requesting a fixed bitrate b^d.

Output: the routing and spectrum allocation (RSA) for each restored demand $d \in D$.

Objective: maximize the total restored bitrate.

As previously discussed, the bitrate squeezing and multipath restoration (BATIDO) problem can be faced using three different approaches:

1) single path restoration, where the total requested bitrate is either restored using a single lightpath or blocked,
2) bitrate squeezing restoration, where part of the originally requested bitrate is restored whilst the rest is blocked,
3) multipath restoration with bitrate squeezing, where several restoration lightpaths can be established to recover partially or totally one single demand.

An Integer Linear Programming (ILP)-based model, which includes these schemes, is presented next.

9.2.2 ILP Formulation

The ILP model is based on the *arc-path* formulation for RSA introduced in Chapter 3, where a set of routes are computed beforehand for each demand (excluding the failed fiber link). It is worth highlighting that the term lightpath includes both the physical route and the allocated slot. The characteristics of the considered modulation format are also embedded in the input data.

The following sets and parameters have been defined:

Topology

N Set of optical nodes, index n.
E Set of fiber links, index e.
K Set of precomputed routes excluding the failed fiber link, index k.
P Set of SBVTs, index p.
$N(k)$ Subset of optical nodes that are source or destination of route k.
$K(n)$ Subset of routes which source or destination optical node is n.
$P(n)$ Subset of SBVTs of optical node n.

h_e^k Equal to 1 if route k uses link e, 0 otherwise.

b^k Maximum bitrate (in Gb/s) that route k can convey (due to physical impairments).

fF_p^n Number of free modulators in SBVT p of optical node n. If $p \notin P(n)$, fF_p^n is equal to 0.

b_p^n Unreserved capacity of SBVT p of optical node n in Gb/s.

Spectrum

S Set of frequency slices available in each link, index s.

u_{es} Equal to 1 if the slice s in fiber link e is free, 0 otherwise. To compute this parameter, only nonfailed lightpaths are considered.

C Set of slots, index c. Each slot c contains a subset of contiguous slices.

l_{cs} Equal to 1 if slot c uses slice s, 0 otherwise.

b_c Bitrate capacity of slice c in Gb/s.

Failed Demands

D Set of failed optical demands to be restored, index d.

b^d Bitrate requested by demand d in Gb/s.

B Largest bitrate of demands, that is, $B = \max\{b^d, d \in D\}$

$K(d)$ Subset of routes for demand d (fulfilling the reachability constraint).

$C(d)$ Subset of feasible slots for demand d to restore some amount of bitrate in the range $(0, b^d]$.

sQ Equal to 1 if squeezing is used, 0 otherwise.

mP Equal to 1 if multipath is used, 0 otherwise.

mk^d Maximum number of lightpaths that can be used to restore demand d, when multipath is selected. Note that $mk^d = 1$ if $mP = 0$; otherwise $mk^d \geq 1$.

The decision variables are:

x_c^{dk} Binary. Equal to 1 if demand d uses route k and slot c for restoration, 0 otherwise.

x_{cp}^{kn} Binary. Equal to 1 if restoration lightpath with route k and slot c uses SBVT p in optical node n, 0 otherwise.

y^d Positive real. Restored bitrate for demand d.

y_c^k Positive real. Bitrate conveyed by restoration lightpath with route k and slot c.

y_{cp}^{kn} Positive real. Bitrate conveyed by restoration lightpath with route k and slot c using SBVT p in optical node n.

w^d Binary. Equal to 1 if demand d is restored (total or partially), 0 otherwise.

z^d Positive integer accounting for the number of lightpaths used to restore demand d.

The ILP formulation for the BATIDO problem is as follows:

$$(\text{BATIDO}) \quad \text{Max} \quad \Phi = \sum_{d \in D} y^d \tag{9.1}$$

subject to:

$$\sum_{k \in K(d)} \sum_{c \in C(d)} x_c^{dk} \le mP \cdot \left(mk^d - 1 \right) + 1 \quad \forall d \in D \tag{9.2}$$

$$\sum_{k \in K(d)} \sum_{c \in C(d)} x_c^{dk} = z^d \quad \forall d \in D \tag{9.3}$$

$$z^d \le mk^d \cdot w^d \quad \forall d \in D \tag{9.4}$$

$$y^d \le \sum_{k \in K(d)} \sum_{c \in C(d)} b_c \cdot x_c^{dk} \quad \forall d \in D \tag{9.5}$$

$$y^d \le b^d \quad \forall d \in D \tag{9.6}$$

$$B \cdot sQ + y^d \ge b^d \cdot w^d \quad \forall d \in D \tag{9.7}$$

$$\sum_{d \in D} x_c^{dk} \le \sum_{p \in P(n)} x_{cp}^{kn} \quad \forall k \in K,\ c \in C,\ n \in N(k) \tag{9.8}$$

$$\sum_{p \in P(n)} x_{cp}^{kn} \le 1 \quad \forall k \in K,\ c \in C,\ n \in N(k) \tag{9.9}$$

$$\sum_{k \in K} \sum_{c \in C} x_{cp}^{kn} \le fF_p^n \quad \forall n \in N,\ p \in P(n) \tag{9.10}$$

$$\sum_{d \in D} b_c \cdot x_c^{dk} \le y_c^k \quad \forall k \in K,\ c \in C \tag{9.11}$$

$$y_c^k \le b^k \quad \forall k \in K,\ c \in C \tag{9.12}$$

$$y_{cp}^{kn} \le b^k \cdot x_{cp}^{kn} \quad \forall n \in N,\ p \in P(n),\ k \in K(n),\ c \in C \tag{9.13}$$

$$y_{cp}^{kn} \le y_c^k \quad \forall n \in N,\ p \in P(n),\ k \in K(n),\ c \in C \tag{9.14}$$

$$y_{cp}^{kn} \geq y_c^k - b^k \cdot \left(1 - x_{cp}^{kn}\right) \quad \forall n \in N, \ p \in P(n), \ k \in K(n), \ c \in C \quad (9.15)$$

$$\sum_{k \in K} \sum_{c \in C} y_{cp}^{kn} \leq b_p^n \quad \forall n \in N, \ p \in P(n) \quad (9.16)$$

$$\sum_{d \in D} \sum_{k \in K(d)} \sum_{c \in C(d)} h_e^k \cdot l_{cs} \cdot x_c^{dk} \leq u_{es} \quad \forall e \in E, \ s \in S \quad (9.17)$$

Equation (9.1) maximizes the total restored bitrate Φ. Constraint (9.2) accounts the number of lightpaths assigned to each failed demand. Constraint (9.3) stores the number of lightpaths used to restore each demand and constraint (9.4) limits that number to the maximum value allowed while keeping track of the restored demands. Constraint (9.5) accounts the restored bitrate of each demand, which is limited by the demand's bitrate in constraint (9.6). In case bitrate squeezing is not allowed, constraint (9.7) ensures that all its bitrate is restored or blocked. Note that the large value of B allows relaxing this constraint when bitrate squeezing is allowed.

Constraints (9.8)–(9.17) deal with lightpath allocation. Constraints (9.8) and (9.9) assign one SBVT at each end of the lightpath, whilst constraint (9.10) ensures that the available number of modulators in each SBVT is not exceeded. Constraint (9.11) accounts for the bitrate conveyed by a lightpath, whereas constraint (9.12) makes sure that the maximum bitrate for the assigned route k (b^k) is not exceeded. Constraints (9.13)–(9.15) force y_{cp}^{kn} to take the value of $x_{cp}^{kn} \ldots y_c^k$, and constraint (9.16) limits the total bitrate that can be allocated in each SBVT. Constraint (9.17) guarantees that not more than one lightpath uses each slice in each link.

Regarding the size of the BATIDO problem, the number of variables is $O(|D| \cdot |K| \cdot |C| + |N| \cdot |P| \cdot |K| \cdot |C|)$ and the number of constraints is $O(|D| + |N| \cdot |P| \cdot |K| \cdot |C| + |E| \cdot |S|)$. As an example, the number of variables and constraints for the Telefonica (TEL) network and the scenario presented in Section 9.2.4 is 10^9 for both variables and constraints.

Although the ILP model can be solved for small instances, its exact solving becomes impractical for realistic backbone networks under appreciable load, even using commercial solvers, for example, CPLEX [CPLEX]. Hence, aiming at providing near-optimal solutions for the BATIDO problem within the stringent required times (e.g., hundreds of milliseconds), we present next a heuristic algorithm to solve it.

9.2.3 Heuristic Algorithm

To cope with the required computation time constraints, we propose a simple but efficient heuristic algorithm, based on the bulk RSA algorithm presented in Chapter 5, which generates a number of randomized solutions (Table 9.1).

Table 9.1 Heuristic algorithm pseudo-code.

INPUT: N, E, D, mP, sQ, maxIterations
OUTPUT: bestΦ, bestSol

```
 1: bestΦ ← 0
 2: bestSol ← ∅
 3: for i = 1…maxIterations do
 4:    tempΦ ← 0
 5:    tempSol ← ∅
 6:    shuffle(D)
 7:    for each d ∈ D do
 8:       Kd ← kSP(G, d)
 9:       sort(Kd, SLOT_WIDTH, DEC)
10:       z^d ← 0
11:       tempB ← 0
12:       for each k ∈ Kd do
13:          z^d++
14:          c ← getLargerSlot(k, b^d)
15:          tempB ← tempB + getBitrate(c)
16:          if tempB ≥ b^d then
17:             allocate(k, c)
18:             tempΦ ← tempΦ + b^d
19:             tempSol ← tempSol U {d, k, c}
20:             break
21:          else
22:             if mP then
23:                allocate(k, c)
24:                tempΦ ← tempΦ + tempB
25:                tempSol ← tempSol U {d, k, c}
26:                if z^d = mk^d then
27:                   break
28:             else if sQ then
29:                allocate(k, c)
30:                tempΦ ← tempΦ + tempB
31:                tempSol ← tempSol U {d, k, c}
32:                break
33:             else
34:                break
35:    if tempΦ > bestΦ then
36:       bestΦ ← tempΦ
37:       bestSol ← tempSol
38:    resetAllocations(tempSol)
39: return <bestΦ, Sol>
```

The algorithm includes performing a fixed number of iterations (*maxIterations*) in which one solution is built from a randomly sorted set of demands (lines 3–6 in Table 9.1). Next, k-shortest routes are computed for each demand (lines 7–8), which are afterward sorted in decreasing slot width order (line 9). For each route, the larger available slot (considering both continuous free slices and available resources at end nodes) is selected, and the restored bitrate is updated (lines 12–15).

In case the demand can be totally restored, the corresponding lightpath is allocated on graph G to prevent these resources from being used by subsequent restoration lightpaths. The lightpath is finally added to the solution being built (lines 16–20), and the algorithm continues with the next demand. In case the demand can be only partially restored, we need to consider two options. Firstly, if multipath is allowed, the lightpath is allocated, added to the solution, and new lightpaths are considered provided that the number of lightpaths already used does not exceed the given limit (lines 22–27). Secondly, if bitrate squeezing is allowed, the lightpath is allocated and added to the solution (lines 28–32). Otherwise, the demand cannot be restored, and it is blocked (lines 33–34).

Once a solution is built, its total restored bitrate is compared against the best solution obtained so far, which is updated as long as Φ is improved (lines 35–37). Finally, before building new solutions, all the allocated resources are released from graph G (line 38) and the best solution is eventually returned (line 39).

The proposed heuristic performs a constant number of iterations (*maxIterations*); on each iteration one permutation, as well as one k-shortest paths computation, based on Yen's algorithm [Ye71], and one sorting for each demand are performed. Therefore, the time complexity of the heuristic is polynomial and can be expressed as $O(maxIterations \cdot |D| \cdot |K| \cdot |N| \cdot (|E| + |N| \cdot \log(|N|)))$.

The performance of the proposed heuristic was compared against the optimal solution obtained from solving the ILP model for small instances. In all the tests performed, the optimal solution was found within running times of few milliseconds, in contrast to more than 1 h needed to find the optimal solution with CPLEX. Consequently, we use the proposed algorithm to solve the instances presented in the next section.

9.2.4 Numerical Results

In this section, we focus on studying the performance of the restoration schemes. To that end, we consider the 30-node and 56-link TEL, and the 22-node and 35-link BT topologies; each node location is equipped with one single optical node.

For evaluation purposes, we developed an ad hoc event-driven simulator in OMNeT++ [OMNet]. A dynamic network environment was simulated where incoming connection requests arrive at the system following a Poisson process and are sequentially served without prior knowledge of future incoming connection requests. The holding time of the connection requests is exponentially distributed with a mean value equal to 2 h. Source/destination pairs are randomly chosen with equal probability (uniform distribution) among all nodes. Different values of the offered network load are created by changing the interarrival rate while keeping the mean holding time constant. We assume that no retrial is performed; if a request cannot be served, it is immediately blocked. Regarding the optical spectrum, the total width was fixed to 4 THz.

Besides, optical link failures follow a Poisson process with a Mean Time to Failure (MTTF) equal to 50 h, and link failures are randomly chosen with equal probability. We consider that the link is repaired immediately after the restoration process ends.

In our experiments, the bitrate of each connection request was selected considering 80% of the connections were of 100 Gb/s and the other 20% of 400 Gb/s. Note that each point in the results is the average of 10 runs with 150,000 connection requests each.

Figure 9.2 plots blocking probability as a function of the offered load for the restoration approaches and network topologies considered. Note that offered loads have been limited to those unleashing blocking probability in the meaningful range [0.1–5%]. For the sake of comparability between topologies, offered loads have been normalized to the value of the highest load.

As shown, all three approaches behave similarly, that is, whatever the restoration approach is selected the blocking probability for provisioning remains unchanged. Note that the considered MTTF value is large enough to reduce the probability of two lightpaths being affected by two consecutive failures and hence virtually all the restored lightpaths have been torn down when a new failure occurs.

When we analyze the results for aggregated restorability, that is, combined values for 100 and 400 Gb/s lightpaths (Figure 9.3), we observe that multipath and bitrate squeezing approaches restore almost all the failed bitrate, in contrast to the single path (nonsplit) one. Certainly, aggregated restorability using multipath and bitrate squeezing is higher than 95%, even for the highest considered loads, remarkably higher than that obtained using the nonsplit approach, whose values range in the interval [60–80%].

To get insight into the performance of the different approaches, Figure 9.4 focuses on the restorability for 400 Gb/s lightpaths. The performance of multipath and bitrate squeezing is noticeably divergent: the multipath approach shows significantly better performance than that of the bitrate squeezing. The rationale behind that behavior is that the multipath approach implements bitrate squeezing and additionally adds the ability to use several lightpaths to restore one single demand.

To appreciate the way the multipath approach works, let us evaluate the distribution of lightpaths actually used for restoration. Note that since 400 Gb/s demands can be restored using any combination of 400, 200, and 100 Gb/s lightpaths. Thus, one, two, three, or four different lightpaths can be used by the

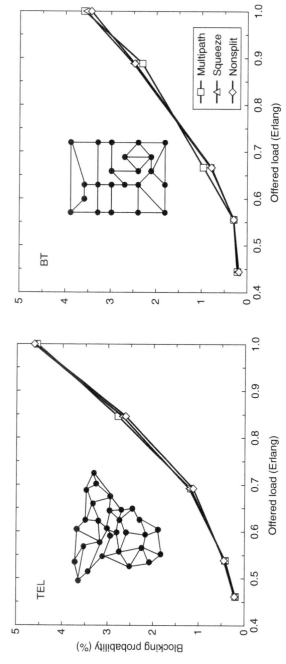

Figure 9.2 Blocking probability against offered load.

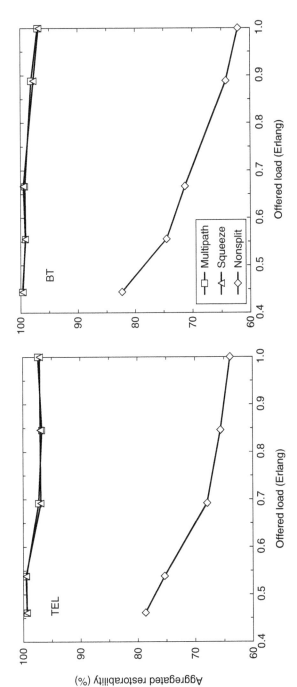

Figure 9.3 Restorability against offered load.

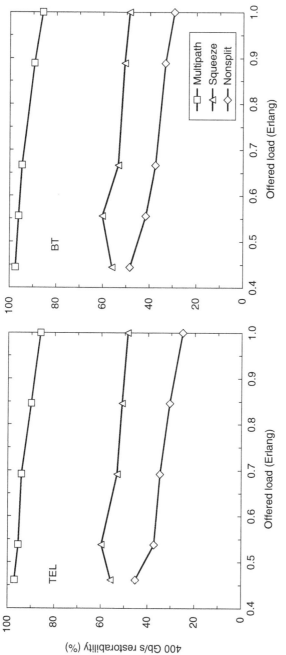

Figure 9.4 400 Gb/s connections' restorability against offered load.

restoration approach. Then, Figure 9.5 depicts the number of demands restored using z^d lightpaths and its average value as a function of the offered load. Figure 9.5 shows an upward trend of z^d average that clearly is as a consequence of the lower probability of finding a single lightpath with enough spectral resources for each demand under higher loads. That can be clearly observed analyzing the distribution of z^d values; the number of demands restored using one single lightpath decreases as the offered load increases, whereas the number of demands restored using more than one lightpath increases with the load.

Consequently, as the offered load grows, the multipath approach takes advantage of using subconnections to maximize the total restored bitrate.

9.3 Modulation Format-Aware Restoration

In this section, we extend the BATIDO problem by selecting the appropriate modulation format among those available in the installed SBVTs for the optical connections being restored; we call this the modulation format-aware restoration (MF-Restoration) problem. The problem thus consists of computing the Routing, Modulation, and Spectrum Allocation (RMSA), as well as of selecting the SBVTs for the set of connections affected by a failure in the optical network. An example of such modulation format awareness is illustrated for the 14-node Spanish topology depicted in Figure 9.6a. Table 9.2 shows the spectral efficiency and reachability of the considered modulation formats; each of them is assumed to support a set of bitrates depending on the number of subcarrier modules configured, as summarized in Table 9.3. For instance, a 300 Gb/s demand requires configuring three subcarrier modules with DP-QPSK or two subcarrier modules with DP-QAM8.

Three connections are established in Figure 9.6b; their main characteristics before the failure are given in Table 9.4. In the event of a failure in link 4–8, connections are restored; Table 9.4 shows the characteristics of the restored connections where main changes are highlighted. Three different cases can be identified as a result of longer restoration routes (in km) and/or resources availability: (i) the lower spectral efficiency of the new modulation format entailed using more frequency slices to serve all the requested bitrate ($lsp1$); (ii) although the number of slices was kept invariant, the restored bitrate was squeezed ($lsp2$); and (iii) the number of slices was reduced because resource availability and the bitrate were squeezed ($lsp3$).

An additional problem is the availability of transponders in the end nodes of each demand. Although many parameters can be used to model SBVTs, in this section, according to the architecture in [Sa15], we use just two: (i) total SBVT capacity, for example, 400 Gb/s, and (ii) the number of subcarrier modules in the transponder, for example, four. We assume that any other parameter is fixed, for example, the list of modulation formats; the baud rate of each

(a)

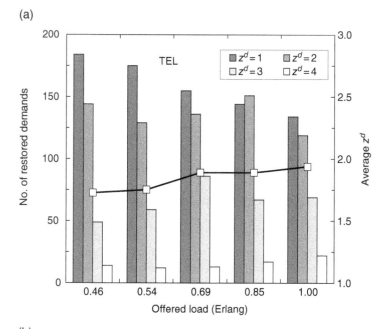

(b)

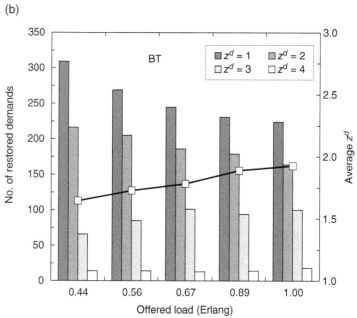

Figure 9.5 Distribution and average z^d values for restored 400 Gb/s demands. (a) TEL and (b) BT topologies.

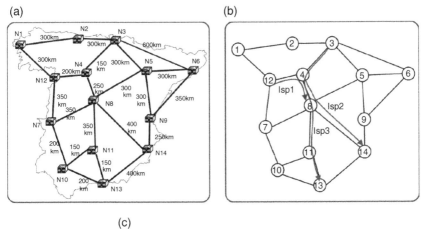

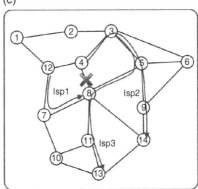

Figure 9.6 Network topology with link distances (a). Example of the network before failure (b) and after restoration (c).

Table 9.2 Modulation formats.

Modulation format	Spectral efficiency (bit/symbol)	Reach (km)
DP-QPSK	4	3000
DP-QAM8	6	1000
DP-QAM16	8	650

Table 9.3 Bitrates per modulation format.

No. of subcarrier modules	DP-QPSK	DP-QAM8	DP-QAM16
1	100 Gb/s	150 Gb/s	200 Gb/s
2	200 Gb/s	300 Gb/s	400 Gb/s
3	300 Gb/s	—	—
4	400 Gb/s	—	—

Table 9.4 Connection configuration.

LSP	Bitrate (Gb/s)	Length (km)	Hops	Modulation format	No. of slices
Before failure					
lsp1	200	450	2	DP-QAM16	2
lsp2	400	650	2	DP-QAM16	4
lsp3	300	900	4	DP-QAM8	4
After restoration					
lsp1	200	700	2	DP-QAM8	4
lsp2	300	1000	4	DP-QAM8	4
lsp3	100	1100	4	DP-QPSK	2

subcarrier module, for example, 25 Gbaud/s; the central frequency granularity, for example, 6.25 GHz; and so on.

Furthermore, we assume that each subcarrier module can be configured independently of the rest of the modules in terms of modulation format, slot width, and central frequency. Because the different subcarrier modules in the same SBVT are combined optically, frequency slices can be used by one single subcarrier at the most, which would entail changing connections from its original SBVT to a different one during the restoration process.

The next section focuses on formally stating the restoration problem under the earlier assumptions.

9.3.1 The MF-Restoration Problem

The MF-Restoration problem can be stated as follows:

Given:
- a graph $G(N, E)$ representing the network topology. The set N includes the set of optical nodes and the set of packet nodes. Packet nodes are equipped with SBVTs and connected to optical nodes.
- the availability of every frequency slice in the optical spectrum of every link in E;
- the characteristics and availability of the installed SBVTs, including available capacity and subcarrier modules;
- a set F of modulation formats supported by the SBVTs;
- a set E_F of failed links;
- a set D of demands affected by the failure to be restored.

Output: the SBVTs and the RMSA for each demand in D.

Objective: maximize served bitrate while minimizing the used optical resources.

An algorithm to solve the MF-Restoration problem is presented next.

9.3.2 Algorithm for MF-Restoration

As introduced in the previous section, we assume that SBVTs are installed in the packet nodes, so optical connections start/end in those nodes thus enabling them to use any of the SBVT modules available there. Hence, one optical link needs to be considered connecting every SBVT to an optical node. As a result of the large number of links to be added, the size of the resulting graph might entail long computation times. For this very reason, we extend the algorithm for RMSA presented in Chapter 5 and create a per-demand simplified auxiliary graph containing only the optical nodes and the optical links connecting two optical nodes having free resources (e.g., frequency slot with a minimum spectral width of $2 \times 12.5 = 25$ GHz) to compute k-shortest paths. Moreover, we also precompute the set of candidate source–destination pairs of SBVTs connected to the source/destination routers having free resources (e.g., minimum free spectral width of 25 GHz and one transmitter/receiver subcarrier module at least).

Table 9.5 presents the pseudo-code of the proposed algorithm to solve the RMSA problem with SBVT availability. Once the auxiliary graph is created for the given demand d (line 2 in Table 9.5) and the candidate pairs of source–destination SBVTs are found (line 3), a set of shortest paths is precomputed (line 4); each path includes its physical route k (sequence of hops) and the width of the largest continuous slot in that route, n_k. Then, the precomputed paths are combined with each pair of source–destination SBVTs (lines 5–7), and the width of largest slot available is checked (lines 8–9). Next, the best modulation format is selected from set F, which is ordered by its spectral efficiency provided that the reach works for the length of the route (lines 10–15). Finally, the path is accepted as long as the included SBVTs have enough resources available (subcarrier modules and capacity) (lines 16–18). The set of paths satisfying the previous constraints, if any, is sorted first by its bitrate and second by the length of its route (line 20) and the best path is selected (line 21). A slot of the proper width is selected, and the computed lightpath is eventually returned (line 22).

Once the extended RMSA algorithm has been introduced, we focus on the MF-Restoration problem.

To solve the MF-Restoration problem, we propose the heuristic algorithm presented in Table 9.6. The algorithm first deallocates the demands in D (line 2 in Table 9.6) and removes the failed links in E_F from G (line 3). Next, *maxIter* solutions are generated; at each iteration, a new solution S is created first preallocating the demands in D (line 5) and then expanding the frequency slots of the allocated demands (line 6). Next, the fitness of S is computed, and the best solution found so far (*bestS*) is updated if S improves it (lines 7–8); *bestS* is eventually returned (line 9). Note that since D contains failed demands, they are discarded, that is, they are not reallocated at the end of the algorithm. This

Table 9.5 Compute RMSA with SBVT availability algorithm.

INPUT: $G(N, E)$, d, $maxSlotWidth$, bw
OUTPUT: $<d, k, c_k, f_k, b_k>$

```
 1: Q' ← ∅
 2: G' ← createAuxGraph(G, d)
 3: TP ← findCandidateSBVTPairs(G, d.src, d.dst)
 4: Q = {<k, n_k>} ← kSP(G', d)
 5: for each q ∈ Q do
 6:   for each tp ∈ TP do
 7:     q' ← combine(tp, q)
 8:     if q'.n_k < minSlotWidth then continue
 9:     q'.n_k ← min(q'.n_k, maxSlotWidth)
10:     q'.f_k ← 0
11:     for each f ∈ F do
12:       if len(q'.k) > len(f) then continue
13:       q'.f_k ← f
14:       q'.b_k ← min (computeBitrate(f.m, q'.n_k), bw)
15:     if q'.f_k = 0 then continue
16:     if not TPFeasible(tp.src, q'.b_k) then continue
17:     if not TPFeasible(tp.dst, q'.b_k) then continue
18:     Q' ← Q' ∪ {q'}
19: if Q' = ∅ then return ∅
20: sort(Q', <q.b_k, DESC>, <|q.k|, ASC>)
21: q' = <k, n_k, f_k, b_k> ← first(Q')
22: return selectSlot(d, q')
```

Table 9.6 MF-restoration heuristic algorithm.

INPUT: $G(N, E)$, E_F, D
OUTPUT: $bestS$

```
1: bestS ← ∅
2: for each d in D do deallocate(G, d)
3: E ← E \ E_F
4: for i = 1...maxIter do
5:   <S, G'> ← bulkPreAllocation(G, D)
6:   S ← slotExpansion(G', S)
7:   Compute Φ(S)
8:   if Φ(S) > Φ(bestS) then bestS ← S
9: return bestS
```

Table 9.7 slotExpansion function.

```
INPUT: G(N, E), S
OUTPUT: S
```

```
 1: while true do
 2:    P ← ∅
 3:    for each p in S do
 4:       if p.bₖ = p.d.b_req then continue
 5:       <benefit, k, cₖ, fₖ, bₖ> ← checkExpansion(G', p)
 6:       if benefit ≤ 0 then break
 7:       P ← P U {<p, benefit, k, cₖ, fₖ, bₖ>)}
 8:    if P = ∅ then break
 9:    sort (P, benefit)
10:    benefit ← first(P).benefit
11:    for each p in P do
12:       if benefit < p.benefit then break
13:       p ← expandSlot(G', p, k, cₖ, fₖ, bₖ)
14: return S
```

---

two-step restoration algorithm maximizes the number of restored demands and fairly shares out the remaining resources among the demands.

The *bulkPreAllocation* function is similar to the bulk RSA algorithm presented in Chapter 5; it generates a number of solutions, each restoring all the demands in a random order using the RMSA algorithm in Table 9.5 with a minimum slot width and eventually returning the solution maximizing the restored demands while minimizing the resources used.

The *slotExpansion* function, shown in Table 9.7, tries to iteratively expand the frequency slot initially assigned to each restored demand. A demand expansion tries to assign enough subcarrier modules to the demand to increase its bitrate; recall that discrete bitrate capacities were assumed, thus avoiding infrautilization of subcarrier modules. First, the *benefit* of expanding each restored demand with a bitrate lower than the requested one is computed as a function of the extra bitrate served per extra subcarrier module (lines 3–5). If the expansion is feasible, that candidate expansion is added to P (lines 6–7). After evaluating the demands, if no candidate expansion is found, the loop finishes (line 8), and solution S is returned (line 14); otherwise, all candidate expansions with the same benefit are applied (lines 9–13) and next iteration is started.

9.3.3 Protocol Extensions and Proposed Workflows

In this section, we propose protocol extensions to convey information related to SBVTs, as well as the workflow to implement the MF-Restoration algorithm.

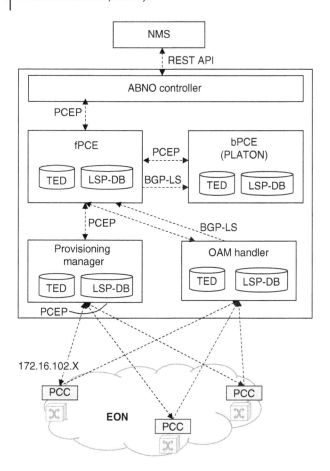

Figure 9.7 Control and management architecture.

Figure 9.7 shows the considered ABNO-based architecture described in Chapter 4. The Network Management System (NMS) issues operation requests to the Representational State Transfer Application Programming Interface (REST API) northbound interface of the ABNO controller. The ABNO components communicate among them by means of the Path Computation Element Communication Protocol (PCEP) and Border Gateway Protocol with Link State Distribution extension (BGP-LS) protocols. BGP-LS is used to synchronize the local Traffic Engineering Database (TEDs) of fPCE and bPCE. PCEP is used for delegating computations by means of PCReq and PCRep messages, while PCRpt messages are used for LSP-DB synchronization. Likewise, Path Computation Notification (*PCNtf*) messages are used to issue immediate asynchronous notifications between ABNO components, such as link failures.

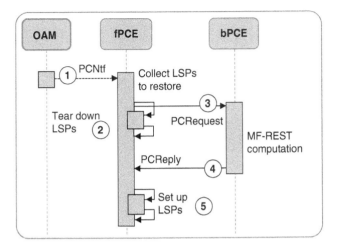

Figure 9.8 Proposed workflow.

Protocol extensions for BGP-LS are needed to convey information related to SBVTs and for PCEP to specify which subcarrier modules and modulation formats have to be used for a specific LSP. These extensions can be based on those proposed in [Ma15] for Open Shortest Path First for Traffic Engineering (OSPF-TE) and Resource Reservation Protocol for Traffic Engineering (RSVP-TE), respectively, so we refer the reader to that reference for details. Specifically, the new Explicit Transponder Control (ETC) subobject for Route Object (RO) objects (e.g., Explicit Route Object (ERO)) was proposed.

The workflow shown in Figure 9.8 has been defined to implement the MF-Restoration algorithm in the ABNO architecture. It is triggered by the OAM Handler after it receives failure notifications from the data plane and correlates them to locate the failed element(s). Next, it issues an asynchronous PCNtf message to the fPCE identifying the failed elements, for example, a fiber link (message 1 in Figure 9.8).

Upon reception of the PCNtf message identifying the failed elements, the fPCE collects every LSP supported by these elements, tears failed LSPs down (message 2), and concurrently delegates the computation of the restoration paths to the bPCE by means of a PCReq message. The PCReq message (message 3) includes an Exclude Route Object (XRO) object with the failed elements for the bPCE to exclude those elements from the path computation, a SVEC object to jointly compute the restoration paths of the set of LSPs, and an Objective Function (OF) object specifying the optimization algorithm, MF-Restoration in this case.

The bPCE computes the solution and sends it back to the fPCE in a PCRep message (message 4) that contains, for each LSP, either an ERO object with the updated LSP's resource allocation and a BANDWIDTH object with the new,

No.	Time	Source	Destin	Proto	Info
① 8	0.001	OAMH	fPCE	PCEP	Notification (PCNtf)
② 12	0.001	fPCE	PM	PCEP	Path Computation LSP Initiate (PCInitiate)
③ 13	0.001	fPCE	bPCE	PCEP	Path Computation Request (PCReq)
② 18	0.046	PM	fPCE	PCEP	Path Computation LSP State Report (PCRpt)
④ 21	0.063	bPCE	fPCE	PCEP	Path Computation Reply (PCRep)
② 24	0.098	PM	fPCE	PCEP	Path Computation LSP State Report (PCRpt)
28	0.172	PM	fPCE	PCEP	Path Computation LSP State Report (PCRpt)
32	0.178	fPCE	PM	PCEP	Path Computation LSP Initiate (PCInitiate)
⑤ 34	0.225	PM	fPCE	PCEP	Path Computation LSP State Report (PCRpt)
38	0.301	PM	fPCE	PCEP	Path Computation LSP State Report (PCRpt)
42	0.377	PM	fPCE	PCEP	Path Computation LSP State Report (PCRpt)

Figure 9.9 PCEP exchange messages for the MF-RESTORATION workflow.

possibly squeezed, bitrate of that LSP or a NO-PATH object if the LSP cannot be restored. The fPCE is in charge of implementing the solution in the network through the provisioning manager (PM); thus, a PCInit message (messages 5) is issued to the PM to signal the setup of the LSPs. Finally, when the signaling of the LSPs is completed, the PM issues PCRpt messages to the fPCE confirming that the LSPs have been set up.

9.3.4 Experimental Assessment

In this section, we present the experimental setup reproducing the architecture shown in Figure 9.7, where the MF-Restoration algorithm and workflow were assessed. All modules were implemented in C++ for Linux.

The NMS uses a REST API to issue the demand set up and tear down requests to the ABNO controller, which translates them into PCInit messages and forwards them to the fPCE. The rest of the ABNO components use PCEP and BGP-LS protocols with the extensions described in Section 9.3.3.

We configured the 14-node Spanish network topology shown in Figure 9.6; each network link has a spectrum width of 4 THz, and each location has been equipped with an optical node and a router with 10 SBVTs; each SBVT consists of 10 transmitter and 10 receiver subcarrier modules, a switching capacity per SBVT of 1 Tb/s, and the modulation formats characterized in Tables 9.2 and 9.3.

Figure 9.9 shows the meaningful PCEP messages exchanged during the operation of the MF-Restoration workflow, while Figure 9.10 shows an extract of the PCRep message (4) sent by the bPCE containing the restored path and the ETC subobjects in the ERO object for lsp2.

9.4 Recovering Anycast Connections

Anycast optical connections were introduced in Chapter 7 in the context of a content distribution scenario in the telecom cloud to connect a given metro datacenter (DC) to any DC in a set of core DCs that store a specific content.

```
▸ RP object
▾ EXPLICIT ROUTE object (ERO)
    Object Class: EXPLICIT ROUTE OBJECT (ERO) (7)
    0001 .... = Object Type: 1
  ▸ Flags
    Object Length: 444
  ▸ SUBOBJECT: Unnumbered Interface ID: 172.16.102.104:18
  ▸ SUBOBJECT: Label Control         Tx SubTransponders
  ▸ SUBOBJECT: Explicit Transponder Control
  ▸ SUBOBJECT: Unnumbered Interface ID: 172.16.102.104:3
  ▸ SUBOBJECT: Label Control
  ▸ SUBOBJECT: Unnumbered Interface ID: 172.16.102.103:5
  ▸ SUBOBJECT: Label Control
  ▸ SUBOBJECT: Unnumbered Interface ID: 172.16.102.105:9
  ▸ SUBOBJECT: Label Control
  ▸ SUBOBJECT: Unnumbered Interface ID: 172.16.102.109:14
  ▸ SUBOBJECT: Label Control
  ▸ SUBOBJECT: Unnumbered Interface ID: 172.16.102.114:28
  ▸ SUBOBJECT: Label Control         Rx SubTransponders
  ▸ SUBOBJECT: Explicit Transponder Control
▸ BANDWIDTH object
```

Figure 9.10 New route for lsp2 in message (4).

Let us assume that core DCs are permanently synchronized among them. Since network failures might affect those anycast connections, recovery mechanisms need to be proposed to reconnect metro DCs to the same core DC or to any other DC in the set.

Figure 9.11 presents an example with three core DCs (labeled as 10, 11, and 12) and two metro DCs (labeled as 1 and 6); we assume that a Layer 2 (L2) switch connects each DC to the optical transport network. When a metro DC needs to synchronize its local contents with those in the core DCs, it establishes an anycast optical connection to any of the core DCs (see Chapter 7). For example, in Figure 9.11a, two anycast connections (orange lines) are established: 1–10 and 6–11. In the event of a failure, anycast connections can be affected; in Figure 9.11b a link failure has torn down connection 6–11 that has been recovered by creating the anycast connection 6–12; note that the destination core DC has changed.

9.4.1 ILP Formulations and Algorithm

The anycast connection recovery (AC_RECOVER) problem aims at recovering the anycast connections affected by a network failure involving several nodes and/or links. As for the anycast problem presented in Chapter 7, our approach for solving this problem is to create an augmented graph where a *dummy* node and links connecting it to every core DC are added.

(a)

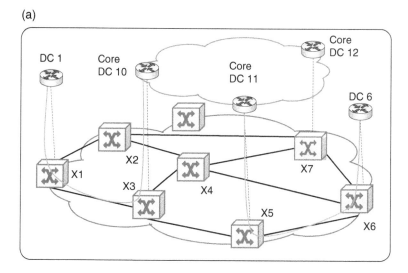

(b)

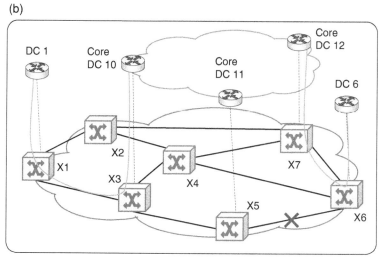

Figure 9.11 Anycast recovery. Anycast connections before (a) and after (b) a link failure.

The AC_RECOVER problem can be stated as follows:

Given:

- the network topology represented by a graph $G(N, L)$, where the failed elements (optical links or nodes) have been removed,
- the number of available slices η_l in each link l,
- the set D with the failed anycast connections. Each anycast connection d is defined by the tuple $< o_d, T, \chi >$, where o_d represents the source node, T the set of destination nodes, and χ the number of slices required.

Output: the route for the bidirectional LSPs used to recover the failed anycast connections.

Objective: maximize the recovered anycast demands.

The next sets, parameters, and decision variables have been defined:

Topology

N set of all nodes in the network, that is, nodes in the optical network and Ethernet switches in the DCs, index n.

L set of optical links connecting two nodes, index l.

δ_{nl} 1 if link l is incident to node n; 0 otherwise.

η number of available spectrum slices in link l.

D set of optical anycast demands to be recovered, index d.

o_d source node for demand d.

T set of core DCs (destinations).

χ number of slices required for each optical connection.

Decision Variables

x_{dl} binary. 1 if anycast connection for demand d is supported by optical link l; 0 otherwise.

y_d binary. 1 if demand d is recovered; 0 otherwise.

As mentioned earlier, to solve the AC_RECOVER problem we create a dummy node and connect it to every core DC. Then, the ILP formulation, assuming the graph is undirected, is as follows; we assume an opaque optical core network, so contiguity and continuity constraints are relaxed.

$$\left(\text{AC\_RECOVER}\right) \quad \max \quad \sum_{d \in D} y_d \tag{9.18}$$

subject to:

$$\sum_{l \in L} \delta_{nl} \cdot x_{dl} = y_d \quad \forall d \in D, n \in \{o_d, dummy\} \tag{9.19}$$

$$\sum_{l \in L} \delta_{nl} \cdot x_{dl} \leq 2 \quad \forall d \in D, n \in N \backslash \{o_d, dummy\} \tag{9.20}$$

$$\sum_{\substack{l' \in L \\ l' \neq l}} \delta_{nl'} \cdot x_{dl'} \geq \delta_{nl} \cdot x_{dl} \quad \forall d \in D, n \in N \backslash \{o_d, dummy\}, l \in L \tag{9.21}$$

$$\sum_{d \in D} \chi \cdot x_{dl} \leq \eta \quad \forall l \in L \tag{9.22}$$

The ILP is a node-arc formulation (see Chapter 2) to set up the underlying lightpaths. The objective function (9.18) maximizes the number of recovered

Table 9.8 AC-RECOVER heuristic algorithm.

`INPUT: G(N, L), maxIter, D, T, χ` `OUTPUT: LSPs`

`1: bestSolution ← ∅`
`2: for it = 1...maxIter do`
`3:   solution ← ∅`
`4:   shuffle(D)`
`5:   for each d in D do`
`6:     p ← AC-RSA (G(N, L), d.o, T, χ)`
`7:     if p ≠ ∅ then`
`8:       solution ← solution ∪ {p}`
`9:   bestSolution ← Select_Best(bestSolution, solution)`
`10: return bestSolution`

anycast connections. Constraints (9.19)–(9.21) route the affected anycast demands through the optical network, while constraint (9.22) ensures that optical resources used in a link do not exceed its capacity.

The AC-RECOVER problem needs to be solved in real time, for example, less than 1 s. In view of that, a heuristic algorithm, based on the bulk-path computation algorithm presented in Chapter 5, is proposed; Table 9.8 lists the pseudo-code. The heuristic uses the AnyCast RSA (AC-RSA) algorithm proposed in Chapter 7 to find an anycast optical connection between a selected metro DC and one of the core DCs. Solutions are generated by randomly sorting the anycast demand list and keeping the best solution, which is eventually returned.

9.4.2 Proposed Workflow

In this section, we focus on implementing the AC_RECOVER problem by deploying the proposed algorithm on the ABNO architecture. We assume that a bPCE is responsible for dealing with computationally intensive algorithms, such as recovery algorithms.

The AC_RECOVER workflow, presented in Figure 9.12, is initiated by the DC resource managers, issuing anycast recovery requests to the Application Service Orchestrator (ASO) when their services are affected by a failure. As a result, the ASO might receive multiple recovery requests. To accelerate the recovery process, after receiving a number of recovery requests, the ASO can infer the source of the failure and find the affected services [Ca14].

ASO issues a request to the controller for anycast recovery (labeled as 1 in Figure 9.12), which sends a PCReq message to fPCE (2). Anycast recovery is delegated to the bPCE, so the fPCE forwards the message to the bPCE (3) that runs the AC-RECOVER algorithm. Once a solution has been found, the bPCE sends a

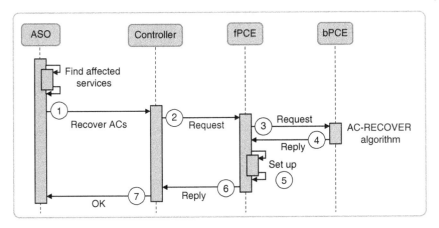

Figure 9.12 AC_RECOVER workflow.

PCRep message (4) to the fPCE. Next, the fPCE issues a request to the PM to implement the lightpaths needed for recovery (5). The PM sends a PCRpt message after each lightpath has been set up. When all the PCRpt messages are received, the fPCE informs the controller (6), which in turn informs the ASO (7). The ASO updates its service database and notifies the cloud resource manager.

9.4.3 Validation

In this section, illustrative simulation results evaluating the proposed algorithm over a realistic network topology are first presented and next the feasibility of the proposed workflow is experimentally assessed.

9.4.3.1 Simulation Results

The anycast recovery algorithm presented in the previous section has been evaluated on an ad hoc event-driven simulator based on OMNeT++. To that end, the 24-node 43-link Large Carrier US topology is used (see Chapter 7), where metro DCs were placed in every node location and up to six core DCs were strategically placed in the network. Each DC is connected to the optical transport network through an L2 switch equipped with 100 Gb/s transponders.

A dynamic network environment was simulated, where core DCs are initially created and interconnected, and then 100 Gb/s optical anycast connection requests arrive following a Poisson process. The holding time of the anycast connection requests is exponentially distributed with a mean value equal to 1 h. As the source of each anycast request, the metro DC is randomly chosen with equal probability (uniform distribution) among all metro DCs. Different loads are created by changing the arrival rate while keeping the mean holding time constant.

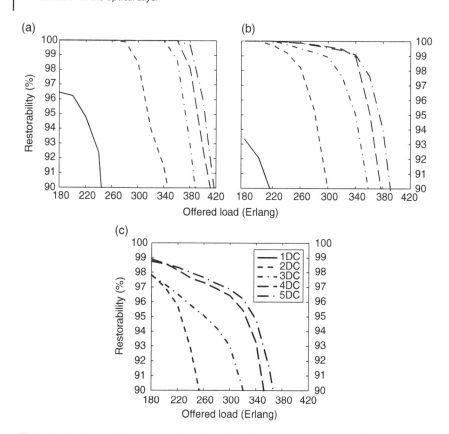

Figure 9.13 Restorability against offered load for different numbers of core DCs. (a) Single link, (b) single node, and (c) 2 adjacent nodes.

Let us analyze the restorability of the proposed algorithm considering single optical link failures, single optical node failures, and double adjacent optical node failures. As illustrated in Figure 9.13, increasing the number of core DCs increases restorability significantly; for example, for an offered load of 300 Erlang, a 99% restorability can be obtained using two core DCs under single link failure scenarios, while three core DCs have to be deployed for single node failures to obtain an equivalent restorability. In the two adjacent node failures scenario under the same offered load, only when five core DCs are deployed, 97% of restorability can be obtained.

Table 9.9 summarizes the increments of traffic load for 90% of restorability with respect to one core DC. For instance, by deploying three core DCs, the traffic load can be doubled under the two adjacent node failures scenario while maintaining the rate of restorability of affected demands.

Table 9.9 Load increment (%) with respect to 1 DC for 90% restorability.

	Failure type		
	1 link	1 node	2 adjacent nodes
2 DCs	41.8	38.9	64.3
3 DCs	59.0	65.7	107.8
4 DCs	68.4	74.5	128.6
5 DCs	70.9	81.0	137.7

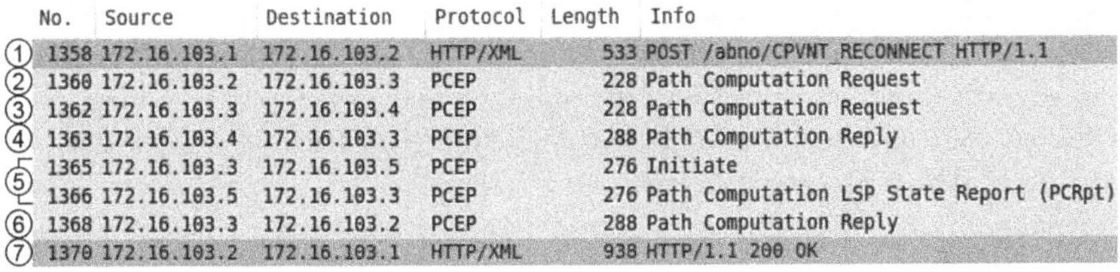

```
No.   Source          Destination    Protocol  Length  Info
①  1358 172.16.103.1   172.16.103.2   HTTP/XML      533 POST /abno/CPVNT_RECONNECT HTTP/1.1
②  1360 172.16.103.2   172.16.103.3   PCEP          228 Path Computation Request
③  1362 172.16.103.3   172.16.103.4   PCEP          228 Path Computation Request
④  1363 172.16.103.4   172.16.103.3   PCEP          288 Path Computation Reply
⑤  1365 172.16.103.3   172.16.103.5   PCEP          276 Initiate
   1366 172.16.103.5   172.16.103.3   PCEP          276 Path Computation LSP State Report (PCRpt)
⑥  1368 172.16.103.3   172.16.103.2   PCEP          288 Path Computation Reply
⑦  1370 172.16.103.2   172.16.103.1   HTTP/XML      938 HTTP/1.1 200 OK
```

Figure 9.14 AC_RECOVER message list.

9.4.3.2 Experimental Assessment

Once the performance of the AC_RECOVER algorithm has been studied, let us now focus on the experimental validation of the proposed workflow. ASO issues XML-encoded messages to the controller through the HTTP REST API. Control and management modules run in the IP subnetwork 172.16.103.X. Specifically, ASO runs in .1, the controller in .2, fPCE in .3, bPCE in .4, and PM in .5. An emulated data plane was deployed for the experiments.

Figure 9.14 shows the relevant messages for the AC_RECOVER workflow. The control plane contribution to the M2C anycast recovery workflow processing time was less than 3 ms, including messages exchange and algorithms.

9.5 Conclusions

An effective restoration scheme enabling multipath recovery and bitrate squeezing in EONs was reviewed. The restoration scheme exploits the advanced flexible capabilities provided by SBVTs, which support the adaptation of connection parameters in terms of the number of subcarrier modules, bitrate, transmission parameters, and reserved spectrum resources.

The BATIDO problem was stated to efficiently recover connections from network failures by exploiting limited portions of spectrum resources along multiple routes; an ILP model and a heuristic algorithm were proposed. Illustrative results obtained by simulations showed that the proposed recovery scheme could even double the percentage of restored bitrate with respect to single path restoration, where no squeezing or multipath are exploited.

Next, the BATIDO problem was extended considering both different modulation formats and SBVT availability. The modulation format-aware algorithm, named as MF-Restoration, that maximizes the number of restored demands affected by a network failure, was proposed. The algorithm squeezes the original bitrate of the demands only when enough resources are not available.

A workflow was proposed enabling the implementation of MF-Restoration in an ABNO-based architecture. An OAM Handler component was responsible for collecting and correlating the network failure events and triggering the workflow. The required protocol extensions for PCEP and BGP-LS enabling the advertisement and control of the ingress and egress SBVT modules in the network were also proposed. The workflow implementing the MF-Restoration algorithm was experimentally assessed.

Finally, the problem of anycast connection recovery (AC_RECOVER) was faced. The problem was formally stated, mathematically formulated, and a heuristic algorithm was eventually proposed for solving it in real time. Exhaustive simulations were carried out to evaluate the performance of the proposed algorithm. The results showed significant improvement in the network restorability for different types of failures when the number of core DCs (destinations) was increased.

To implement the AC_RECOVER problem, a workflow was proposed and experimentally assessed using an ABNO-based control plane architecture. The ASO module was responsible for service management and issues requests to the ABNO. As a result of the complexity of the problem and the stringent time in which it needs to be solved, a bPCE was used.

10

After-Failure-Repair Optimization

Lluís Gifre[1], Filippo Cugini[2], Marc Ruiz[1] and Luis Velasco[1]

[1] Universitat Politècnica de Catalunya, Barcelona, Spain
[2] CNIT, Pisa, Italy

As discussed in Chapter 8, capacity usage in dynamic Elastic Optical Networks (EONs) may not be optimal due to the permanent process of setting up and tearing down of connections, which might lead to spectrum fragmentation and, as a result, to increased connection blocking. On top of this, a restoration mechanism that is triggered in reaction to a link (or node) failure restores the affected lightpaths (see Chapter 9); eventually, when the link is repaired, and its capacity becomes available for new connections, the unbalance between lightly and heavily loaded links increases, thus further decreasing the probability of finding optical paths with continuous and contiguous spectrum for future connection requests. In this chapter, we study the effects of reoptimizing the network after a link failure has been repaired (namely, the *AFRO* (after-failure-repair optimization) problem), as an effective way for both reducing and balancing capacity usage and, by these means, for improving network performance. Illustrative numerical results show that AFRO allows to significantly decrease the request blocking probability in realistic dynamic network scenarios. Besides, traffic disruptions resulting from lightpath rerouting are practically negligible. From the recovery perspective, using multiple paths (named as *subconnections*) to restore every connection affected by a failure might increase connection restorability. However, the multipath restoration scheme might result in poor resource utilization entailing a lesser grade of service. In view of that, the AFRO problem is extended to the so-called multipath AFRO (MP-AFRO) to reduce the subconnections count by aggregating those belonging to the same original connection and rerouting the resulting connection to release spectral resources.

Provisioning, Recovery, and In-operation Planning in Elastic Optical Networks,
First Edition. Luis Velasco and Marc Ruiz.
© 2017 John Wiley & Sons, Inc. Published 2017 by John Wiley & Sons, Inc.

Table of Contents and Tracks' Itineraries

10.1 Introduction

In the context of dynamic networks, the Application-Based Network Operations (ABNO) [ABNO] architecture can be used to control dynamic operations such as solving the routing and spectrum allocation (RSA) problem for incoming connection requests (Chapter 5), performing in-operation network planning, for example, spectrum defragmentation (see Chapter 8), and restoring light-paths to deal with network failures (Chapter 9).

In the case of a link (cable) failure, bulk path computation (see Chapter 5) can be used to increase the number of restored lightpaths without dramatically increasing restoration times, which are kept in the order of hundreds of milliseconds. Note that between the restoration instant and the failure repair instant, the network operates with reduced available capacity, thus increasing the blocking probability of new connection requests. Once the link is repaired, the overall available capacity of the network increases. However, the difference in terms of free available capacity between the highest and the lowest loaded links becomes a problem when solving RSA for new connection requests because of the difficulty of finding routes with a sufficient number of continuous and contiguous frequency slices and therefore the connection blocking probability remains high even after the failure repair.

In order to address this problem, in this chapter, we focus on a dynamic network scenario in which optical connections are rerouted during network operation. The reoptimization is performed after a link has been repaired from a failure—which we refer to as *after-failure-repair optimization* (AFRO)—and it concerns rerouting of some existing lightpaths with the objective to improve network performance. The effectiveness of AFRO relies on the fact that lightpaths are rerouted from highly loaded links to the recently restored link that results in a more balanced network load and, by these means, increases the possibility to accept further connection requests.

It is clear that, even without the application of AFRO, the imbalance in capacity between the repaired link and other network links will be mitigated after some time as old lightpaths are torn down, and new ones are established. Therefore, in very dynamic scenarios in which the connection holding time (HT) is much smaller than the mean time between two consecutive link failures, the gains from applying AFRO, in terms of the improved network performance, seem to lose its importance. Moreover, in such scenarios, the need to suspend the connection setup/release operations during network reoptimization and the possible traffic disruptions during lightpath rerouting may deteriorate network performance. However, if the connection HT increases, the benefits from applying reoptimization increases as well.

Thus, when the connections remain for days, weeks, or even months (which is more than the usual duration in certain real-world scenarios), the application of AFRO allows a rapid improvement in the balancing of network capacity. The reoptimization not only increases the possibility to admit following connection requests, but it also improves the effectiveness of the lightpath restoration algorithm used after link failures. In order to illustrate it, let us consider a national network with a total of about 10,000 km of fiber and a link failure rate of 2.72×10^{-3} per km per year [Vr05], which corresponds to about 30 failures in a year. If we assume this number and that the connections remain active for weeks or months, the probability that a connection is affected by more than one failure (and therefore by more than one restoration) is high. The accumulation of several restorations without the ability to reoptimize the network can decrease the effectiveness of the restoration mechanisms. Therefore, AFRO can be considered as a problem of network reoptimization that aims at improving load distribution so as to reduce the blocking of incoming connection requests and improve the efficiency of the restoration algorithm. Hence, AFRO solving times as stringent as hundreds of milliseconds are required.

10.2   The AFRO Problem

In order to describe the problem addressed in this chapter, in Figure 10.1 we show a process that a dynamic network undergoes after being subject to a certain link failure. Complementing Figure 10.1, Figure 10.2 illustrates an example of the network states at different time instants. Here, we consider a dynamic network that serves a set of lightpaths under normal operation (denoted as state *t0*). Specifically, a number of connections, namely, *P1–P5*, are shown in Figure 10.2a before the event of a fiber cut in link 6–7. The route of connections *P1* and *P2* is represented in the network topology, and the spectrum usage is shown in the diagram, where the spectrum allocation for all five established connections is specified.

After a failure in link 6–7, the restoration mechanism finds an alternative route for the lightpaths affected by the fiber cut (state *t1*). In our example, connection *P2* is rerouted by the restoration algorithm in Figure 10.2b.

Once the solution of the restoration algorithm is implemented, the network operates without the failed link until it is restored (6–12 h are usually needed to repair a fiber link), which is denoted as state *t2*. For the sake of a clear presentation, we assume that *t1* = *t2* in Figure 10.2, that is, neither setups nor teardowns of lightpaths occur during the period of link restoration. At this point, we can see that after the link is repaired, the network capacity increases; its performance, however, is not necessarily improved.

With the aim to improve network performance and make use of resources in the repaired link we consider that lightpaths are rerouted after the link is repaired (i.e., in state *t2*). To this end, we solve the AFRO problem, which gets the network state information as input, in particular those concerning RSA of lightpaths established in the network, and returns a list of rerouted lightpaths with their new routes and spectrum allocation. Since AFRO tries to make use of the recently restored link capacity, new routes are forced to use the repaired link; the lightpaths that cannot be routed over the restored link are kept without changes and are not considered for reoptimization. This constraint allows reducing the amount of lightpath reallocations in the network significantly. After solving AFRO and implementing its solution (state *t3*), the network is ready to operate, and its performance is expected to improve.

Figure 10.1 Dynamic evolution of network states (*t*) with link failure and link repair events.

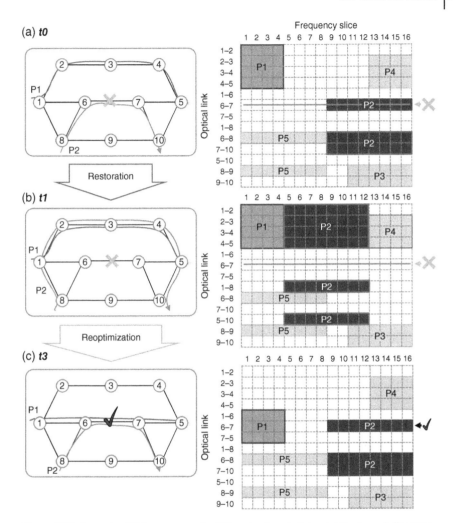

Figure 10.2 Example of link occupancy in different network states. (a) Route and allocation of lightpaths under normal operation, (b) after restoration of a cut on in link 6–7, and (c) after reoptimization.

Note that during the AFRO processing time, the process of establishing new connection requests is stopped, as in other dynamic network reoptimization operations. For this very reason, the time needed for solving AFRO and implementing network changes should be short in order to reduce the delay of that process. Indeed, to coordinate with other operations and to reduce even more the impact over new incoming traffic, AFRO might be scheduled in a time slot with less dynamic activity (e.g., night hours) and therefore not applied immediately after the failure repair.

10.2.1 Problem Statement

The AFRO problem can be formally stated as follows:

Given:
- an Elastic Optical Networks (EONs) represented by a graph $G(N, E)$, where N is the set of optical nodes and E is the set of fiber links,
- a set D of (traffic) demands,
- an optical spectrum (given at each link), which is divided into a set S of frequency slices,
- a set of lightpaths currently established in the network,
- a link $e^r \in E$, representing the link recently repaired that was previously unavailable.

Output: a subset of demands from set D, to be rerouted and, for each of these demands, a new lightpath over the network such that it uses the repaired link e^r, aiming at utilizing recently restored link capacity.

Objective: to optimize the usage of spectrum in the network.

10.2.2 Optimization Algorithm

In Figure 10.3 we present the main processing steps to solve AFRO and implement the solution found. Starting from the set of lightpaths established in $t2$, some of them are selected for reoptimization. In particular, at this step, we decide for which demands it is potentially worthwhile to reroute their lightpaths (the preselection of these demands is described later in this section). Next, for the selected subset of demands, the AFRO problem is solved, and a reoptimization solution (i.e., a set of lightpaths for the selected demands) is obtained. The generation of new lightpaths is traffic disruption-aware (see Section 10.2.5). Finally, in a postprocessing step, lightpaths are arranged appropriately in an order in which the lightpath rerouting will be implemented in the network. The aim of this ordering is to minimize traffic disruptions that might occur during the rerouting. The lightpath ordering procedure is also described in Section 10.2.5.

The first issue that we have to face while solving the AFRO problem is to define a set of demands to be taken into account while reoptimizing a network. Here, two extremities exist: on the one hand, we can consider all the demands served in a network. This approach assures us that we do not miss an optimal solution. However, it keeps a size of the problem large, thus making it difficult

Figure 10.3 Processing steps of the AFRO algorithm.

Table 10.1 Find candidate demands algorithm.

INPUT: $G(N, E)$, D, e^r
OUTPUT: D'

1: $D' \leftarrow \emptyset$
2: **for each** $d \in D$ **do**
3:   $K \leftarrow \text{KSP}(G, d, |E(d.p)|)$
4:   **for each** $p \in K$ **do**
5:     **if** $e^r \in E(p)$ **then**
6:       $D' \leftarrow D' \cup \{d\}$
7:       **break**
8: **return** D'

to solve; on the other hand, we can improve the tractability by limiting a set of considered demands and sacrificing some parts of the feasibility region. This leads us to the second extremity, which is selecting one single demand in the beginning and solving a rather simple problem of rerouting a single lightpath.

In our research, we have developed a procedure that places itself between the two extremities described earlier.

It consists of considering all the demands in which the current lightpath is not shorter than the shortest path that includes the repaired link. An algorithm to find the candidate demands is presented in Table 10.1. The algorithm receives as input the network represented by graph G, the set of demands currently being served, and the repaired link e^r, and it finds the number of shortest paths, the length of which in hops does not exceed the length of the current lightpath.

10.2.3 ILP Formulation

In this section, we focus on modeling the AFRO problem. As already stated, the AFRO problem consists of deciding, for each demand $d \in D$, whether it should be rerouted to a new lightpath, and, if so, in determining a route over the network that uses the recently repaired link e^r. We assume that regeneration requirements are met for all potential paths and that only one link can fail and be repaired at a time. However, the proposed method is not restricted to solely single link failure scenarios and can be used to handle multiple failures, that is, when e^r is not just a single link but a set of recently repaired links.

The Integer Linear Programming (ILP) model is based on the *arc-path* formulation for routing, modulation, and spectrum allocation (RMSA) (see Chapter 3), where a set of routes is precomputed for each candidate demand to be reoptimized.

The following sets and parameters have been defined:

E Set of network links, index e.

S Set of slices in the spectrum of each link, index s.

D Set of candidate demands, index d. Each demand d represented by tuple $<s_d, t_d, b_d, p_d>$, where s_d is the source node, t_d is the target node, b_d is the requested bitrate, and p_d is the current path.

$K(d)$ Set of precomputed paths for demand d, index p.

$C(d)$ Set of slots for demand d, index c.

M Set of modulation formats, index m.

R Set of pairs in reach-modulation format, index r.

δ_{pe} 1 if path p contains link e; 0 otherwise.

α_{es} 1 if slice s in link e is used by any already established connection not in D; 0 otherwise.

γ_{cs} 1 if slot c contains slice s; 0 otherwise.

$b(c, r)$ Maximum bitrate that can be conveyed using slot c with the modulation format given by pair r.

$len(r)$ Reachability of pair r (in km).

$len(p)$ Length of path p (in km).

$h(p)$ Length of path p (in hops).

Decision Variables

x_{pc} Binary. 1 if path $p \in P(d)$ and slot $c \in C(d)$ are used to serve demand d; 0 otherwise.

y_{pcr} Binary. 1 if pair r and slot $c \in C(d)$ are used to serve demand d using path $p \in P(d)$; 0 otherwise.

The $K(d)$ set is computed using equation (7.1) as the union between the current path and the set of all shortest paths between s_d and t_d, using the repaired link e^r.

$$K(d) = \{p_d\} \cup \left\{ p \in KSP(s_d, t_d), e^r \in E(p) \right\} \quad \forall d \in D \tag{10.1}$$

The ILP formulation for the AFRO problem is as follows:

$$\left(\text{AFRO}\right) \quad \min \quad \Phi = \sum_{d \in D} \sum_{p \in K(d)} \sum_{c \in C(d)} h(p) \cdot x_{pc} \tag{10.2}$$

subject to:

$$\sum_{p \in K(d)} \sum_{c \in C(d)} x_{pc} = 1 \quad \forall d \in D \tag{10.3}$$

$$\sum_{d \in D} \sum_{p \in K(d)} \sum_{c \in C(d)} \delta_{ke} \cdot \gamma_{cs} \cdot x_{pc} \leq (1 - \alpha_{es}) \quad \forall e \in E, s \in S \tag{10.4}$$

$$\sum_{p \in K(d)} b_d \cdot x_{pc} \leq \sum_{r \in R} b(c,r) \cdot y_{pcr} \quad \forall d \in D, c \in C(d) \tag{10.5}$$

$$\sum_{p \in K(d)} \sum_{c \in C(d)} len(p) \cdot x_{pc} \leq \sum_{r \in R} \sum_{c \in C(d)} len(r) \cdot y_{pcr} \quad \forall d \in D \tag{10.6}$$

$$\sum_{r \in R} \sum_{p \in K(d)} \sum_{c \in C(d)} y_{pcr} = 1 \quad \forall d \in D \tag{10.7}$$

The objective function in equation (10.2) minimizes the sum of all slices used on the links. Constraint (10.3) assigns a lightpath (either new or the current one) to a demand. Constraint (10.4) ensures that any single slice is used to convey one demand at the most, provided that it is not already used by other demands not in D. Constraints (10.5)–(10.7) deal with modulation formats. Constraint (10.5) guarantees that using slot c with modulation format given by pair r is enough for serving all the requested bitrate. Constraint (10.6) limits the length of the used path to the reachability associated to pair r. Finally, constraint (10.7) selects one pair r for each demand.

The size of this formulation for the AFRO problem is $O(|D| \cdot |K(d)| \cdot |C(d)| \cdot |R|))$ variables and $O(|D| \cdot |C(d)| + |E| \cdot |S|)$ constraints.

Although the ILP model can be solved in a short period of time, for example, minutes, using a commercial solver such as CPLEX, during AFRO computation new connections could arrive, which need to be queued and delayed until the AFRO solution is implemented in the network. Aiming at reducing provisioning delay, while providing a good trade-off between complexity and optimality, a heuristic algorithm is presented next to solve AFRO in the stringent required times (e.g., <1 s).

10.2.4 Heuristic Algorithm

The AFRO algorithm (Table 10.2) is based on the bulk-path computation algorithm introduced in Chapter 5. The algorithm receives as input the network represented by $G(N, E)$, the candidate list of demands to be reoptimized D, the repaired link e^r, and the maximum number of iterations $maxIter$ to be run; it returns the ordered set $bestS$ containing the best solution found, where the lightpath to serve each demand is specified.

All the demands are first deallocated from the original graph G (line 2 in Table 10.2) and then the algorithm performs a number ($maxIter$) of iterations, where the set D is served in a random order (lines 3–5). At each iteration, a copy of the original graph is stored into an auxiliary graph G', so that every subsequent operation is performed over G'. Each iteration consists of serving

Table 10.2 AFRO heuristic algorithm.

INPUT: $G(N, E)$, D, e^r, maxIter
OUTPUT: bestS

```
 1: bestS ← D
 2: for each d ∈ D do deallocate(G, d)
 3: for i = 1…maxIter do
 4:   S ← ∅; G' ← G; feasible ← true
 5:   sort(D, random)
 6:   for each d ∈ D do
 7:     <p, m, c> ← get_RMSA(G', e^r, d)
 8:     if p = ∅ then
 9:       feasible ← false; break
10:     allocate(G', d, <p, m, c>)
11:     S ← S ∪ {d}
12:   if NOT feasible then
13:     for each d ∈ S do deallocate(G', d)
14:     continue
15:   Compute Φ(S)
16:   for each d ∈ D do deallocate(G', d.SC')
17:   if Φ(S) > Φ(bestS) then bestS ← S
18: for each d ∈ D do allocate(G, d)
19: return bestS
```

---

the demands ensuring the requested bitrate through the shortest path that it either is the original path or includes e^r in its route (lines 6–11). A solution is feasible only if every demand can be served, so in case that one demand cannot be served, the solution is discarded (lines 8–9); otherwise, the resources selected are reserved in G' (line 10).

After a complete solution is obtained, its fitness value is computed (line 15), and the resources used in the solution are released from G' (line 16). The fitness value of the just obtained solution is compared to that of the best solution found so far (the incumbent) and stored provided that it is feasible and its fitness value is better than that of the incumbent (line 17). The state of G is restored, and the incumbent solution is eventually returned (lines 18–19).

10.2.5 Disruption Considerations

One of the potential problems that may occur during network reconfiguration is disruption of traffic on rerouted lightpaths. In order to minimize this effect, in the optimization algorithm we apply two techniques: first, when finding lightpaths and, second, in a postprocessing phase. These techniques are described here.

A vast majority of disruptions result from a situation when a newly rerouted lightpath cannot be established before the one it replaces is torn down. The reason behind this fact is that both of them are using some common slices in one or more links. We can prevent such situations from happening by not generating lightpaths that overlap with the lightpaths they are supposed to replace, in terms of route and spectrum allocation. However, in order to make the problem less complex, our excluding rule is slightly harsher; in our experiments, we decided to generate lightpaths that use no common slices with the current lightpath. However, they can use links utilized by the current lightpath. Such an approach does not take into consideration feasible lightpaths that share one or more slices with the current lightpath but are link disjoint. Although less exact than the model approach described earlier, this approach can be easily adapted by the applied RMSA algorithm.

The postprocessing procedure aims at ordering and grouping the lightpaths that have to be rerouted in such a way that all the changes can be performed in a minimum number of bundles without disruption. We define a bundle without disruption as a subset of reroutings that can be performed simultaneously (i.e., in parallel) and by means of a *make-before-break* procedure [RFC3209]. The make-before-break procedure consists of establishing the new lightpath and rerouting the traffic from the old lightpath to the new one before tearing down the old lightpath. It can be assumed that the signaling related to the setup and teardown of n lightpaths in a bundle can be performed in parallel so that the time required to reroute all lightpaths in a bundle corresponds to the time necessary to reroute just one (the longest) lightpath in the bundle.

However, it may happen that the make-before-break strategy cannot be applied to all the rerouted lightpaths especially if it is necessary to remove some existing lightpaths to establish some others. In such case, it will lead to traffic disruptions on the affected lightpaths. Therefore, the process of ordering and grouping the lightpaths into bundles should aim at reducing such situations. For solutions without disruptions or with the same number of disruptions, the secondary objective is to minimize the number of bundles to minimize the time required to perform lightpath rerouting.

In order to address the objectives discussed earlier, we implemented a simple iterative heuristic algorithm in which, given a sequence of lightpaths to be rerouted, we create and fill the bundles by including into them the lightpaths in the order in which they appear in the sequence. Therefore, in the first iteration, the first bundle of lightpaths is created by checking all lightpaths in the sequence and including into the bundle such lightpaths that can be rerouted without disruptions (i.e., with the make-before-break strategy). After that, the first bundle is closed, and a new one is open. Now, in consecutive iterations, the procedure is repeated for the rest of lightpaths until there are no lightpaths left, or for the rest of lightpaths it is not possible to perform their rerouting with the make-before-break strategy. In the latter case, we assume that such lightpaths will be disrupted.

Finally, in order to diversify the set of solutions, several random sequences of lightpaths are generated, and the heuristic discussed earlier is run for all of them returning the best solution found.

10.2.6 Performance Evaluation

In this section, we evaluate the performance of a dynamic EON operating with AFRO. To this end, we apply the AFRO framework in a dynamic network scenario and evaluate it by means of network simulations. Eventually, we study the impact of applying AFRO only for a selected set of network links, instead of using it in all network links, with the aim to reduce the AFRO operational complexity.

10.2.6.1 Scenario

To evaluate the performance of AFRO, we use two representative core network topologies: the 22-node British Telecom (BT) and the 28-node European (EN) networks; the details of the networks are presented in Table 10.3. We consider the fiber links with the spectrum width equal to 2 THz and assume a link failure rate of 2.72×10^{-3} per km per year [Vr05]. Taking this into account and considering the total length of fiber links in BT and EN (as presented in Table 10.3), the average number of cuts per year is equal to 14 and 70, respectively. Accordingly, the mean time to failure (MTTF) is equal to 625 and 125 h, respectively, for BT and EN; we assume that the mean time to repair (MTTR) is the same for both networks and equal to 12 h. The AFRO ILP problem is solved using the CPLEX v.12.5 optimizer [CPLEX], and simulations are run on an ad hoc event-driven simulator implemented in OMNeT++ [OMNet], on a 2.4 GHz Quad-core machine with 8 GB RAM.

Connection requests are generated following a Poisson process, and lightpaths are torn down after an exponentially distributed HT. The source and destination nodes are uniformly chosen. We consider that the requested bitrate can be equal to 40, 100, or 400 Gbit/s under traffic profile TP-6 (see traffic profiles in Chapter 5). Such traffic profile combines several bitrates and takes advantage of high spectrum granularity of EONs. Assuming the quadrature phase shift keying (QPSK) with spectrum efficiency equal to 2 bit/s/Hz is the applied modulation format, the number of slices requested for each

Table 10.3 Network parameters.

Topology	Nodes	Links	Total link length (km)	Average no. of cuts per year	MTTF (h)	MTTR (h)
BT	22	35	5,145	~14	625	12
EN	28	41	25,625	~70	125	12

connection type is 4, 8, and 32, respectively. One of the performance metrics that we use in the evaluation is the blocking probability (*Pb*) of incoming connection requests weighted by the requested bitrate.

10.2.6.2 Dynamic Performance of AFRO

In Figure 10.4 we show the results of *Pb* as a function of the offered load, which is normalized to the load for which *Pb* is equal to 1% in a network operating without failures (i.e., for $MTTF = \infty$); the results are obtained for different values of HT. We can see that the application of AFRO leads to lower values of *Pb* in all studied scenarios when compared to a network without AFRO. Moreover, the relative gain in terms of *Pb* after applying AFRO increases with traffic load, as well as with higher values of HT. It can be concluded that if a network has to guarantee certain target level of *Pb* (e.g., *Pb* = 1%), the application of AFRO allows increasing the network load. Finally, we can see that the difference between the cases with AFRO and without AFRO is more prominent in the EN topology than in the BT topology.

In order to evaluate the gain in supported traffic load, in Figure 10.5 we present the results obtained for different values of HT at *Pb* = 1%. Again, we can see that the higher the value of HT, the higher the gain in network load after applying AFRO. Also, in order to achieve a 10% gain, the connection HT can be shorter in EN (HT ~ 1.5 months) than in BT (HT ~ 4 months).

Table 10.4 presents averaged results of the number of rerouted lightpaths, the number of lightpath disruptions, and the number of lightpath bundles. Recall that lightpath disruption occurs when the lightpath cannot be rerouted with the make-before-break mechanism. The number of bundles corresponds to the signaling overhead of lightpaths rerouting. We can see that both the number of lightpaths selected to be rerouted, which is a result of the optimization algorithm, and the signaling overhead are not affected by using the proposed traffic disruption-aware mechanisms. At the same time, the very small number of disrupted lightpaths obtained leads to a virtually hitless network reoptimization.

Finally, in Table 10.4 we also evaluate the impact of AFRO on the effectiveness of the restoration mechanism used in our network. In particular, we can see that in both BT and EN topologies the percentage of restored connections in a network operating with AFRO is not only preserved, but it is even slightly better than in a network without AFRO.

10.2.6.3 Solving AFRO after Selected Link Repairs

The last set of experiments concerns network scenarios in which AFRO is run only after the repair of a selected set of links, instead of using it for all network links. In particular, in this study AFRO is applied for the links which are most loaded during normal network operation.

In Figure 10.6, we present the performance of AFRO, in terms of weighted *Pb*, for the BT topology and HT = 9 months, assuming a different percentage of

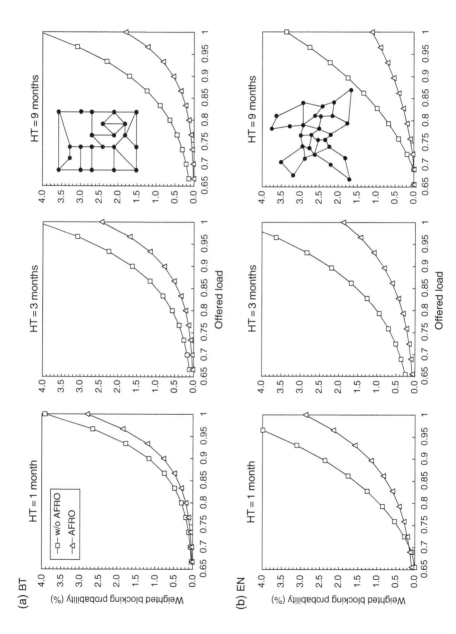

Figure 10.4 Weighted blocking probability versus normalized network load: (a) BT and (b) EN topologies.

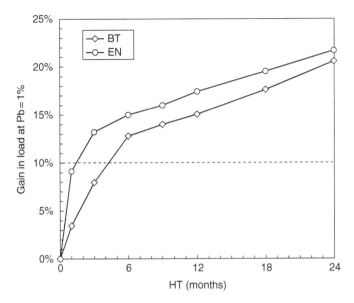

Figure 10.5 Gain in supported load at $Pb = 1\%$ versus connection HT.

Table 10.4 Disruption and restorability analysis.

Network	Rerouted lightpaths	Number of disruptions	Number of bundles	% of restored connections	
				w/o AFRO	AFRO
BT	14.19	0.47	2.49	99.58	99.65
EON	13.51	0.26	2.52	99.48	99.52

network links considered for reoptimization (denoted as ρ). Note that AFRO is not applied if $\rho = 0\%$ and it is always used if $\rho = 100\%$; for instance, for $\rho = 40\%$ it is applied for 40% of (most loaded) network links. Apart from that, we calculate the AFRO relative gain, which represents the improvement in Pb with respect to the scenario without using AFRO.

In Figure 10.6, we can see that it is not necessary to run AFRO for all network links in order to obtain its maximal efficiency. For instance, if we restrict AFRO to be applied after the repair of links which are among 60% of the most loaded links, the AFRO relative gain is already equal to 100%. Moreover, it is enough to use it for only 20% of links, so that to achieve a 50% reduction of Pb. These results indicate that AFRO can be applied with lower frequency without affecting its effectiveness significantly, thus reducing its operational complexity.

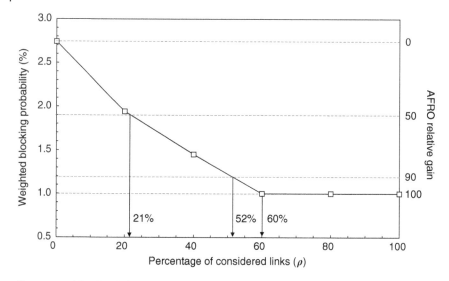

Figure 10.6 Weighted *Pb* and AFRO relative gain versus percentage of links considered for reoptimization.

10.3 Restoration and AFRO with Multiple Paths

Using multiple subconnections to restore every connection affected by a failure might increase connection restorability. In this section, we extend the AFRO problem to deal with demands that have been restored using multiple subconnections (named as multipath AFRO—MP-AFRO), so as to reduce the subconnections count by aggregating those belonging to the same original connection and rerouting the resulting connection to release spectral resources.

For illustrative purposes, as in Figure 10.2, Figure 10.7 reproduces three snapshots with the state of a small EON where several connections are currently established. The link 6–7 has failed in Figure 10.7a; multipath restoration has been applied in Figure 10.7b and connection *P2* has been split into two parallel subconnections *P2a* and *P2b* squeezing the total conveyed bitrate to fit into the available spectrum. Failed link 6–7 has been repaired in Figure 10.7c and reoptimization has been performed by solving MP-AFRO, so subconnections have been merged back, and bitrates have been expanded to the originally requested values.

In view of the example, it is clear that multipath restoration allows increasing restorability, in particular when enough contiguous spectrum cannot be found along a single path, as happened when restoring *P2*. Nonetheless, this benefit is at the cost of an increased resource usage, not only as a result of using (not shortest) parallel routes and squeezing the total conveyed connection's bitrate but also because the spectral efficiency is degraded when connections

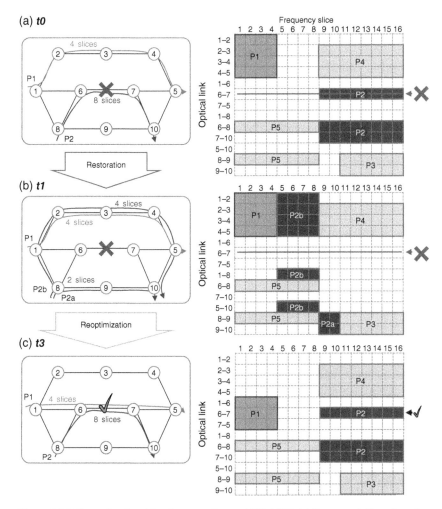

Figure 10.7 Example of multipath restoration and MP-AFRO. (a) Route and allocation of lightpaths under normal operation, (b) after restoration of a cut on in link 6–7, and (c) after re-optimization.

are split. For instance, a 400 Gb/s aggregated flow can be conveyed on one single 100 GHz connection or on four parallel 37.5 GHz subconnections thus using 50% more spectral resources even in the case of being routed through the same links.

For this very reason, resource utilization can be improved by applying MP-AFRO, that is, by rerouting established connections on shorter routes by merging parallel subconnections to achieve better spectrum efficiency and expanding the conveyed connections' bitrates to its original ones. Figure 10.7c illustrates an example of such reoptimization, where connection *P1* has been rerouted using a shorter route that includes the repaired link, whilst subconnections *P2a* and *P2b* have been merged on a single connection conveying the originally requested bandwidth.

10.3.1 Problem Statement

In the MP-AFRO problem, the objective is to reduce the number of subconnections resulting from the restoration process, as an indirect way to optimize the spectrum in the network. In addition, since the original connection bitrate could be squeezed during restoration, the originally requested bitrate should be served back. With these considerations, the MP-AFRO problem can be formally stated as follows:

Given:
- an EON represented by a graph $G(N, E)$, where N is the set of optical nodes and E is the set of fiber links,
- a set D of (traffic) demands,
- an optical spectrum (given at each link) which is divided into a set S of frequency slices,
- a set of lightpaths currently established in the network,
- a link $e^r \in E$, representing the link recently repaired that was previously unavailable.

Output: the route and spectrum allocation for demands in D, merging subconnections serving each original connection.

Objective: maximize the bitrate served while minimizing the total number of subconnections used to convey the traffic served.

10.3.2 MILP Formulation

The Mixed Integer Linear Programming (MILP) model for the MP-AFRO problem extends that for the AFRO one in Section 10.2.3. In addition to the sets and parameters already introduced for the AFRO formulation, we define the following:

$P(d)$	Set of subconnections being used to serve d.
β	Objective function weight.
$b(p, c)$	Maximum bitrate that can be conveyed through path p using slot c.

Regarding decision variables, new variables are defined:

z_{pc}	Positive Real. Served bitrate for demand d through path p and slot c.
w_d	Positive Integer. Total number of subconnections used to serve demand d.

The MILP formulation for the MP-AFRO problem is as follows:

$$\left(\text{MP-AFRO}\right) \qquad \max \quad \Phi = \sum_{d \in D} \left(\frac{\beta}{b_d} \sum_{p \in K(d)} \sum_{c \in C(d)} z_{pc} - \frac{1}{\left|P(d)\right|} w_d \right) \qquad (10.8)$$

subject to:

$$\sum_{p \in K(d)} \sum_{c \in C(d)} x_{pc} = w_d \quad \forall d \in D \qquad (10.9)$$

Constraint (10.4) from the AFRO ILP

$$w_d \le \left|P(d)\right| \quad \forall d \in D \qquad (10.10)$$

$$\sum_{p \in K(d)} \sum_{c \in C(d)} z_{pc} \le b_d \quad \forall d \in D \qquad (10.11)$$

$$z_{pc} \le b_d \cdot x_{pc} \quad \forall d \in D, p \in K(d), c \in C(d) \qquad (10.12)$$

$$z_{pc} \le \sum_{r \in R} b\left(c,r\right) \cdot y_{pcr} \quad \forall d \in D, p \in K(d), c \in C(d) \qquad (10.13)$$

$$\sum_{c \in C(d)} len\left(p\right) \cdot x_{pc} \le \sum_{r \in R} \sum_{c \in C(d)} len\left(r\right) \cdot y_{pcr} \quad \forall d \in D, p \in K(d) \qquad (10.14)$$

$$\sum_{r \in R} \sum_{c \in C(d)} y_{pcr} = 1 \quad \forall d \in D, p \in K(d) \qquad (10.15)$$

The objective function in equation (10.8) maximizes the total served bitrate while minimizing the number of subconnections used to serve that bitrate.

Constraint (10.9) accounts for the number of subconnections used to serve each demand (note the difference with constraint (10.3) in the AFRO ILP). Constraint (10.10) guarantees that the subconnections count to serve any specific demand is not increased. Constraint (10.11) assures that served bitrate does not exceed the demand's requested bitrate. Constraint (10.12) sets to zero the bitrate of unused subconnections.

Constraints (10.13)–(10.15) work similarly as constraint (10.5)–(10.7) in the AFRO ILP, although here they apply to every subconnection individually. Constraint (10.13) limits the bitrate conveyed by any specific subconnection to the maximum associated to pair r, whereas constraint (10.14) limits the length of each used subconnection to the reachability associated to pair r. Finally, constraint (10.15) selects one pair r for each subconnection.

The size of the MILP formulation for the MP-AFRO problem is $O(|D| \cdot |K(d)| \cdot |C(d)| \cdot |R|)$ variables and $O(|D| \cdot |K(d)| \cdot |C(d)| + |E| \cdot |S|)$ constraints. In view of its size, a heuristic algorithm is presented next to solve MP-AFRO in the stringent required times.

10.3.3 Heuristic Algorithm

The MP-AFRO algorithm is shown in Table 10.5. The algorithm maximizes the amount of bitrate that is served (note that only part of the original bitrate could be restored) while minimizing the number of subconnections per

Table 10.5 MP-AFRO heuristic algorithm.

INPUT: $G(N, E)$, D, e^r, $maxIter$
OUTPUT: $bestS$

```
 1: bestS ← D
 2: for each d ∈ D do deallocate(G, d.SC)
 3: for i = 1…maxIter do
 4:    S ← ∅; G' ← G; feasible ← true
 5:    sort(D, random)
 6:    for each d ∈ D do
 7:       d.SC' ← ∅; d.reoptBw ← 0
 8:       {<p, m, cₚ>} ← getMP_RMSA(G', eʳ, d, d.servedBw)
 9:       if b({<p, m, cₚ>}) < d.servedBw then
10:          feasible ← false; break
11:       d.reoptBw ← b({<p, m, cₚ>}); d.SC' ← {<p, m, cₚ>}
12:       allocate(G', d.SC'); S ← S ∪ {d}
13:    if NOT feasible then
14:       for each d ∈ S do deallocate(G', d.SC')
15:       continue
16:    for each d ∈ D do
17:       if d.reoptBw < d.bw then
18:          {<p, m, cₚ>} ← expandSlots(d, d.bw, G')
19:          if b({<p, m, cₚ>}) > d.reoptBw then
20:             d.reoptBw ← b({<p, m, cₚ>})
21:             d.SC' ← {<p, m, cₚ>}; continue
22:          if d.reoptBw < d.bw AND |d.SC'| < |d.SC| then
23:             {<p, m, cₚ>} ← getMP_RMSA(G', eʳ, d, d.bw - d.reoptBw)
24:             if b({<p, m, cₚ>}) = 0 then break
25:             d.reoptBw ← d.reoptBw + b({<p, m, cₚ>})
26:             d.SC' ← d.SC' ∪ {<p, m, cₚ>}
27:             allocate(G', d.SC')
28:    Compute Φ(S)
29:    for each d ∈ D do deallocate(G', d.SC')
30:    if Φ(S) > Φ(bestS) then bestS ← S
31: for each d ∈ D do allocate(G, d.SC)
32: return bestS
```

---

demand being reoptimized. The algorithm extends that for the AFRO problem presented in Section 10.2.4.

All the subconnections of each demand are first deallocated from the original graph G (line 2 in Table 10.5) and then the algorithm performs a number (*maxIter*) of iterations, where the set D is served in a random order (lines 3–5). Each iteration consists of two steps performed sequentially. In the first step (lines 6–12), the set of demands is served ensuring the currently served bitrate in the hope of finding a shorter route. The *getMP_RMSA* function computes the set of subconnections consisting of the routes with the highest available capacity, using equation (10.16), and the minimum cost and selects those to serve the required bitrate (line 8).

$$c_p(d) = \left\{ c^* \; \middle| \; \begin{array}{l} b(p,c^*) \geq b(p,c) \quad \forall c \in C(d) \\ \alpha_{es} = 0 \quad \forall e \in p, s \in S(c) \end{array} \right\} \tag{10.16}$$

A solution is feasible only if every demand obtains at least the same bitrate as the current allocation, so the solution is discarded in case the required bitrate cannot be served (lines 9–10); otherwise, the resources selected are reserved in the auxiliary graph G' (line 12). In the second step, the bitrate of the demands is increased first by allocating wider slots along the same routes (lines 17–21) and then by adding more subconnections (lines 22–27).

After a complete solution is obtained, its fitness value is computed (line 28). Next, the used resources in the solution are released from the auxiliary graph (line 29). The fitness value of the just obtained solution is compared to that of the best solution found so far (line 30) and stored provided that it is feasible and its fitness value is better than that of the incumbent. The state of G is restored, and the best solution is eventually returned (lines 31–32).

10.3.4 MP-AFRO Performance Evaluation

To evaluate the performance of MP-AFRO, we used two representative core network topologies: the 30-node Spanish Telefonica (TEL) and the 28-node EN networks. We considered the fiber links with the spectrum width equal to 2 THz.

The MP-AFRO heuristic was developed in C++ and integrated into an ad hoc event-driven simulator based on OMNeT++. Connection requests are generated following a Poisson process and are torn down after an exponentially distributed connection HT with mean equal to 6 months. The source and destination nodes are randomly chosen using the uniform distribution.

As in AFRO, the bitrate of each connection request was randomly selected considering traffic profile TP-6 (see traffic profiles in Chapter 5); we assumed a link failure rate of 2.72×10^{-3} failures per km per year and considered that the link is repaired immediately after the restoration process ends. Note that every point in the results is the average of 10 runs with 100,000 connection requests each.

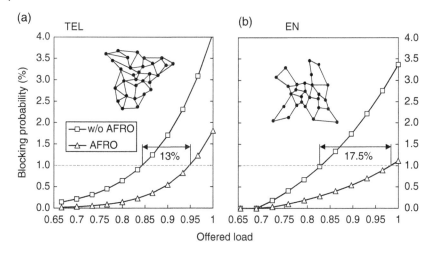

Figure 10.8 Blocking probability (*Pb*) against offered load: (a) TEL and (b) EN topologies.

Aiming at comparing the performance when MP-AFRO is used, Figure 10.8 plots *Pb* against the offered load with and without applying MP-AFRO. *Pb* is weighted using the connections' bitrate, and the load is normalized to the largest load unleashing *Pb* < 4%. As shown, gains above 13% can be obtained using MP-AFRO.

10.4 Experimental Validation

In this section, we propose a workflow to implement AFRO and MP-AFRO problems. The proposed workflow running the MP-AFRO algorithm is experimentally assessed in a test bed.

10.4.1 Proposed Reoptimization Workflow

For the workflow, we assume that an operator in the Network Management System (NMS) triggers the MP-AFRO workflow after a link has been repaired. To that end, the NMS issues a service request toward the ABNO Controller (see Chapter 4). Figure 10.9 reproduces the reoptimization workflow involving ABNO components and its execution flow diagram (Figure 10.10). When the request from the NMS arrives at the ABNO controller, reoptimization is requested by sending a PCReq message (labeled as 1 in Figure 10.9) to the front-end PCE (fPCE). Upon receiving the request, the fPCE collects relevant data to be sent to the back-end PCE (bPCE) in the form of a PCReq message containing a Global Concurrent Optimization (GCO) request (3). In light of the AFRO and MP-AFRO problem statements and assuming that the network topology and

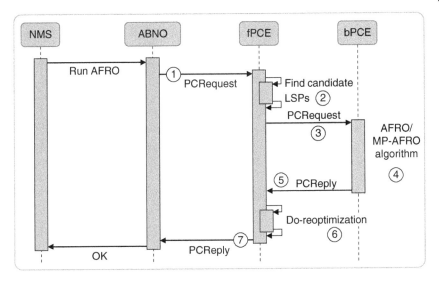

Figure 10.9 Proposed workflow.

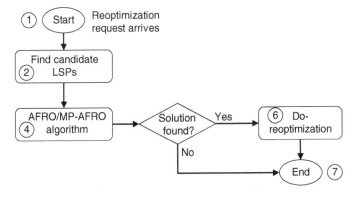

Figure 10.10 Reoptimization flow diagram.

the current state of the resources have been synchronized, data to be included in the GCO request include the set of established Label Switched Paths (LSPs) that are a candidate for reoptimization (2) (*D*) that can be computed using the algorithm in Table 10.1.

Data regarding the candidate connections is sent in a PCReq message (3) containing the GCO request. Each candidate connection is sent as an individual request, identified by an Request Parameters (RP) object; the end points and original bandwidth are also included using the END-POINTS and the BANDWIDTH objects, respectively. In addition, for each individual subconnection related to the original connection, its current route and spectrum allocation are specified using an Record Route Object (RRO) object that interleaves the route's nodes with its spectrum allocation. The sequence of nodes is encoded using the Unnumbered Interface subobjects encoding each node and the corresponding outgoing port identifiers to reach the next node, and an IPv4 Prefix subobject to identify the destination node. The spectrum allocation is encoded using the Label Control subobject that contains the tuple <*n, m*> that unambiguously defines any frequency slot, where *n* is the central frequency index (positive, negative, or 0) of the selected frequency slot from a reference frequency (193.1 THz), whereas *m* is the slot width in the number of slices at each side of the central frequency. Aiming at finding an optimal solution for the entire problem, individual requests are grouped together using a Synchronization Vector (SVEC) object.

The desired network-wide GCO related criterion, such as "AFRO" or "MP-AFRO," is specified by means of the Objective Function (OF) object. Finally, the repaired link must be used for reoptimization. To that end, we extended current standards adding an IRO (Include Route Object) object that identifies those repaired links that should be included in the new routes in a symmetric way with respect to the XRO (Exclude Route Object) object that specifies the links that should be excluded from route being computed.

Upon receiving the PCReq, the bPCE runs the specified heuristic algorithm. The solution of the selected problem is encoded in a PCRep (labeled as 5 in Figure 10.9); each individual request is replied specifying the bitrate that could be served and the route and spectrum allocation of the connections related to each request in a list of ERO objects. Note that the solution might entail merging several existing subconnections to create one or more new connections. The Order type-length-value (TLV) needs to be included in RP objects to indicate the order in which the solution needs to be implemented in the data plane.

Upon receiving the PCRep message with the solution from the bPCE, the fPCE runs the *Do-Re-Optimization* algorithm (labeled as 6 in Figure 10.9), listed in Table 10.6, to update the connections in the network's data plane. The algorithm sorts *D* in the same order as solution *S* (line 1 in Table 10.6). Next, each demand *d* in *D* is sequentially processed. The possibly updated demand

Table 10.6 Do-reoptimization algorithm.

INPUT: *D, S*
1: sort(*D, S.order*)
2: **for each** *d* ∈ *D* **do**
3:    *ds* ← getDemand(*S, d*)
4:    **if** *d* ≠ *ds* **then** update(*d, ds*)

ds in S corresponding to the current demand d in D is taken (line 3) by matching them using the identifier stored in its related RP object. If d differs from ds, the bPCE has found a better set of subconnections for the demand, so the fPCE should send the corresponding PCUpd/PCInit messages toward source Generalized Multi-Protocol Label Switching (GMPLS) controllers (line 4) to update/implement it.

When the solution is completely implemented, that is, when the corresponding PCRpt messages are received from the nodes, the fPCE replies the completion of the requested operation to the ABNO controller using a PCRep message (labeled as 7 in Figure 10.9), which eventually informs the NMS.

10.4.2 Experimental Assessment

For the experiments, we consider the network topology illustrated in Figure 10.7b. The data plane includes programmable Wavelength Selective Switches (WSSs) and to complete the topology node emulators are additionally deployed. Nodes are handled by colocated GMPLS controllers running the Resource Reservation Protocol for Traffic Engineering (RSVP-TE) with flexgrid extensions. The controllers run a proprietary configuration tool for automatic filter reshaping with a resolution of 1 GHz. Controllers communicate with the fPCE by means of PCEP through Gigabit Ethernet interfaces. The controllers operate on an EON test bed derived from [Cu12] that includes 100 Gb/s polarization multiplexed (PM) QPSK optical signals and bandwidth-variable cross-connects.

Figure 10.11 depicts the test-bed setup where experiments have been carried out. A link failure has already triggered a multipath restoration to reroute the failed connections splitting them into multiple subconnections per demand. After the failed link is repaired, the operator in the NMS decides to trigger the MP-AFRO algorithm to improve network efficiency, so the NMS requests the computation to the ABNO controller. The ABNO controller, in turn, sends the PCReq message (message 1 in Figure 10.12) to the fPCE containing the repaired link identification by means of an IRO object, and an OF object to specify the algorithm to be executed, in this experiment the MP-AFRO.

When the fPCE receives the ABNO request, it runs the *Find Candidate Demands* algorithm using as repaired link the one encoded in the received IRO object, to obtain the candidate list of demands to be reoptimized, and a PCReq to be sent to the bPCE is composed.

This PCReq message encodes the candidate demands using RP objects to identify them, an END-POINTS object defining its source and target nodes, a BANDWIDTH object to specify the requested bandwidth, and one RRO object for each subconnection being used by the demand. Included candidate demands are grouped by means of a SVEC object so that they are reoptimized jointly. The received OF and IRO objects are also included in the PCReq message to inform to the bPCE which algorithm should be executed and which link has been repaired, respectively (message 3).

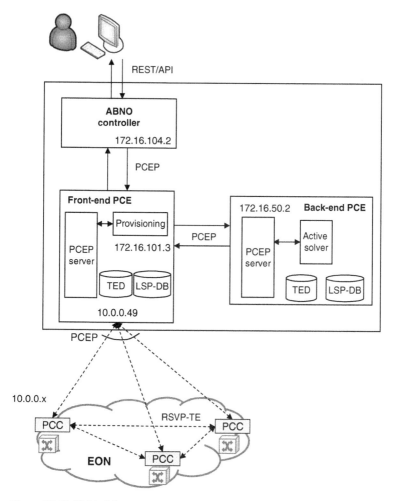

Figure 10.11 Field trial setup.

No.	Time	Source	Destination	Info
❶17	8.097681	172.16.104.2	172.16.101.3	PATH COMPUTATION REQUEST
❸18	8.097845	172.16.101.3	172.16.50.2	PATH COMPUTATION REQUEST
❺19	8.199351	172.16.50.2	172.16.101.3	PATH COMPUTATION REPLY
20	8.220174	10.0.0.49	10.0.0.1	PATH COMPUTATION UPDATE
❻21	8.223918	10.0.0.1	10.0.0.49	PATH COMPUTATION REPORT
22	11.281405	10.0.0.49	10.0.0.8	PATH COMPUTATION UPDATE
23	11.285044	10.0.0.8	10.0.0.49	PATH COMPUTATION REPORT
❼24	16.639926	172.16.101.3	172.16.104.2	PATH COMPUTATION REPLY

Figure 10.12 PCEP messages exchanged.

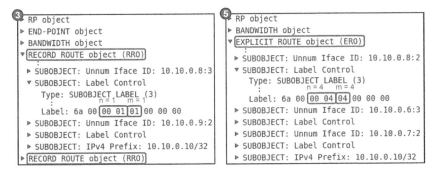

❸ RP object
‣ END-POINT object
‣ BANDWIDTH object
▼ RECORD ROUTE object (RRO)
⋮
  ‣ SUBOBJECT: Unnum Iface ID: 10.10.0.8:3
  ▼ SUBOBJECT: Label Control
      Type: SUBOBJECT LABEL (3)
      ⋮ $n = 1$ $m = 1$
      Label: 6a 00 00 01 01 00 00 00
  ‣ SUBOBJECT: Unnum Iface ID: 10.10.0.9:2
  ‣ SUBOBJECT: Label Control
  ‣ SUBOBJECT: IPv4 Prefix: 10.10.0.10/32
‣ RECORD ROUTE object (RRO)

❺ RP object
‣ BANDWIDTH object
▼ EXPLICIT ROUTE object (ERO)
⋮
  ‣ SUBOBJECT: Unnum Iface ID: 10.10.0.8:2
  ▼ SUBOBJECT: Label Control
      Type: SUBOBJECT LABEL (3)
      ⋮ $n = 4$ $m = 4$
      Label: 6a 00 00 04 04 00 00 00
  ‣ SUBOBJECT: Unnum Iface ID: 10.10.0.6:3
  ‣ SUBOBJECT: Label Control
  ‣ SUBOBJECT: Unnum Iface ID: 10.10.0.7:2
  ‣ SUBOBJECT: Label Control
  ‣ SUBOBJECT: IPv4 Prefix: 10.10.0.10/32

Figure 10.13 Detail of PCE Communication Protocol (PCEP) messages (3) and (5).

Figure 10.13 details the PCReq message (3) received by the bPCE highlighting the RRO objects encoding the original subconnections. The first RRO object and one of the Label Control subobjects have been expanded to identify the original subconnection's route and frequency slot. Objects in Figure 10.13 correspond to demand *P2a* and *P2b* in Figure 10.7b. *P2a* traverses nodes 8–9–10 using frequency slot $<n = 1, m = 1>$, and *P2b* traverses nodes 8–1–2–3–4–5–10 using frequency slot $<n = -2, m = 2>$.

After receiving the PCReq message, the bPCE computes the MP-AFRO algorithm and responds using a PCRep message (message 5) containing the solution found. For each demand, the received RP object is duplicated and included in the PCRep message, also adding a BANDWIDTH object specifying the served bandwidth, and one ERO object for each subconnection to be established to serve the demand. ERO objects have the same format as the RRO ones.

Figure 10.13 details also the PCRep message (5) replied by the bPCE, highlighting the ERO objects encoding the reoptimized subconnections. In contrast to message (3), objects in message (5) correspond to demand *P2*, the result of merging *P2a* and *P2b* in Figure 10.7b. The new computed connection traverses nodes 8–6–7–10 using frequency slot $<n = 4, m = 4>$; the resulting connection restores the initially failed connection using the repaired link.

Upon PCRep message arrival, the fPCE starts updating connections by sending PCUpd messages toward the involved GMPLS controllers (IPs: 10.0.0.X). Data plane nodes are located in network 10.10.0.X.

Figure 10.14 shows the optical spectrum on links 6–8 and 1–8 before and after the reoptimization is performed. PCRpt responses are generated when the requested actions have been performed (message 6 in Figure 10.12). When the solution computed by the bPCE has been completely deployed, the fPCE confirms to the ABNO controller the end of the requested reoptimization by sending a PCRep message (7).

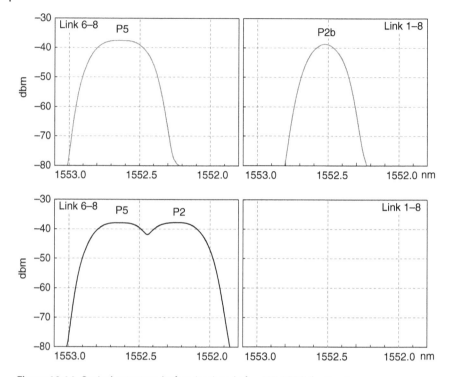

Figure 10.14 Optical spectrum before (top) and after MP-AFRO (bottom).

The overall reoptimization was successfully completed in around 8 s, including circa 150 ms of pure control plane contribution (i.e., messages exchange and path computation algorithms at fPCE and bPCE). Note that around 1 s is required to tune each WSS.

10.5 Conclusions

In this chapter, we presented AFRO, an online reoptimization problem useful for improving traffic performance of an EON subject to link failures and employ dynamic lightpath restoration. Specifically, when a failure is repaired, AFRO considers rerouting the traffic from some existing lightpaths to new lightpaths established through the repaired link thus using the repaired capacity. Through this, a better distribution of network load and, consequently, a decrease in the connection blocking probability can be achieved.

The AFRO problem was defined by means of an arc-path ILP formulation. Due to the AFRO requirement to use the repaired link by new lightpaths and in order to avoid solving large problem instances, that is, with a large set of precomputed lightpaths, a heuristic algorithm was devised in the form of a time-wise acceptable procedure to obtain near-optimal solutions. Moreover, some extensions to minimize traffic disruptions when migrating from old to new lightpaths were proposed.

Numerical results obtained for a national and an EN network showed that AFRO reduces the amount of used capacity with respect to the network state before applying AFRO. Additionally, the proposed heuristic algorithm provides quality solutions in execution times acceptable for real dynamic scenarios. The effectiveness of AFRO in improving the performance of a dynamic network in operation is demonstrated by two main results: (i) a significant blocking probability reduction (>10%) for reasonable values of MTTF, MTTR, and HT is achieved; (ii) application of AFRO allows improving the performance of the dynamic restoration algorithm by increasing the restorability of connections affected by a failure. Finally, it was shown that the operational overhead of AFRO can be reduced by using its mechanism only for a selected set of links (~60% of all links) without a negative impact on its performance.

Next, AFRO was extended to the case where multipath restoration is used to increase restorability of affected connections. After the link is repaired, the cost of multipath restoration in terms of worse resource utilization and spectral efficiency needs to be reduced by solving the MP-AFRO problem; MP-AFRO aggregates multiple subconnections serving a single demand using shorter routes, thus releasing spectrum resources that can now be used to convey new connection requests. The MP-AFRO was modeled as an ILP formulation, and a heuristic algorithm was devised to find good feasible solutions in practical computation times.

The performance of MP-AFRO was evaluated on an ad hoc network simulator and experimentally demonstrated on a test bed.

Part IV

Multilayer Networks

11

Virtual Network Topology Design and Reconfiguration

Fernando Morales, Marc Ruiz and Luis Velasco

Universitat Politècnica de Catalunya, Barcelona, Spain

The introduction of new services requiring large and dynamic bitrate connectivity can cause changes in the direction of the traffic in metro and even core network segments along the day. As introduced in Chapter 4, this leads to large overprovisioning in statically managed virtual network topologies (VNTs) designed to cope with the traffic forecast. To reduce expenses while ensuring the required grade of service, in this chapter we propose the VNT reconfiguration based on real traffic measurements (VENTURE) approach; it regularly reconfigures the VNT for the next period, thus adapting the topology to both the current and the expected traffic volume and direction. The reconfiguration problem is modeled mathematically, and a heuristic is proposed to solve it in practical times. To support VENTURE, we propose a workflow to be realized on the Application-Based Network Operation (ABNO) architecture. Exhaustive simulation results together with the experimental assessment of the proposed architecture are presented.

Table of Contents and Tracks' Itineraries

Provisioning, Recovery, and In-operation Planning in Elastic Optical Networks,
First Edition. Luis Velasco and Marc Ruiz.
© 2017 John Wiley & Sons, Inc. Published 2017 by John Wiley & Sons, Inc.

11.1 Introduction

A multilayer network integrating both Multi-Protocol Label Switching (MPLS) and Elastic Optical Networks (EONs) is seen as a cost- and energy-efficient solution for deploying next-generation transport networks, leveraging the bandwidth flexibility and coarse transport capacity provided by MPLS and EONs, respectively. This leads to an efficient use of the network resources exploiting Traffic Engineering (TE) strategies, where multiple packet circuits are aggregated over virtual links (vlinks) supported by connections on the underlying EON.

In this context, static virtual network topologies (VNTs), where MPLS nodes are connected through vlinks supported by static connections in the optical layer, have been commonly deployed with enough capacity to cope with the *off-net* traffic forecast. However, the introduction of cloud infrastructure in telecom operators' networks to create the telecom cloud [Ve15] facilitates the introduction of new types of service (e.g., live TV and video distribution [Ru16.1]), which require large bitrate connectivity and cause changes in the direction of the traffic along the day in metro and even core network segments. This, together with the overall traffic increment that operators' networks are needing to deal with year after year, entails that static packet network topologies were largely overprovisioned, thus increasing total cost of ownership (TCO). In view of that, network operators are looking for more efficient architectures able to reduce TCO, while providing the required grade of service. To that end, the VNT needs to be dynamically adapted not only to variations in volume but also to changes in the direction of the traffic.

To support VNT dynamicity, the Application-Based Network Operation (ABNO) architecture [ABNO] includes, among others (see Chapter 4 for details), (i) a Virtual Network Topology Manager (VNTM) in charge of reconfiguring on demand the VNT; (ii) a Path Computation Element (PCE) to compute the path for new Label Switched Paths (LSPs) on the Traffic Engineering Database (TED); and (iii) the Operations, Administration, and Maintenance (OAM) Handler to receive notifications and measured traffic counters.

To automate VNT adaptability, traffic can be measured in the packet nodes and counters need to be accessible by the OAM handler. In particular, the disaggregated traffic volume forwarded by each MPLS node to every other destination node, that is, origin–destination (OD) traffic, should be available. In addition, notifications can also be triggered when the used vlink capacity reaches some configured threshold (e.g., 90%). In fact, threshold-triggered VNT reconfiguration, where the OAM handler receives notifications from the control plane and reroutes individual MPLS paths (Packet-Switch Capable Label-Switched Paths (PSC LSPs)) in a reactive manner, was proposed in [Ag15]. However, the number of transponders to be installed tends to be as high as in a static VNT since dimensioning must be done to cope with the maximum of daily traffic forecast during a planning period (e.g., 1 year).

In fact, because of its benefits, VNT reconfiguration has been widely studied in the literature. Authors in [Ag09] proposed a centralized path reallocation module running periodically aiming at minimizing the number of used transponders. To follow traffic changes, authors in [Ge03] proposed to add/remove one single lightpath each time the VNT is reconfigured.

The approach presented in this chapter is different, since it targets at adapting the VNT based on measuring traffic to reduce the number of transponders to be deployed in scenarios were traffic variations are as a result of changes in the traffic directionality.

The next section presents examples of the VNT design and reconfiguration options.

11.2 VNT Design and Reconfiguration Options

As mentioned, several approaches to design and dynamically reconfigure the VNT can be devised. In this section we review two of them: (i) the *static VNT design* and (ii) the *threshold-based VNT capacity reconfiguration*.

In the static VNT design, the topology is designed and dimensioned to cope with the maximum of daily traffic forecast for every *od* pair during a planning period. The resulting topology is thus capable of supporting the traffic at any time during that period provided a perfect traffic forecast. Figure 11.1 presents an example for a seven-node VNT, where the capacity of every vlink supports the maximum daily traffic volume. For illustrative purposes, the plot in Figure 11.1a shows the variability in link 6–1 that needs to be dimensioned with a capacity of 200 Gb/s and Figure 11.1b shows the resulting VNT with the capacity of every vlink. It is clear, in view of Figure 11.1a, that the main drawback of the static VNT design is overprovisioning since most of the available capacity in the VNT will remain underutilized along the day.

To reduce capacity overprovisioning, the capacity of the vlinks can be adapted over time instead of allocating a constant amount of resources. Let us assume that the capacity of the existing vlinks can be increased and decreased to follow the traffic variations, but no new vlinks can be created or removed,

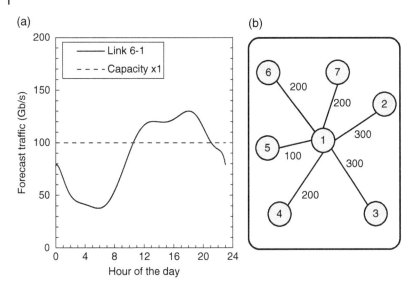

Figure 11.1 Static VNT design. (a) Traffic variability in link 6–1 and (b) VNT design.

hence keeping the topology invariant. In this approach, traffic can be measured at the MPLS routers and when the amount of traffic through a vlink reaches some threshold (e.g., 90%) the network controller can increase the capacity of such vlink by setting up a parallel lightpath between the two MPLS routers; conversely, unused capacity can be released by tearing down lightpaths.

The example in Figure 11.2a presents measured traffic data captured during the last 2 h in node 6, where the three plots represent: (i) traffic from that node to every other node in the VNT (labeled as 6->*N*); (ii) traffic from node 6 to node 7 (6->7); and (iii) traffic from node 6 to every other node except to node 7 (6->*N*\{7}).

Figure 11.2b shows the initial VNT where every vlink is supported by a 100 Gb/s lightpath in the underlying optical layer; the MPLS path for *od* 6–7 is also shown. A 90% threshold is configured and, in the event of threshold violation, the capacity of some vlinks is increased. In our example, two threshold violations for vlinks 1–6 and 1–7 are received, so the VNT capacity is updated (Figure 11.2c). It is worth noting that the MPLS path for *od* 6->7 is not affected by the VNT reconfiguration. As shown in the example, the threshold-based reconfiguration is able to adapt the VNT capacity to traffic changes, so resources in the optical layer are allocated only when vlinks need to increase their capacity. However, the same number of transponders as in the static VNT design approach need to be installed in the MPLS routers; for instance, in the example in Figure 11.2 two transponders are installed in routers 6 and 7 and another four in router 1 reserved for vlinks 1–6 and 1–7.

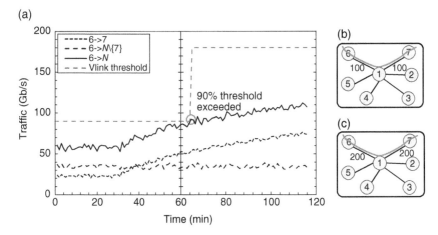

Figure 11.2 Threshold-based VNT capacity reconfiguration. (a) Measured traffic data in node 6; (b) initial VNT design; and updated VNT.

Let us assume now that, instead of measuring vlink capacity usage, OD traffic is measured in the routers. Indeed, analyzing the plots in Figure 11.2a we realize that traffic 6->7 is responsible for the registered traffic increment. In this case, let us assume that new vlinks can be created/removed in addition to increasing the capacity of the existing ones, so the VNT actually changes. We propose an approach where OD traffic has been measured and analyzed, and the current VNT is periodically reconfigured according to the foreseen traffic for the next period. An example following this approach is illustrated in Figure 11.3, where the maximum traffic value for *od* 6->7 is predicted at $t = 60$ for the next hour (e.g., 90 Gb/s). Then, a new vlink between nodes 6 and 7 can be created by establishing a lightpath on the optical layer and traffic 6->7 rerouted (Figure 11.3b). Note that this solution saves two transponders

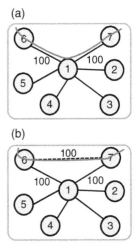

Figure 11.3 VNT reconfiguration: (a) initial VNT design and (b) updated VNT.

to be installed in router 1 compared to the previous approaches. It is clear that this reduction will happen when the amount of traffic is large enough. In particular, when the amount of traffic exceeds the capacity of the installed transponders (e.g., 100 Gb/s) direct vlinks can be created for part of that traffic, while the residual part could be routed through a different IP/MPLS path.

In order to adapt the VNT we assume that, for every OD pair, meaningful values are predicted (e.g., the maximum bitrate for the next hour) and used to adapt the VNT to meet the future traffic matrix, assuming that every OD traffic can be conveyed through two IP/MPLS paths at the most. We call this approach as VENTURE.

We next face the static VNT design problem to be solved off-line and the VNT reconfiguration (VENTURE) problem to be solved in the VNTM. For both problems, they are firstly formally stated, and Integer Linear Programming (ILP) formulations are presented. In light of the complexity of the VENTURE problem and the stringent time in which it needs to be solved, a heuristic algorithm is devised. Since VENTURE runs online, a workflow to implement VNT reconfiguration in the ABNO architecture is eventually proposed.

11.3 Static VNT Design

This section faces the static VNT design problem and provides a simplified ILP formulation, where the optical layer is assumed with enough capacity, so VNT design is just constrained by the availability of transponders.

11.3.1 The VNT Design Problem

The VNT design can be formally stated as follows:

Given:

- A graph $G(N, E)$, where N is the set of routers and E the set of feasible vlinks that can be created.
- The OD matrix with the maximum traffic for each origin-to-destination pair o. The tuple $<s_o, t_o, b_o>$ is defined for every o in OD, where s_o and t_o specify source and destination nodes for the od pair o and b_o the bitrate.
- Every od pair can be served following no more than K different flows.
- The sets $P^+(n)$ and $P^-(n)$ with the transponders available that can be used for outgoing (transmission) and incoming (reception), respectively, for each node n in N.
- The capacity B of every transponder.

Find: a VNT topology represented by graph $G^*(N, E^*)$, where $E^* \subseteq E$.

Objective: Minimize the number of transponders used.

11.3.2 MILP Formulation

The following sets and parameters are defined:

N Set of nodes, index n.
E Set of vlinks, index e.
$E^+(n)$ Subset of vlinks leaving from node n.

$E^-(n)$ Subset of vlinks arriving to node n.

$P^+(n)$ Subset of transmission transponders in node n.

$P^-(n)$ Subset of reception transponders in node n.

$P(n)$ Subset of transponders in node n. $P(n) = P^+(n) \cup P^-(n)$.

B Capacity of every transponder.

OD Set of origin–destination pairs, index o. Every o is defined by the tuple $<s_o, t_o, b_o>$.

K Maximum number of flows to serve every od pair.

The decision variables are:

x_p Binary, 1 if transponder p is used; 0 otherwise.

x_{pe} Binary, 1 if transponder p is used to support vlink e; 0 otherwise.

x_{ok} Integer, fraction of bitrate of pair o served through path k.

x_{oke} Integer, fraction of bitrate of pair o served through path k using vlink e.

z_{oke} Binary, 1 if pair o is routed using path k through vlink e; 0 otherwise.

y_n Integer , number of transponders used at router n.

The static VNT design problem is modeled using the following node-arc formulation (see Chapter 2):

$$\left(VNT\ DESIGN\right)\quad \min\ \sum_{n\in N} y_n \tag{11.1}$$

subject to:

$$\sum_{e\in E^+(n)} z_{oke} - \sum_{e\in E^-(n)} z_{oke} = \begin{cases} 1 & \forall o \in OD, k=1\ldots K, n=s_o \\ 0 & \forall o \in OD, k=1\ldots K, n\in N \setminus \left\{s_o, t_o\right\} \\ -1 & \forall o \in OD, k=1\ldots K, n=t_o \end{cases} \tag{11.2}$$

$$x_{oke} \leq b_o \cdot z_{oke} \quad \forall o \in OD, k=1\ldots K, e\in E \tag{11.3}$$

$$x_{oke} \leq x_{ok} \quad \forall o \in OD, k=1\ldots K, e\in E \tag{11.4}$$

$$x_{ok} - b_o \cdot \left(1 - z_{oke}\right) \leq x_{oke} \quad \forall o \in OD, k=1\ldots K, e\in E \tag{11.5}$$

$$\sum_{k=1}^{K} x_{ok} = b_o \quad \forall o \in OD \tag{11.6}$$

$$\sum_{o\in OD}\sum_{k=1}^{K} x_{oke} \leq B \cdot \sum_{p\in P^+(i)} x_{pe} \quad \forall e = \left(i,j\right) \in E \mid i,j \in N \tag{11.7}$$

$$\sum_{o \in OD} \sum_{k=1}^{K} x_{oke} \leq B \cdot \sum_{p \in P^-(j)} x_{pe} \quad \forall e = (i,j) \in E \mid i, j \in N \tag{11.8}$$

$$\sum_{e \in E^+(n)} x_{pe} \leq x_p \quad \forall n \in N, p \in P^+(n) \tag{11.9}$$

$$\sum_{e \in E^-(n)} x_{pe} \leq x_p \quad \forall n \in N, p \in P^-(n) \tag{11.10}$$

$$\sum_{p \in P^+(n)} x_p \leq y_n \quad \forall n \in N \tag{11.11}$$

$$\sum_{p \in P^-(n)} x_p \leq y_n \quad \forall n \in N \tag{11.12}$$

The multi-objective cost function (11.1) minimizes the number of transponders that are used to create the VNT.

The network flow constraints in (11.2) define paths on the topology for every OD pair. Each of these paths has a continuous capacity assignment along its route, as imposed by constraints (11.3)–(11.5). In addition, the sum of their capacities must be b_o as stated in (11.6).

Constraints (11.7)–(11.12) deal with transponder equipment. Constraints (11.7) and (11.8) assign transmission and reception transponders to vlinks, respectively, to support the capacitated paths. Constraints (11.9) and (11.10) prevent from assigning one transponder to multiple vlinks. Finally, constraints (11.11) and (11.12) compute the maximum between the number of transponders used for transmission and reception at every node, which represents the number of transponders to be installed in every router.

The size of the proposed formulation is $O(|N|^4 + |P| \cdot |N|^2)$ variables and $O(|N|^4 + |P| \cdot |N|)$ constraints. As an example, the size of the formulation for the network instance with $K = 2$ and 14 nodes presented in Section 11.5.1 is of 2×10^5 variables and 10^5 constraints.

11.4 VNT Reconfiguration Based on Traffic Measures

11.4.1 The VENTURE Problem

Given:

1) The current VNT represented by a graph $G(N, E')$, N being the set of routers and $E' \subseteq E$ the set of current vlinks. Set E is the set of all possible vlinks connecting pairs of routers.
2) The sets $P^+(n)$ and $P^-(n)$ with the transponders available that can be used for transmission and reception, respectively, for each node n in N; every transponder with capacity B.

3) The current traffic matrix D.
4) The traffic matrix OD for the next period. The bitrate b_o of the OD pair o must be served following one single path. Only in the case that b_o is enough to fill transponders with an amount over a given boundary usage *tbu*, the bitrate of pair o can be split into two flows and served through different paths.

Output: The reconfigured VNT $G^*(N, E^*)$, where $E^* \subseteq E$, and the paths for the traffic on G^*.

Objective: Maximize the served traffic for the next period, whilst minimizing the total number of transponders used.

11.4.2   ILP Formulation

Note from the problem statement that both the current and the predicted traffic matrices must be served. Consequently, we generate an input traffic matrix OD, where every pair o is the maximum of both the current and the predicted traffic. In addition, a parameter k_o will be used to specify whether pair o can be served using two paths.

The following sets and parameters are defined:

Topology

N	Set of routers, index n.
E	Set of all possible vlinks, index e.
$E^+(n)$	Subset of E with vlinks leaving from router n.
$E^-(n)$	Subset of E with vlinks arriving at router n.

Traffic

OD	Set of origin–destination pairs, index o. Every o is defined by the tuple $<s_o, t_o, d_o, b_o>$, where s_o and t_o specify the source and target nodes, d_o the currently served bitrate, and b_o the maximum of current and predicted bitrate to serve for pair o, respectively.
k_o	Maximum number of paths to serve pair o; $k_o = 2$ if $b_o \geq tbu$; $k_o = 1$ otherwise.

Equipment

P	Set of transponders, index p. Every transponder consists of one transmitter (tx) and one receiver (rx).
$P^+(n)$	Subset of tx transponders in router n.
$P^-(n)$	Subset of rx transponders in router n.
$P(n)$	Subset of transponders in n. $P(n) = P^+(n) \cup P^-(n)$.
B	Capacity of every transponder.

The decision variables are:

x_p Binary, 1 if transponder p is used; 0 otherwise.

x_{pe} Binary, 1 if transponder p is used to support vlink e; 0 otherwise.

x_{ok} Integer, fraction of bitrate of pair o served through path k.

x_{oke} Integer, fraction of bitrate of pair o served through path k using vlink e.

z_{oke} Binary, 1 if pair o is routed using path k through vlink e; 0 otherwise.

y_n Integer, number of transponders used at router n.

v_o Integer, fraction of unserved bitrate of pair o.

Then, the proposed ILP formulation is as follows:

$$\text{(VENTURE)} \quad \min \quad (P+1) \cdot \sum_{o \in OD} v_o + \sum_{n \in N} y_n \tag{11.13}$$

subject to:

$$\sum_{e \in E^+(n)} z_{oke} - \sum_{e \in E^-(n)} z_{oke} = \begin{cases} 1 & \forall o \in OD, k=1\ldots k_o, \ n=s_o \\ 0 & \forall o \in OD, k=1\ldots k_o, \ n \in N \setminus \{s_o, t_o\} \\ -1 & \forall o \in OD, k=1\ldots k_o, \ n=t_o \end{cases} \tag{11.14}$$

$$x_{oke} \leq b_o \cdot z_{oke} \quad \forall o \in OD, \ k=1\ldots k_o, \ e \in E \tag{11.15}$$

$$x_{oke} \leq x_{ok} \quad \forall o \in OD, k=1\ldots k_o, \ e \in E \tag{11.16}$$

$$x_{ok} - b_o \cdot (1 - z_{oke}) \leq x_{oke} \quad \forall o \in OD, \ k=1\ldots k_o, \ e \in E \tag{11.17}$$

$$\sum_{k=1}^{k_o} x_{ok} + v_o \geq b_o \quad \forall o \in OD \tag{11.18}$$

$$\sum_{k=1}^{k_o} x_{ok} \geq d_o \quad \forall o \in OD \tag{11.19}$$

$$\sum_{o \in OD} \sum_{k=1}^{k_o} x_{oke} \leq B \cdot \sum_{p \in P^+(i)} x_{pe} \quad \forall e = (i,j) \in E \,|\, i,j \in N \tag{11.20}$$

$$\sum_{o \in OD} \sum_{k=1}^{k_o} x_{oke} \leq B \cdot \sum_{p \in P^-(j)} x_{pe} \quad \forall e = (i,j) \in E \,|\, i,j \in N \tag{11.21}$$

$$\sum_{e \in E^+(n)} x_{pe} \leq x_p \quad \forall n \in N, p \in P^+(n) \tag{11.22}$$

$$\sum_{e \in E^-(n)} x_{pe} \leq x_p \quad \forall n \in N, p \in P^-(n) \tag{11.23}$$

$$\sum_{p \in P^+(n)} x_p \leq y_n \quad \forall n \in N \tag{11.24}$$

$$\sum_{p \in P^-(n)} x_p \leq y_n \quad \forall n \in N \tag{11.25}$$

The multi-objective cost function (11.13) minimizes both unserved traffic and used transponders, where the highest cost corresponds to the first term.

The constraints of this formulation are similar to those of the static VNT design problem. However, there are some differences that are worth noting. The network flow constraints in (11.14) define paths on the topology for every OD pair. Each of these paths has a continuous capacity assignment along its route, as imposed by constraints (11.15)–(11.17). Notwithstanding, constraint (11.18) allows serving only a fraction of the total capacity b_o of every OD pair; that fraction has to include at least the currently served bitrate d_o as stated in constraint (11.19). The equipment constraints remain the same in this formulation.

The size of the proposed formulation is $O(|N|^4 + |P| \cdot |N|^2)$ variables and $O(|N|^4 + |P| \cdot |N|)$ constraints, the same order of magnitude than the static VNT design problem. As a result, solving the proposed formulation becomes impractical for realistic scenarios even using commercial solvers; in our tests, solving times were longer than 10 h. Consequently, we developed a heuristic algorithm that provides much better trade-off between optimality and complexity.

11.4.3 Heuristic Algorithm

We devise a heuristic algorithm to solve the VENTURE problem consisting of three phases; the pseudo-code is presented in Table 11.1. After deallocating current traffic and releasing used resources (lines 2–5 in Table 11.1), bitrate b_o of OD pairs is split into two different flows and stored in set Q: flow g_o carries bitrate enough to fill transponders with an amount over tbu and flow l_o carries the remaining bitrate; these flows will be routed through different paths (lines 6–7). Next, the first two phases focus on routing every flow g_o through the direct vlink connecting source and destination routers (lines 8–9); set F stores the path of the flows. The first phase selects those flows for which a direct vlink already exists in the current VNT and the second phase does the same for the rest of flows, thus creating new direct vlinks. After these two phases, the residual bitrate u_o is checked and stored in set U. If all traffic has already been served, the algorithm ends (lines 10–12); otherwise, OD pairs are sorted by the amount of unserved bitrate and the third phase eventually routes the unserved

Table 11.1 Main algorithm.

INPUT: $G(N, E')$, D, OD, B, tbu
OUTPUT: G^*, F

```
 1: Q ← ∅, U ← ∅
 2: for each d ∈ D do dealloc(G, d)
 3: for each e ∈ E' do
 4:     setCapacity(e, 0)
 5:     releaseTransponders(e)
 6: for each o ∈ OD do
 7:     Q ← Q U {<o, g_o, l_o> = splitOD(o, B, tbu)}
 8: <Q, F'> ← PhaseI(G, Q)
 9: <G', Q, F'''> ← PhaseII(G, Q)
10: for each q ∈ Q do
11:     U ← U U {<o, u_o = g_o + l_o>}
12: if U = ∅ then return <G', F' U F''>
13: <G*, F'''> ← PhaseIII(G', U, thr)
14: if F''' = ∅ then return INFEASIBLE
15: return <G*, F' U F'' U F'''>
```

---

bitrate by possibly increasing the capacity of existing vlinks or by adding new ones (line 13). The reconfigured VNT and the new routing are eventually returned (line 15).

The algorithm for the first phase is detailed in Table 11.2. The original uncapacitated topology is used to route flows g_o through an existing direct vlink (line 3 in Table 11.2). The number of transponders to be allocated in the end routers of the direct vlink is computed as the minimum between the number of transponders needed to allocate g_o and the number of unused transponders (lines 4–7); those transponders are allocated to add capacity to the direct vlink (line 8), and the shortest path is computed on the resulting VNT (line 9). In case a path is found (i.e., capacity was added to the direct vlink), the path is allocated, and the amount of served bitrate reduced from the one requested (lines 10–13). The updated set Q and the found paths stored in set F are eventually returned (line 14).

The second phase is similar to the first phase, but for flows g_o through nonexisting direct vlinks. New capacitated direct vlinks are thus added to the topology to support those flows.

In the third phase, the current topology is extended to a full mesh topology by adding uncapacitated vlinks (lines 2–4 in Table 11.3). Next, a randomized routing procedure is run for a given number of iterations (lines 6–31); at every iteration, a constructive phase is started by cloning the initial extended topology and the unserved bitrate, this being the latter randomly sorted to give

Table 11.2 Phase I algorithm.

INPUT: $G(N, E)$, Q
OUTPUT: Q, F

```
 1:  F ← ∅
 2:  for each q = <o, g_o, l_o> ∈ Q do
 3:     if e = (s_o, t_o) ∉ E OR g_o = 0 then continue
 4:     n_o ← ceil (g_o / B)
 5:     n_s ← getNumUnusedTransponders (s_o, P⁺)
 6:     n_t ← getNumUnusedTransponders (t_o, P⁻)
 7:     n ← min{n_o, n_s, n_t}
 8:     allocateTransponders(e, n)
 9:     f ← SP(G, o, g_o)
10:     if f ≠ ∅ then
11:        allocate(G, f)
12:        F ← F U {f}
13:        g_o ← g_o - f.b
14:  return <Q, F>
```

---

higher priority to flows with higher current remaining traffic (and secondly predicted remaining traffic) (lines 7–12). Those flows with unserved bitrate are routed using one single path (lines 13–16). Aiming at minimizing the number of used transponders, link metrics are set proportional to the number of transponders needed to allocate the remaining capacity of the current flow (line 15). In case the path capacity does not serve the remaining bitrate, we check whether the capacity of its links can be increased using the available resources, that is, transponders in the end nodes (lines 17–19). Finally, the path is allocated, and the remaining bitrate of the flow is updated (lines 20–22). In case all the current bitrate d_o is served, the iteration cost is updated by penalizing the remaining unserved traffic (lines 23–24); otherwise, the iteration ends as infeasible (lines 25–27).

Once a solution has been built, a local search procedure is executed (line 28) aiming at finding a local minimum. The best topology and the found paths are returned as a final solution (lines 29–33).

The local search procedure tries to reduce the total number of used transponders during the constructive phase. Since the number of transponders used in a node is computed as the maximum between transmission and reception, this procedure focuses on releasing transponders that actively contribute to that maximum at every node. The algorithm is detailed in Table 11.4, where all current vlinks in the VNT are processed (lines 2–20). The vlink with transponders most actively contributing to the cost function

Table 11.3 PhaseII algorithm.

INPUT: $G(N, E)$, U, thr
OUTPUT: $G^*(N, E^*)$, F

```
 1: G*(N, E*) ← G(N, E); F ← ∅
 2: for each e = (i, j) ∉ E | i, j ∈ N, i ≠ j do
 3:   E* ← E* U {e}
 4:   setCapacity(e, 0)
 5: bestCost ← +∞
 6: for ite = 1…maxIter do
 7:   iteUnserved ← 0
 8:   Gite ← G
 9:   Uite ← U
10:   Fite ← ∅
11:   for each u ∈ Uite do u.order ← rand(0, 1) * uo
12:   sort(Uite, u.order, DESC)
13:   for each u ∈ Uite do
14:     if u.uo = 0 then continue
15:     updateMetrics(Gite, u.uo)
16:     f ← SP(Gite, u.o, u.uo)
17:     if f.b < u.uo AND canIncreaseCap(f, Gite, u.uo) then
18:       increaseCap(f, Eite, u.uo)
19:       f.b ← u.uo
20:     Fite ← Fite U {f}
21:     allocate(Gite, f)
22:     u.uo ← u.uo - f.b
23:     if u.uo ≤ bo-do then
24:         iteCost ← iteCost + u.uo * (|P| + 1)
25:     else
26:         iteCost ← +∞
27:         break
28:   <Gite, Fite> ← doLocalSearch(Gite, Fite)
29:   iteCost ← iteCost + numUsedTransponders(Gite)
30:   if iteCost < bestCost then
31:     bestCost ← iteCost
32:     <G*, F> ← <Gite, Fite>
33: return <G*, F>
```

Table 11.4 Local search procedure.

INPUT: $G(N, E)$, F

OUTPUT: $G^*(N, E^*)$, F^*

```
 1:  G* ← G; F* ← F; E_rem ← E
 2:  while E_rem ≠ ∅ do
 3:    for each e ∈ E_rem do
 4:      e.balance ← computeTransponderBalance (e)
 5:    sort(E_rem, e.balance, DESC)
 6:    e ← removeFirst(E_rem)
 7:    F_e ← getPaths(e)
 8:    <G_aux, F_aux> ← <G*, F*>
 9:    release(G_aux, F_e)
10:    E_aux ← E_aux \ {e}
11:    sort(F_e, f.b, DESC)
12:    allRerouted ← true
13:    for each f ∈ F_e do
14:      recomputeZeroCostLinks(G_aux)
15:      f' ← SP(G_aux, o, f.b)
16:      if f'.b < f.b then
17:        allRerouted ← false; break
18:      allocate(G_aux, f')
19:      F_aux ← (F_aux \ {f}) U {f'}
20:    if allRerouted then <G*, F*> ← <G_aux, F_aux>
21:  return <G*, F*>
```

Table 11.5 Time complexity of the algorithms.

Phase I/II	Phase III (per iter)	Local search
$O(\lvert N \rvert^2 \cdot RSA)$	$O(\lvert N \rvert^2 \cdot (\lvert E \rvert \cdot \log \lvert N \rvert + RSA))$	$O(\lvert E \rvert^2 \cdot (\log \lvert E \rvert + \lvert Fe \rvert \cdot \log \lvert N \rvert))$
<1 ms	298 ms	234 ms

is selected along with the set of paths routed through it (lines 6–7). This set is released from the VNT and sorted with respect to the bitrate (lines 8–11). Next, the set of paths is rerouted by possibly using new vlinks at zero objective cost (lines 13–15). In the case of a feasible solution, the VNT is updated with these changes (line 20).

Table 11.5 presents the time complexity of the proposed algorithms, where *routing and spectrum allocation* (RSA) is the worst case time complexity to

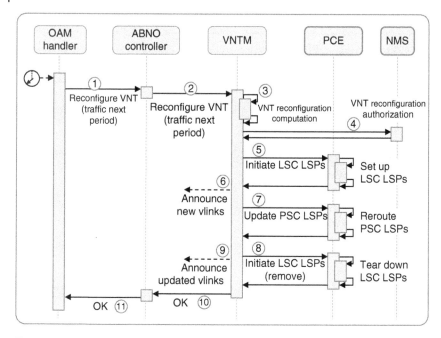

Figure 11.4 Proposed workflow.

find a feasible lightpath to support every direct vlink. For illustrative purposes, the computation time for the scenario with 14 nodes presented in Section 11.5.1 is also provided.

11.4.4 Proposed Workflow

Figure 11.4 presents the proposed workflow, where the OAM handler periodically decides whether the VNT should be updated based on the expected traffic for the next period. In the case of VNT reconfiguration, the OAM handler initiates the workflow by issuing a request to the ABNO controller that includes the traffic for the next period together with some other parameters to facilitate VNT computation (message 1 in Figure 11.4). Since the VNTM is in charge of computing the new VNT, the ABNO controller forwards the request to that module (2).

The VNTM computes the new VNT with the predicted traffic matrix received from the OAM handler (3). Continuing with our seven-node VNT example, let us assume that the new VNT involves adding the new virtual link 6–7 and reducing the capacity of some other vlinks. The solution is first notified to the Network Management System (NMS) (4) and then its implementation is divided into a sequence to avoid traffic disruption as anticipated earlier; firstly, lightpath (Lambda-Switch Capable Label-Switched Path (LSC LSP)) 6–7 is created (5) and the new vlink is advertised (6); next, PSC LSPs are rerouted (7) (a *make-before-break* strategy to avoid disruption can be implemented) and unused capacity in vlinks 1–3 and 1–4 removed by tearing down the underlying LSC LSPs (8); new vlinks' capacity is advertised (9). Upon VNT reconfiguration completion, VNTM sends a reply to the ABNO controller (10), which eventually replies to the OAM handler (11).

The next few sections focus, first, on validating the VENTURE approach through simulation and, then, the proposed workflow is experimentally assessed.

11.5 Results

11.5.1 Simulation Results

For evaluation purposes, we implemented an event-driven simulator in OMNeT++. To measure the effect of volumetric and directional changes in traffic, we implemented generators that inject traffic following two predefined traffic profiles, named as *Users* and Datacenter-to-Datacenter (*DC2DC*) (see average daily evolution in Figure 11.5). In addition, some random values around the average value are usually observed in real traffic. As a consequence, a function ε_t representing random variable traffic is also added. We assume that ε_t follows a 0-centered normal (Gaussian) probability distribution, that is,

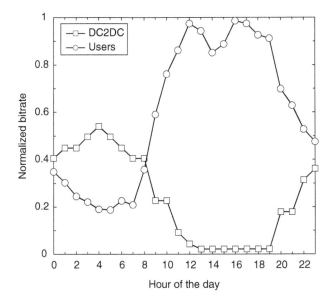

Figure 11.5 Average daily traffic profiles used in the simulation.

$\varepsilon_t \sim N(0, \sigma^2)$ where σ represents the standard deviation. Thus, the daily traffic profile of every OD pair can be defined as $Y_{OD}(t) = \alpha \cdot f(t) + \varepsilon_t$, where function $f(t)$ represents the average traffic profile and α is a scaling factor in Mb/s.

Finally, the set of nodes was divided into two subsets to generate changes in the direction of the traffic; ODs with one of the nodes as the destination in the first subset follow the *Users* profile, while the others follow the *DC2DC* one. We consider a scenario where a maximum of 26×100 Gb/s transponders per node is equipped. With such configuration, the static and threshold-based approaches are applied to a full-mesh 14-node VNT, where the initial capacity of each vlink ranges from 100 to 200 Gb/s.

We compare the effect in the unserved traffic and the number of used transponders under the static VNT design, under the threshold-based approach (we assumed 90% threshold) that runs continuously and under the VENTURE one that it is triggered at fixed intervals of 1 h.

Figure 11.6 presents the obtained blocking probability for the range of loads considered; values for both the static and the threshold-based approaches are omitted since they yield zero blocking probability. In the case of the VENTURE approach, Figure 11.6a plots the average and maximum hourly blocking along a day. We observe that, for a wide range of traffic loads, the maximum blocking probability is below 0.24%, while that on average is virtually zero. Figure 11.6b and c analyze the evolution of blocking probability during the day for the lowest and highest load, respectively. We observe that small peaks of blocking probability appear related to abrupt changes in the injected traffic and last for a couple of hours at the most, which is the time that VENTURE takes in fully adapting the VNT to traffic changes with the specific configuration selected.

Figure 11.7 focuses on the use of transponders. Figure 11.7a plots the maximum transponder usage as a function of the load for each approach. Both the static and the threshold-based approaches show a constant transponder usage for loads lower than 0.5, a usage that is increased from that load up. For low loads, the capacity of vlinks in the fully meshed VNT is 100 Gb/s in both cases, and it is increased to 200 Gb/s for high loads under the static approach. The threshold-based approach, however, is able to manage the use of transponders by flexibly using available transponders to increment the capacity of vlinks running out of capacity; this way it achieves transponder savings up to 11% as compared to the static VNT approach.

Interestingly, transponder usage scales linearly with the load with VENTURE. Compared to the threshold-based approach, VENTURE obtains savings between 8 and 42%.

Figure 11.7b and c focus on the use of transponders along the day for the lowest and highest loads under the three approaches. Apart from the constant transponder usage in the static approach, we show the different usages of the threshold-based and the VENTURE approaches. In particular, we observe how the VENTURE approach is able to remarkably reduce up to 45% transponder

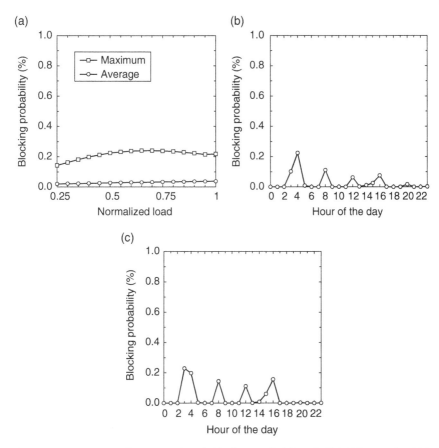

Figure 11.6 Average and maximum hourly blocking probability of VENTURE versus load (a). Blocking probability along 1 day and for normalized loads 0.48 (b) and 1.0 (c).

usage at some hours, mainly when the DC2DC traffic profile is dominant. On the other hand, in those hours when Users traffic profiles dominate, transponder usage under VENTURE still outperforms that of the threshold-based approach.

In conclusion, the VENTURE approach maximizes the overall utilization of available transponders in two different ways: (i) by reconfiguring the virtual topology to follow traffic direction changes, and (ii) by increasing the capacity of vlinks when the traffic increases.

11.5.2   Experimental Assessment

Experiments were carried out to assess the proposed workflow. An HTTP Representational State Transfer Application Programming Interface (REST API) interface was implemented between the OAM handler and the ABNO

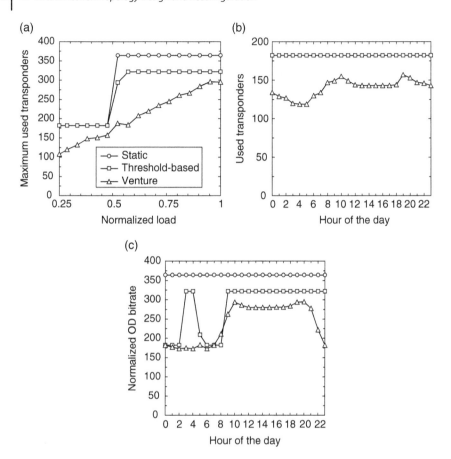

Figure 11.7 Maximum used transponders versus load (a). Used transponders along 1 day and for normalized loads 0.48 (b) and 1.0 (c).

controller and from it to VNTM, so as to report the predicted traffic matrix. PCEP was used between VNTM, PCE, and the provisioning manager. Finally, Border Gateway Protocol with Link State Distribution (BGP-LS) was used to synchronize TEDs. In particular, VNTM is in charge of advertising topological changes in the VNT, including vlink creation and releasing, as well as updating vlink capacity changes.

Figure 11.8 shows the meaningful messages exchanged between ABNO modules. For the sake of clarity, messages are identified following the workflow in Figure 11.4. The OAM handler sends a REST API request to the ABNO controller (message 1) containing the expected traffic matrix for the next period. The details of that message are presented in Figure 11.9. After receiving the traffic matrix, the VNTM computes the optimal VNT and issues requests

No.	Time	Source	Destin	Protocol	Info
(1) 465	*REF*	OAMHandler	ABNOCtrl	HTTP/XML	POST /ctrl/VNTReconfig HTTP/1.0
(2) 468	0.000	ABNOCtrl	VNTManager	HTTP/XML	POST /vntm/VNTReconfig HTTP/1.0
(4) 471	0.028	VNTManager	NMS	HTTP/XML	POST /nms/VNTReconfig HTTP/1.0
475	0.030	NMS	VNTManager	HTTP/XML	HTTP/1.1 200 OK
(5) 478	0.030	VNTManager	PCE	PCEP	Path Computation LSP Initiate (PCInitiate)
483	0.076	PCE	VNTManager	PCEP	Path Computation LSP State Report (PCRpt)
(6) 484	0.077	VNTManager	PCE	BGP	UPDATE Message
486	0.082	VNTManager	PCE	BGP	UPDATE Message
(7) 495	0.097	VNTManager	PCE	PCEP	Path Computation LSP Update Request (PCUpd)
497	0.104	PCE	VNTManager	PCEP	Path Computation LSP State Report (PCRpt)
(8) 498	0.104	VNTManager	PCE	PCEP	Path Computation LSP Initiate (PCInitiate)
499	0.104	VNTManager	PCE	PCEP	Path Computation LSP Initiate (PCInitiate)
(9) 500	0.104	VNTManager	PCE	BGP	UPDATE Message
503	0.109	VNTManager	PCE	BGP	UPDATE Message
(8) 509	0.167	PCE	VNTManager	PCEP	Path Computation LSP State Report (PCRpt)
511	0.216	PCE	VNTManager	PCEP	Path Computation LSP State Report (PCRpt)
(10) 513	0.217	VNTManager	ABNOCtrl	HTTP/XML	HTTP/1.0 200 OK
(11) 515	0.217	ABNOCtrl	OAMHandler	HTTP/XML	HTTP/1.0 200 OK

Figure 11.8 Exchanged messages for VNT reconfiguration.

Figure 11.9 Message (1) details.

```
▼<VNTReconfig>
  ▼<Matrix
    name="bitRateMbps">
    ▶<Data
    ▶<Data
    ▶<Data
    ▶<Data
    ▶<Data
    ▼<Data
      src="172.16.103.106"
      dst_172_16_103_101="17650"
      dst_172_16_103_102="7600"
      dst_172_16_103_103="3140"
      dst_172_16_103_104="9680"
      dst_172_16_103_105="4060"
      dst_172_16_103_107="72100"/>
    ▶<Data
    </Matrix>
  ▶<Params>
  </VNTReconfig>
```

to the PCE to implement the LSC LSPs supporting the new vlinks, reroute the selected PSC LSPs, and tear down unused LSC LSPs. In addition, VNT changes are advertised to the rest of the ABNO modules.

The signaling process took 217 ms from the instant the OAM handler triggered the workflow.

11.6 Conclusions

Different approaches for multilayer MPLS-over-EON design and reconfiguration have been reviewed. The VNT design uses maximum expected traffic to design the topology. The capacity of vlink in the deployed VNT might remain static or adapted to the traffic load by using a threshold-based mechanism, where new lightpaths are created to add more capacity or torn down when no extra capacity is required. The VNT design problem was formally stated and formulated as an ILP.

Next, a different approach, named as VENTURE, was proposed to adapt the current VNT to expected traffic conditions aiming at minimizing TCO. The approach consists in using expected OD traffic as the input of a reconfiguration problem. The VENTURE reconfiguration problem was formally stated and formulated as an ILP. In view of its complexity for valid short-term solutions, a heuristic algorithm to provide near-optimal solutions in practical computation times was proposed.

Regarding its implementation in the ABNO architecture, the OAM handler has been proposed to store traffic models to support VENTURE and the VNTM uses expected traffic as input to find the optimal VNT by solving the VENTURE reconfiguration problem.

The performance of VENTURE was compared through simulation against the static and the threshold-based approaches. We observed savings between 8 and 42% in the number of transponders to be installed in the routers when the VENTURE approach was applied. In addition, VENTURE is able to deactivate transponders during low traffic hours, thus decreasing the energy consumption and releasing lightpaths from the underlying optical layer, which contribute to costs reduction. Finally, the proposed architecture was experimentally assessed.

12

Recovery in Multilayer Networks

Alberto Castro[1], Lluís Gifre[2], Ricardo Martínez[3], Marc Ruiz[2] and Luis Velasco[2]

[1] University California Davis, Davis, CA, USA
[2] Universitat Politècnica de Catalunya, Barcelona, Spain
[3] Centre Tecnològic de Telecomunicacions de Catalunya, Castelldefels, Spain

In multilayer networks, an optical link failure may cause the disruption of multiple aggregated Multi-Protocol Label Switching (MPLS) Label Switched Paths (LSPs). Thereby, efficient recovery schemes, such as multilayer restoration and survivable virtual network topology (VNT) design, are required.

Regarding restoration, we focus on multilayer MPLS-over-EON (Multi-Protocol Label Switching over Elastic Optical Networks) operated with a Generalized Multi-Protocol Label Switching (GMPLS) control plane, where a centralized Path Computation Element (PCE) is in charge of computing the route of MPLS connection requests. In the event of a failure, the PCE sequentially computes backup paths for the set of failed MPLS paths. Since the Traffic Engineering Database (TED) in the PCE is not updated until an LSP is actually set up, it is very likely that the PCE assigns the same network resources to different backup paths. To improve this, we propose the DYNamic restorAtion in Multilayer MPLS-over-EON (DYNAMO) problem to compute the restoration routes for all the affected MPLS paths together, while taking advantage of sliceable bandwidth variable transponders (SBVTs) to provide additional flexibility for the reconfiguration of optical connections. The DYNAMO problem is modeled as a mathematical programming (MP) formulation. However, due to the high complexity of the problem and the required stringent solving times, a Greedy Randomize Search Procedure (GRASP)-based heuristic (see Chapter 2) is developed. The performance evaluation is comprehensively conducted by comparing the devised heuristic against the traditional sequential restoration.

Provisioning, Recovery, and In-operation Planning in Elastic Optical Networks,
First Edition. Luis Velasco and Marc Ruiz.
© 2017 John Wiley & Sons, Inc. Published 2017 by John Wiley & Sons, Inc.

As for survivable VNT design, we focus on both the problem of designing VNTs to ensure that any single failure in the optical layer will affect one single virtual link (vlink) at the most and on providing a VNT mechanism that reconnects the VNT in the case of vlink failure. The main use case of these survivable VNTs is for datacenter (DC) interconnection, where contents in a number of databases need to be replicated among the DCs. VNT creation and recovery problems are first formally stated and modeled by means of an integer linear programming (ILP) formulation. Solving these models, however, becomes impractical for real-sized scenarios, so heuristic algorithms are also proposed. Workflows to implement the algorithms within the Application-Based Network Operations (ABNO) architecture are proposed and experimentally validated.

Table of Contents and Tracks' Itineraries

12.1 Introduction

Multilayer MPLS-over-EON (Multi-Protocol Label Switching over Elastic Optical Network) networks must be designed to be fault-tolerant allowing a rapid and efficient recovery of disrupted MPLS Label Switched Paths (LSPs) [Ch10]. In such multilayer networks, Multi-Protocol Label Switching (MPLS) LSPs are routed through virtual links (vlink) supported by lightpaths on the Elastic Optical Networks (EONs); in case the multilayer network is operated with a Generalized Multi-Protocol Label Switching (GMPLS) control plane, vlinks are named as Forwarding Adjacency LSPs (FA LSPs). When an LSP crosses the boundary from an upper to a lower layer, it may be nested into a lower-layer FA LSP that crosses the lower layer [RFC5212]. From a signaling perspective, there are two alternatives to establish the lower-layer FA LSP: preprovisioned and triggered. The dynamic preprovisioned establishment of FA LSPs [RFC6107] is an efficient way to create and manage virtual network topologies (VNTs), where the set of lower-layer LSPs can be planned (see Chapter 11). However, the provisioning of end-to-end upper-layer LSPs with dynamic triggering of FA LSPs allows for a dynamic FA LSPs setup if such lower-layer LSPs do not already exist when the upper-layer LSP is being signaled.

In the first part of this chapter, we face the path restoration problem under the assumption of dynamic FA LSPs triggering, which introduces a number of interesting problems that need to be solved. In the second part of the chapter, we face the problem of designing survivable VNTs that are based on FA LSP preprovisioning.

12.2 Path Restoration in GMPLS-Controlled Networks

In this section, we focus on dynamic restoration, where failed MPLS LSPs are dynamically restored by a backup path computed by the Path Computation Element (PCE), assuming the dynamic triggering of FA LSPs.

In the case of failure in the optical layer, the ingress node of each disrupted MPLS LSP notifies the Application-Based Network Operations (ABNO) architecture about that problem, so a path computation request is issued to the PCE. Although multiple MPLS LSPs can be affected by a single failure, each request is sequentially and independently served by the PCE using as input the current state of the network in the Traffic Engineering Database (TED) and the particular path constraints (e.g., exclude failed links) specified in the request message. Since the TED and LSP database (LSP-DB) are not updated between path computations, when the number of concurrent LSPs to be restored grows, the same resources (e.g., link bandwidth, wavelengths and ports) may be assigned to different requests causing resource contention when signaling, and being unable to exploit the grooming opportunities through new created vlinks. Consequently, this PCE architecture attains a poor *restorability* (defined as the ratio between the number of LSPs that are successfully restored and the total number of LSPs to restore) [Cl02].

A solution to increase the restorability at the PCE relies on applying Global Concurrent Optimization (GCO) [RFC5557]. GCO aims at serving the bulk of path requests attaining the optimal solution for the whole network. Specifically, the path requests are served taking into account not only the current state of the network and particular path constraints but also a copy of the TED constructed to consider additional topology and resources (i.e., vlinks) derived/resulting from the establishment of previously served paths within the same bulk. Notice that the latter information will not be reflected in the regular TED until those LSPs are successfully set up. Nevertheless, anticipating the potential network state may achieve a better path computation from a twofold perspective: on the one hand, the connection blocking caused by resource contention among multiple concurrent LSPs may be reduced; on the other hand, the grooming opportunities can be better exploited since vlinks induced by formerly computed paths are reused to route the subsequent paths within the same bulk. To do this, it is necessary to synchronize and delay PCE responses to yield sufficient time to create the optical LSPs associated to specific vlinks, which will then be used by other subsequent MPLS LSPs. The main drawback of this approach, however, stems from the fact that it increases the response time affecting the overall restoration time.

For illustrative purposes, Figure 12.1a shows a simple physical network consisting of five optical cross-connects (OXCs) and four MPLS routers. OXCs are

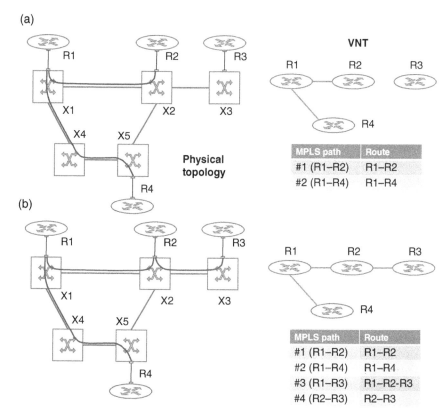

Figure 12.1 Example of multilayer network setup: (a) Initial physical and virtual networks state (a) and (b) Networks state after LSP setup.

connected by bidirectional fiber links, while each MPLS router is connected to the collocated OXC. Finally, two MPLS client paths are already being served. We assume that the bitrate of each path is 1 Gb/s. Two lightpaths were established in the physical topology to support an equal number of vlinks in the VNT, and its route over the VNT is given in the adjacent table.

At this stage, let us assume that a new MPLS request #3 between R1 and R3 needs to be served. After requesting a route to the PCE, it computes R1–R2–R3, where the existing vlink R1–R2 is reused, and a new vlink R2–R3 must be created, which triggers the new lightpath R2–R3 to be established. Later, another MPLS request #4 between R2 and R3 arrives, and it is served through the route R2–R3, using capacity available in vlink R2-R3. Figure 12.1b describes the configuration of both the physical and the virtual topologies once all four MPLS paths have been set up.

Next, a failure in fiber link X1–X2 does cause that affected LSPs (#1 and #3) request a restoration route to the centralized PCE.

In Figure 12.2a, the restoration route has been computed sequentially by the PCE and two parallel lightpaths have been set up to create vlinks used to restore the affected MPLS LSPs. In the example, both lightpaths could be created because enough resources, that is, the optical spectrum in the links and SBVTs in the MPLS routers, were available. Frequently, nonetheless, that is not the case, and *resource contention* may arise. In that regard, note that both restoration routes reuse vlink R1–R4; again, resource contention could arise as a consequence of not enough capacity being available for both MPLS paths in that vlink.

In Figure 12.2b, the PCE has grouped all restoration requests and performed bulk path computation. In the example, the restoration routes of both MPLS paths have been computed, and the bulk path computation algorithm has decided to create vlink R4–R2 using it to restore both affected LSPs, thus reducing the amount of resources used compared to the sequential approach.

However, for the bulk path restoration to work, restoration routes need to be sequenced: one of the routes must be signaled first, so as to trigger actual vlink creation; after waiting enough time, vlink R4–R2 is effectively created and the second route reusing it can be signaled. In our example, LSPs #1 and #3 need vlink R4–R2 to be created. Then, one of the demands is rerouted and the other one must be delayed enough time to allow the vlink R4–R2 to be effectively created. This fact introduces a set of *dependencies* among the demands that must be considered, so as to minimize recovery times.

As a consequence of the efficiency that bulk restoration reaches by reusing vlinks, the number of ports connecting MPLS routers and OXCs can be reduced; let us assume that those ports are SBVTs. Figure 12.3 is intended for illustrating that reduction; Figure 12.3a shows how the MPLS paths were routed before the failure; each router is equipped with a number of client ports where the MPLS paths arrive. One SBVT with four subcarrier modules is connected to each router. For instance, in Figure 12.3a two lightpaths end in the

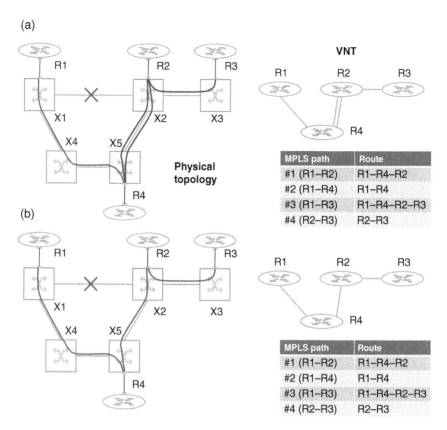

Figure 12.2 Sequential (a) and bulk (b) restoration after a failure in link X1–X2.

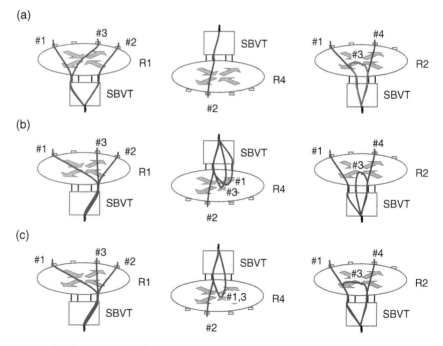

Figure 12.3 Routing of MPLS demands. (a) Before link X1–X2 fails, (b) after sequential restoration, and (c) after bulk restoration.

SBVT and MPLS paths #1 and #3 are groomed together into a single lightpath, whereas path #2 uses a different one. MPLS path #3 is routed through intermediate router R2, so that path enters in router R2 by one of the lightpaths terminating in the SBVT in that router and leaves it using the same SBVT but aggregated with path #4 into lightpath R2–R3.

Figure 12.3b and c show the routing at each router after the restoration has been done sequentially and in bulk, respectively. The main difference can be seen in routers 2 and 4 where more resources in each SBVT have been used in the case of sequential restoration because of the increased number of lightpaths created. Hence, efficiency in the use of the resources can reduce notably the amount of resources needed for the same grade of service, thus reducing costs remarkably when dealing with expensive resources such as SBVTs.

The next section formally states the DYNamic restorAtion in Multilayer MPLS-over-EON (DYNAMO) problem.

12.2.1 The DYNAMO Problem

The problem can be formally stated as follows:

Given:
- a network topology represented by a graph $G_o(N, L)$, N being the set of OXC nodes and L the set of bidirectional fiber links connecting two OXC nodes, excluding failed ones; each link consists of two unidirectional optical fibers,
- a set S of available frequency slices in each link in L,
- the VNT represented by a graph $G_v(V, E)$, V being the subset of N where MPLS routers are placed, and E the set of vlinks defining the connectivity among the MPLS nodes,
- a set D of MPLS demands to be recovered. Each demand d is defined by the tuple $<s_d, t_d, b_d>$, where s_d and t_d represent demand's source and destination MPLS routers, respectively, and b_d its bitrate.

Output:
- the routing of every recovered MPLS demand over the VNT, and
- the route, modulation, and spectrum allocation (RMSA) of the new lightpaths used to support new vlinks.

Objective: maximize the total amount of bitrate recovered whilst minimizing the amount of resources used (i.e., frequency slices and SBVTs) and the total recovery time.

The next section presents the Mathematical Programming (MP) formulation proposed for the DYNAMO problem.

12.2.2 MP Formulation

The mathematical programming model for the DYNAMO problem performs routing in both the optical and the MPLS layers using node-arc formulations for each layer (see Chapter 2). A set of vlinks is precomputed beforehand; each vlink connects two locations with MPLS nodes provided that a feasible optical route can be found. A set of lightpaths is available for each vlink, although its actual RMSA (see Chapter 3) in the optical topology is determined during the resolution of the problem.

The following sets and parameters have been defined:

Optical Topology

N	Set of OXC nodes, index n.
L	Set of fiber links, index l.
$L(n)$	Subset of fiber links incidents to OXC node n.
$len(l)$	Length of fiber link l (km).
R	Set of bitrate-reach pairs (Gb/s, km), index r.
$len(r)$	Reach of a path using bitrate-reach pair r in km.
$b(r)$	Maximum bitrate of a path using bitrate-reach pair r.

Optical Spectrum

S	Set of frequency slices, index s.
C	Set of slots, index c. Each slot c contains a subset of contiguous slices.
h_{cr}	Equal to 1 if slot c uses bitrate-reach pair r.
a^{ls}	Equal to 1 if slice s in fiber link l is being used.
u^{cs}	Equal to 1 if slot c includes slice s.
b_c	Capacity of slot c (Gb/s).

Virtual Topology

V	Set of MPLS routers ($V \subseteq N$), index v ($v = n$ provided that OXC node with index n is physically connected to the MPLS router with index v).
E	Set of virtual links, index e.
$P(v)$	Set of SBVTs in MPLS router v, index p.
$K(e)$	Set of routes to support virtual link e, index k.
$K_1(e)$	Subset of $K(e)$ already deployed in the optical topology.
$K_2(e)$	Subset of $K(e)$ not currently deployed in the optical topology.
$E(v)$	Subset of virtual links incident to MPLS router v.
$N(e)$	Set of end OXC nodes (nodes connected to the correspondent MPLS router) of virtual link e.
b_{ek}	Available capacity in virtual link e using lightpath k (Gb/s).
b_{pv}	Available capacity in SBVT p in MPLS router v (Gb/s).
f_{pv}	Number of lightpaths that can be assigned to SBVT p in MPLS router v.
g_{ekpv}	Equal to 1 if virtual link e using lightpath k ends in SBVT p in MPLS router v.

Demands To Be Recovered

D Set of MPLS demands to be recovered, index d.

$SD(d)$ Set of $\{s_d, t_d\}$ MPLS routers of demand d.

b_d Bitrate of demand d (Gb/s).

The decision variables are:

ω_{dek} Binary. Equal to 1 if demand d is routed through virtual link e using lightpath k; 0 otherwise.

δ_{ek}^c Binary. Equal to 1 if lightpath k of virtual link e uses slot c; 0 otherwise.

λ_{ek}^{lc} Binary. Equal to 1 if lightpath k of virtual link e uses slot c in fiber link l; 0 otherwise.

σ_d Binary. Equal to 1 if demand d is recovered; 0 otherwise.

γ_{pv} Binary. Equal 1 if SBVT p in MPLS node v is allocated; 0 otherwise.

v_{ek}^r Binary. Equal 1 if lightpath k of virtual link e uses bitrate-reach pair r; 0 otherwise.

χ_{dt} Binary. Equal 1 if demand d is established in time interval t; 0 otherwise.

φ_{ekt} Binary. Equal 1 if lightpath k of virtual link e is active in time interval t; 0 otherwise.

C_{ek} Positive integer. Completion time of lightpath k of virtual link e.

C_{max} Positive integer with the total recovery time.

Finally, the mathematical programming formulation for the DYNAMO problem is as follows:

$$(\text{DYNAMO}) \quad \text{maximize} \quad A_1 \cdot \sum_{d \in D} b_d \cdot \sigma_d - A_2 \cdot \sum_{v \in V} \sum_{p \in P(v)} \gamma_{pv} - A_3 \cdot C_{max}$$

$$(12.1)$$

subject to:

$$\sum_{e \in E(v)} \sum_{k \in K(e)} \omega_{dek} = \sigma_d \quad \forall d \in D, v \in SD(d) \tag{12.2}$$

$$\sum_{e \in E(v)} \sum_{k \in K(e)} \omega_{dek} \leq 2 \quad \forall d \in D, v \in \overline{SD(d)} \tag{12.3}$$

$$\sum_{\substack{e' \in E(v) \\ e' \neq e}} \sum_{k \in K(e')} \omega_{de'k} \geq \sum_{k \in K(e)} \omega_{dek} \quad \forall d \in D, v \in \overline{SD(d)}, e \in E(v) \tag{12.4}$$

$$\sum_{d \in D} b_d \cdot \omega_{dek} \leq b_{ek} \quad \forall e \in E, k \in K_1(e) \tag{12.5}$$

$$\sum_{d \in D} b_d \cdot \omega_{dek} \leq \sum_{c \in C} b_c \cdot \delta_{ek}^c \quad \forall e \in E, k \in K_2(e) \tag{12.6}$$

$$\sum_{l \in L(n)} \sum_{c \in C} \lambda_{ek}^{lc} = \sum_{c \in C} \delta_{ek}^c \quad \forall e \in E, k \in K_2(e), n \in N(e) \tag{12.7}$$

$$\sum_{l \in L(n)} \sum_{c \in C} \lambda_{ek}^{lc} \leq 2 \quad \forall e \in E, k \in K_2(e), n \in \overline{N(e)} \tag{12.8}$$

$$\sum_{\substack{l' \in L(n) \\ l' \neq l}} \sum_{c \in C} \lambda_{ek}^{l'c} \geq \sum_{c \in C} \lambda_{ek}^{lc} \quad \forall e \in E, k \in K_2(e), n \in \overline{N(e)}, l \in L(n) \tag{12.9}$$

$$\sum_{l \in L} \lambda_{ek}^{lc} \leq |L| \cdot \delta_{ek}^c \quad \forall e \in E, k \in K_2(e), c \in C \tag{12.10}$$

$$\sum_{c \in C} \delta_{ek}^c \leq 1 \quad \forall e \in E, k \in K_2(e) \tag{12.11}$$

$$\sum_{e \in E} \sum_{k \in K_2(e)} \sum_{c \in C} \lambda_{ek}^{lc} \cdot u^{cs} + a^{ls} \leq 1 \quad \forall l \in L, s \in S \tag{12.12}$$

$$\sum_{e \in E(v)} \sum_{k \in K_2(e)} \sum_{c \in C} \delta_{ek}^c \cdot g_{ekpv} \leq f_{pv} \cdot \gamma_{pv} \quad \forall v \in V, p \in P(v) \tag{12.13}$$

$$\sum_{d \in D} \sum_{e \in E(v)} \sum_{k \in K_2(e)} \omega_{dek} \cdot b_d \cdot g_{ekpv} \leq b_{pv} \quad \forall v \in V, p \in P(v) \tag{12.14}$$

$$\sum_{d \in D} b_d \cdot \omega_{dek} \leq \sum_{r \in R} b(r) \cdot v_{ek}^r \quad \forall e \in E, k \in K_2(e) \tag{12.15}$$

$$\delta_{ek}^c \leq \sum_{r \in R} h_{rc} \cdot v_{ek}^r \quad \forall e \in E, k \in K_2(e), c \in C \tag{12.16}$$

$$\sum_{l \in L} \sum_{c \in C} len(l) \cdot \lambda_{ek}^{lc} \leq \sum_{r \in R} len(r) \cdot v_{ek}^r \quad \forall e \in E, k \in K_2(e) \tag{12.17}$$

$$\sum_{r \in R} v_{ek}^r \leq 1 \quad \forall e \in E, k \in K(e) \tag{12.18}$$

$$\sum_{d \in D} \omega_{dek} \cdot \chi_{dt} \leq 1 + |D| \cdot \sum_{\substack{t' \in T \\ t' < t}} \varphi_{ekt'} \quad \forall e \in E, k \in K_2(e), t \in T \tag{12.19}$$

$$\sum_{d \in D} \sum_{\substack{t' \in T \\ t' \leq t}} \chi_{dt'} \geq \varphi_{ekt} \quad \forall e \in E, k \in K_2(e), t \in T \tag{12.20}$$

$$\sum_{t \in T} \chi_{dt} = \sum_{e \in Ek \in K_2(e)} \sum \omega_{dek} \quad \forall d \in D \tag{12.21}$$

$$C_{ek} = \sum_{t \in T} (1 - \varphi_{ekt}) - |T| \cdot \left(1 - \sum_{c \in C} \delta_{ek}^c\right) \quad \forall e \in E, k \in K_2(e) \tag{12.22}$$

$$C_{\max} \geq C_{ek} \quad \forall e \in E, k \in K_2(e) \tag{12.23}$$

The objective function (12.1) maximizes the total bitrate recovered, whilst minimizing the use of SBVTs and the total restoration time. A_1, A_2, and A_3 are constants.

Constraints (12.2)–(12.4) compute the route and perform aggregation of demands through the VNT. Constraint (12.2) defines whether a demand is restored by selecting a vlink incident to source and destination routers. Constraints (12.3) and (12.4) perform routing and aggregation in the intermediate routers.

Constraints (12.1)–(12.6) allow demands to use existing vlinks or new ones, in which case new lightpaths need to be created. Constraint (12.1) ensures that demands use existing vlinks with enough capacity, whereas constraint (12.6) ensures that a slot with enough capacity is allocated for the amount of bitrate assigned to each new lightpath to be created.

Constraints (12.7)–(12.12) compute the route and spectrum allocation over the optical topology for lightpaths supporting new vlinks. Note that constraints (12.7)–(12.9) compute the route in a similar way as constraints (12.2)–(12.4) do. Constraint (12.10) guarantees that every lightpath uses the same slot along its route. Constraint (12.11) implements the slot continuity constraint ensuring that no more than one slot is allocated to one lightpath. Constraint (12.12) allows each slice in each optical link to be used by only one lightpath provided that it has not been previously used in the network.

Constraints (12.13) and (12.14) ensure that SBVTs capacity is not exceeded. Constraint (12.13) guarantees that the number of new lightpaths assigned to each SBVT does not exceed the given availability. Constraint (12.14) ensures that the bitrate associated to each SBVT does not exceed the given maximum.

Constraints (12.15)–(12.18) take care of the modulation format assignment by the proper selection of a bitrate-reach pair for every lightpath. Constraint (12.15) chooses a pair with enough bitrate for the traffic to be transmitted. In addition, constraint (12.16) ensures that the selected slot is compatible with the chosen bitrate-reach pair, which is mandatory for linking the bitrate capacity constraint at the optical layer in (12.6) with the bitrate capacity constraint at MPLS layer in (12.15). Constraint (12.17) ensures that the reach of that pair works for the length of the lightpath, whereas constraint (12.18) ensures that only one pair is chosen.

Constraints (12.19)–(12.23) sequence the demands assigning demand establishing, and so new vlinks to time intervals. Constraint (12.19) guarantees that each new vlink to be established is triggered by exactly one demand. Once the vlink is established, several demands using it can be set up simultaneously, that is, these latter demands depend upon the former to be established. Constraint (12.20) ensures that a given vlink becomes available once the first demand using it has been established. Constraint (12.21) ensures that each demand with assigned restoration resources is assigned to a time period. Constraint (12.22) accounts for the completion time for each vlink that is established. Note that the completion time of any vlink not to be established is set to zero. Finally, constraint (12.23) computes the total completion time defined as the max function of the completion time for every single vlink.

Note that constraint (12.19) entails multiplying two binary variables, thus converting the mathematical model into a nonlinear one. Notwithstanding, variable multiplication can be easily solved at the expense of introducing additional binary variables and constraints. However, its exact solving becomes impractical for the stringent times required for restoration and, as a result, a heuristic algorithm is needed to provide good near-optimal solutions in the time periods required for recovering.

12.2.3 Heuristic Algorithm

In this section, we propose a Greedy Randomize Search Procedure (GRASP)-based heuristic to solve the DYNAMO problem. Table 12.1 describes the proposed greedy randomized constructive algorithm, where parameter α controls the size of the Restricted Candidate List (RCL). Equation (12.24) is used to quantify the quality of recovering a given demand d, in line with the objective function for the mathematical model.

$$q(d) = A_1 \cdot d.bw - A_2 \cdot d.newResources - A_3 \cdot d.depend \qquad (12.24)$$

During the local search, the demands' routes are changed, so as to try to avoid new vlinks to be created.

The heuristic was validated for really small instances against the MP. To this aim, constraint (12.19) was linearized with additional variables and constraints, thus converting the formulation to integer linear programming (ILP) and allowing CPLEX [CPLEX] to solve instances to optimality. We observed that in all the instances checked, the heuristic provided the optimal solution, that is, the same solution obtained from exactly solving the ILP. In light of this, the heuristic was used to obtain the results presented next.

12.2.4 DYNAMO Numerical Results

The performance of the considered restoration approaches was compared on two national network topologies: the 21-node TEL and the 21-node DT topologies, where each location contained one MPLS router and one OXC.

Table 12.1 Greedy randomized constructive algorithm.

INPUT: $G_o(N, L)$, $G_V(V, E)$, D, α
OUTPUT: Sol

```
 1: Sol ← ∅
 2: Q ← D
 3: while Q ≠ ∅ do
 4:   for each d ∈ Q do
 5:     d.route = shortestPath(G_V, d)
 6:     if d.route = ∅ then
 7:       q(d) ← -INF
 8:     else
 9:       evaluate the quality q(d) using equation (12.24)
10:   q^min ← min{q(d) : d∈Q}
11:   q^max ← max{q(d) : d∈Q}
12:   if q^max = -INF then
13:     break
14:   RCL ← {d∈Q : q(d) ≥ q^max - α(q^max - q^min)}
15:   Select an element d from RCL at random
16:   for each e ∈ d.route do
17:     if not e.isImplemented then
18:       implement(G_o, e)
19:   implement(G_V, d)
20:   Q ← Q \ {d}
21:   Sol ← Sol ∪ {d}
22: return Sol
```

Evaluation of the restoration approaches was performed by using the simulation algorithm presented in Table 12.2. To load the network (line 2), we developed an ad hoc event-driven simulator in OMNeT++ [OMNet]; a dynamic network environment was simulated where incoming MPLS connection requests arrive at the system following a Poisson process and are sequentially served without prior knowledge of future incoming connection requests.

To compute the RMSA of the lightpaths, we used the bulk RMSA algorithm described in Chapter 5. The holding time of the connection requests is exponentially distributed with a mean value equal to 2 h. Source/destination pairs are randomly chosen with equal probability (uniform distribution) among all MPLS nodes. Different values of the offered network load are created by changing the interarrival rate while keeping the mean holding time constant. Finally, note that each point in the results is the average of $10 \cdot |L|$ runs and that sequential and bulk restoration approaches are executed using identical input data.

Table 12.2 Simulation algorithm.

```
 1: for i = 1…10 do
 2:     Load the network to the desired level
 3:     store the state of the network
 4:     for each l ∈ L do
 5:         cut l
 6:         perform sequential restoration
 7:         repair l
 8:         restore network state
 9:         cut l
10:         perform bulk restoration
11:         repair l
12:         restore network state
```

In our experiments, the bitrate of each MPLS path was set to 1 Gb/s, the Quadrature Phase-Shift Keying modulation format was used for the optical signals, the optical spectrum width was set to 1 THz, and each MPLS router was equipped with one SBVT with a number of subcarrier modules ranging from 2 to 5.

To find the appropriate loads, we first run the simulator without cutting links and store the resulting blocking probabilities. Five traffic loads unleashing blocking probabilities ranging from 0.1 to 5% for each of the networks and SBVT capacities considered were found.

Figure 12.4 presents the percentage of unrestorability as a function of the number of demands to restore for the TEL and DT networks. Five plots for both sequential and bulk restoration approaches are presented, one for each of the found traffic loads, where each point corresponds to one SBVT capacity value. As anticipated, the sequential approach produces unrestorability values as high as 27% to above 50% as a function of the traffic load. In contrast, the bulk approach achieves unrestorability values of almost 0%, that is, virtually all demands are restored for even the most stringent traffic load. The behavior is the same for the TEL network as for the DT, as shown in Figure 12.4.

Table 12.3 provides insights on the results for the TEL network using SBVTs with five subcarrier modules. There, the number of MPLS paths to be restored ranges, on average, from 37 to 46 as a function of the load offered to the network. Unrestorability values are given for both the sequential and the bulk approach and two main causes behind unrestorability are detailed:

- no route could be found during path computation, and
- resource contention, that is, resources were already in use during the signaling phase. This gets together frequency slices and existing vlinks that were available in the TED when the route was computed, or SBVTs resources that are actually allocated during lightpaths' setup.

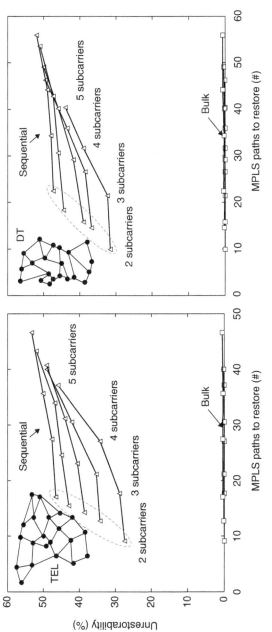

Figure 12.4 Unrestored MPLS flows against the total amount of flows to restore.

Table 12.3 Restoration results for the TEL network using five-subcarrier SBVTs.

Offered load	No. of demands to restore	Unrestorability sequential			Unrestorability bulk		
		Total	Path computation	Resource contention	Total	Path computation	Resource contention
357	37.14	47.72%	0.00%	47.72%	0.02%	0.02%	0.00%
378	39.98	50.10%	0.00%	50.10%	0.13%	0.13%	0.00%
383	40.71	49.11%	0.00%	49.11%	0.13%	0.13%	0.00%
395	43.28	51.95%	0.00%	51.95%	0.21%	0.21%	0.00%
409	46.59	53.21%	0.00%	53.21%	0.56%	0.56%	0.00%

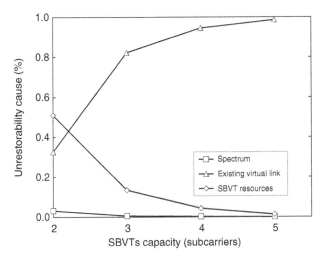

Figure 12.5 Unrestorability by cause for the sequential approach.

As detailed, the reason for the high unrestorability of the sequential approach is resource contention; restoration routes are computed using the state of the resources in the TED. However, as a result of the number of path computation requests arriving at the PCE, the TED becomes immediately outdated. Indeed, the same resources could be assigned to several routes and ports' availability decreases notably, so no new lightpaths could be established. The bulk restoration approach, in contrast, reaches negligible unrestorability values, since network resources are globally optimized. Once restoration routes are computed for a bulk of requests, resource contention disappears completely.

Figure 12.5 explores the causes of unrestorability for the sequential restoration approach when TEL network simulation was run with the load unleashing 1% blocking probability. As shown, when the capacity of the SBVTs is low, the percentage of resource contention as a consequence of the lack of the resources in SBVTs is the dominant cause of unrestorability. However, as soon as higher capacity SBVTs are used, the main cause rapidly changes to contention in the use of existing vlinks' capacity. The same results and conclusions are valid for any other load on both TEL and DT networks.

The advantages of bulk restoration come at the cost of increased restoration times when compared to those of the sequential approach. This is particularly noticeable for high traffic loads where a large number of MPLS paths need to be restored as illustrated in Figure 12.6. Plots for maximum bulk restoration computation times for each network and capacity of the installed SBVTs are presented; an almost linear trend with the amount of paths to restore can be observed.

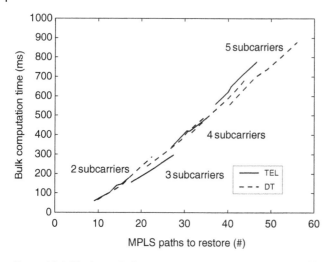

Figure 12.6 Maximum bulk computation times versus number of flows to restore.

Table 12.4 Bulk computation for the TEL network using five-lightpath SBVTs.

Offered load	Dependences			Maximum computation time (ms)
	Avg	Max	Min	
357	0.8	3.0	0.0	560
378	1.0	4.0	0.0	622
383	1.0	3.0	0.0	649
395	1.1	5.0	0.0	706
409	1.4	5.0	0.0	776

Those computation times translate into restoration times that include both sequencing restoration route signaling to allow new vlinks to be created prior to being reused and actual signaling. Thus, the time to restore an MPLS path depends on the depth of the dependencies list for that flow and the length of the route to be signaled.

As described in the DYNAMO mathematical model earlier, dependence depth needs to be minimized, so as to minimize restoration times in bulk restoration, and, as such, it was included in the heuristic algorithm. Table 12.4 presents the dependence depth values (max, min, and average) as a function of the offered load. As shown, the maximum value is only 5, which in turn introduces a considerable delay for the last set of restoration routes to be signaled.

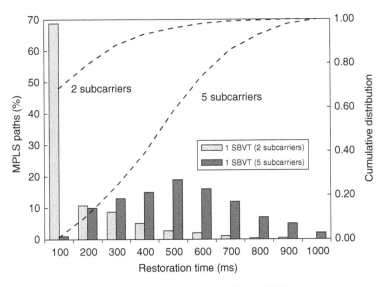

Figure 12.7 Restoration times distribution for different SBVT capacities.

The histograms in Figure 12.7 represent the restoration time distribution when the TEL network was loaded with the medium intensity. Cumulative distributions are also plotted. Signaling times computation were performed using equations and experimental times described in [Ve10]. The main conclusion is that restoration times lower than 1 s can be achieved even when the number of MPLS paths to be restored is as high as 50; indeed, more than 50% of them were restored in less than 500 ms.

12.2.5 PCE Architecture

Figure 12.8 depicts the architecture of the implemented GCO module inside a stateless PCE for solving the DYNAMO problem. For both regular LSP provisioning and restoring disrupted LSPs, the ingress PCC is responsible for sending a path request to the PCE using the PCE Communication Protocol (PCEP) (step 1 in Figure 12.8).

At the PCE, the incoming request is stored in either the provisioning or the restoration queues. An eXclude Route Object (XRO) [RFC5521] received in the path computation request, conveying the failed physical link with the fail bit active, allows the PCE to decide storing it in the restoration queue; otherwise, the request is assumed to be a regular working path provisioning and is stored in a priority queue attended/served by a thread pool. The selected routing algorithm is determined by the objective function (OF) object included in the PCEP request. Such OF must match with one of the algorithms available in the Algorithm API.

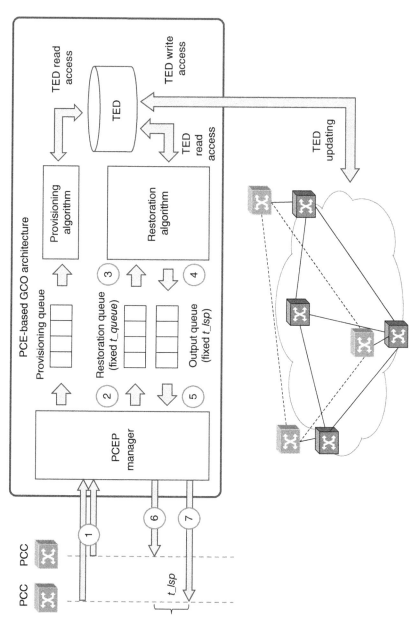

Figure 12.8 Designed and implemented PCE-based GCO architecture.

Path computation requests originating from restoration processes are grouped in a bulk (step 2) and processed after a configurable timer (t_queue, step 3) expires. This allows synchronizing such bulk of backup paths to be computed. Once t_queue elapses, the path computation for regular LSP provisioning is stopped to avoid any interference with the restoration procedure, with incoming requests temporary stored in the provisioning queue until that procedure ends. Next, the selected restoration algorithm available in the Algorithm API is triggered (step 4). Before the restoration algorithm is called, a copy of the TED is created, and both failed physical and logical links specified in the XRO are removed from that copy. In addition, the restoration algorithm receives the set D of pairs of source and destination end points and the required bitrates.

Besides computing the path for each request, the algorithm specifies the ordering, sequence, and timing of the PCEP responses (step 5). In other words, the PCE decides whether a PCEP response is immediately sent back to the PCC or delayed. The latter is forced to occur when the computed path has some dependencies (e.g., creation of a new logical link) with respect to a formerly computed path (e.g., Figure 12.2b). In this situation, the PCEP response is artificially delayed by t_lsp to guarantee that, for instance, the required logical links are actually induced and created (steps 6 and 7).

12.2.6 Experimental Results

In this section, we show the results obtained when using sequential (no GCO) versus bulk PCE GCO restoration. A GMPLS/PCE control plane platform is used for experimentally validating the implemented PCE-based dynamic restoration with GCO for multilayer networks. For the experiments, we used a 12-node European network topology.

We performed some tests to experimentally retrieve the values for t_queue and t_lsp, without excessively penalizing the restoration time. We finally found that, in the test bed, at least 100 ms are needed to guarantee that every PCEP request arrives at the PCE, whereas 150 ms are needed for t_lsp to delay PCEP replies ensuring that LSPs are actually established in the network. For illustrative purposes, let us assume that two MPLS LSPs have been affected by a failure. Exchanged PCEP request/reply messages for multilayer restoration are shown in Figure 12.9. Two path computation requests *PCReq1* and *PCReq2* are received in the PCE. Since those requests contain an XRO object, they are placed into the PCEP restoration queue. After the timer t_queue (started upon the reception of *PCReq1*) expires, the PCE invokes the algorithm in the GCO to solve the DYNAMO problem to simultaneously compute the routes for all the received backup path requests. Next, the PCE sends a response message to those LSPs that can be immediately established,

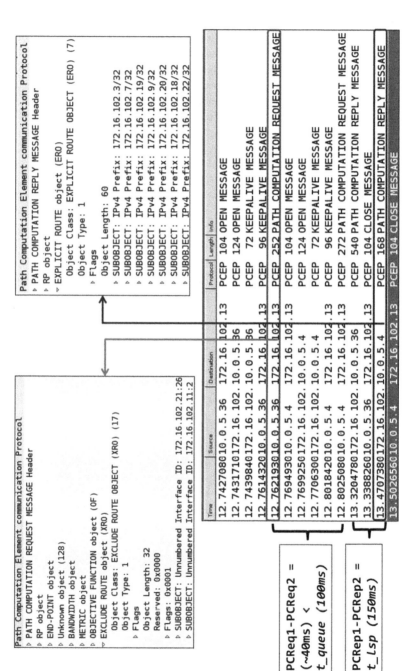

Path Computation Element communication Protocol
▷ PATH COMPUTATION REQUEST MESSAGE Header
▷ RP object
▷ END-POINT object
▷ Unknown object (128)
▷ BANDWIDTH object
▷ METRIC object
▷ OBJECTIVE FUNCTION object (OF)
▽ EXCLUDE ROUTE object (XRO)
  Object Class: EXCLUDE ROUTE OBJECT (XRO) (17)
  Object Type: 1
  ▷ Flags
  Object Length: 32
  Reserved: 0x0000
  ▷ Flags: 0x0001
  ▷ SUBOBJECT: Unnumbered Interface ID: 172.16.102.21:26
  ▷ SUBOBJECT: Unnumbered Interface ID: 172.16.102.11:2

Path Computation Element communication Protocol
▷ PATH COMPUTATION REPLY MESSAGE Header
▷ RP object
▽ EXPLICIT ROUTE object (ERO)
  Object Class: EXPLICIT ROUTE OBJECT (ERO) (7)
  Object Type: 1
  ▷ Flags
  Object Length: 60
  ▷ SUBOBJECT: IPv4 Prefix: 172.16.102.3/32
  ▷ SUBOBJECT: IPv4 Prefix: 172.16.102.7/32
  ▷ SUBOBJECT: IPv4 Prefix: 172.16.102.19/32
  ▷ SUBOBJECT: IPv4 Prefix: 172.16.102.9/32
  ▷ SUBOBJECT: IPv4 Prefix: 172.16.102.20/32
  ▷ SUBOBJECT: IPv4 Prefix: 172.16.102.18/32
  ▷ SUBOBJECT: IPv4 Prefix: 172.16.102.22/32

Time	Source	Destination	Protocol	Length	Info
12.742708	010.0.5.36	172.16.102.13	PCEP	104	OPEN MESSAGE
12.743171	0172.16.102.10.0.5.36		PCEP	124	OPEN MESSAGE
12.743984	0172.16.102.10.0.5.36		PCEP	72	KEEPALIVE MESSAGE
12.761432	010.0.5.36	172.16.102.13	PCEP	96	KEEPALIVE MESSAGE
12.762193	010.0.5.36	172.16.102.13	PCEP	252	PATH COMPUTATION REQUEST MESSAGE
12.769493	010.0.5.4	172.16.102.13	PCEP	104	OPEN MESSAGE
12.769925	0172.16.102.10.0.5.4		PCEP	124	OPEN MESSAGE
12.770630	0172.16.102.10.0.5.4		PCEP	72	KEEPALIVE MESSAGE
12.801842	010.0.5.4	172.16.102.13	PCEP	96	KEEPALIVE MESSAGE
12.802508	010.0.5.4	172.16.102.13	PCEP	272	PATH COMPUTATION REQUEST MESSAGE
13.320478	0172.16.102.10.0.5.36		PCEP	540	PATH COMPUTATION REPLY MESSAGE
13.338826	010.0.5.36	172.16.102.13	PCEP	104	CLOSE MESSAGE
13.470738	0172.16.102.10.0.5.4		PCEP	168	PATH COMPUTATION REPLY MESSAGE
13.502656	010.0.5.4	172.16.102.13	PCEP	104	CLOSE MESSAGE

PCReq1-PCReq2 =
(~40ms) <
t_queue (100ms)

PCRep1-PCRep2 =
t_Lsp (150ms)

Figure 12.9 Messages exchange between PCC and PCE applying PCE-based GCO.

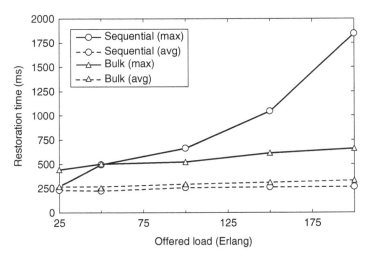

Figure 12.10 Restoration times versus offered load.

since they do not depend on new FA TE links. In Figure 12.9, the PCE sends the response to *PCReq1* and waits a time *t_lsp* before sending the response to *PCReq2*.

Figure 12.10 shows the obtained restoration times as a function of the traffic load. Average times in Figure 12.10 show that the bulk approach increases the restoration time up to 26% with respect to the sequential approach, being faster than 350 ms even under the highest offered load. This is the cost/penalty of the bulk approach, which comes from waiting for *t_queue* (100 ms) to create the bulk request, and *k* times *t_lsp* (150 ms) for those MPLS LSPs which need that *k* FA TE links are previously established (i.e., LSP dependencies). Notwithstanding, in view of the obtained times, the value of *k* is low (not higher than 2 on average). When focusing on the maximum restoration times, they increase remarkably for the sequential approach.

For low and high offered loads in Figure 12.10, Figure 12.11 provides insights on maximum restoration times disaggregating those into their five components: (i) input queue, (ii) computations time, (iii) delay waiting for dependencies, and times for (iv) PCEP and (v) GMPLS signaling. Remarkably high values can be observed for the PCE input queue, which is as a consequence of TED updating messages sent immediately after the first LSP is signaled, thus blocking the TED for the update, as stated in the previous section. In contrast, the increment in the bulk approach is almost linear with the load; starting from 445 ms under low load, it increases to only 662 ms for the highest one. Note that in the bulk approach, TED updating starts after the whole bulk computation is performed.

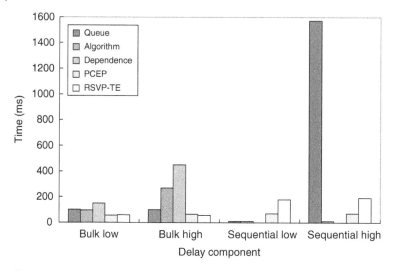

Figure 12.11 Disaggregated maximum values of time components.

12.3 Survivable VNT for DC Synchronization

The telecom cloud infrastructure has been proposed to deal with expected traffic growth coming from services like video distribution. Since those services entail large capacity connections from users to datacenters (DCs), the Telecom Cloud infrastructure can place contents closer to the users by deploying DCs in metro areas, thus reducing the impact of the traffic between users and DCs.

In such scenario, a hierarchical content distribution architecture for the telecom cloud can be devised, where core DCs, placed in geographically distributed locations, are interconnected through permanent VNTs (Figure 12.12). The VNT allows core DCs to be permanently synchronized, that is, any modification performed in the contents in one core DC is propagated to the rest of the core DCs through the VNT. Additionally, metro DCs need to be interconnected with the core DC and metro-to-core optical *anycast* connections (see Chapter 7) can be established periodically for content synchronization. Since network failures might disconnect the VNTs, recovery mechanisms need to be proposed to reconnect both topologies and anycast connections.

This hierarchical architecture brings benefits, including the reduction of traffic in the core, as well as its inherent high availability against DC failures. However, a failure affecting a vlink in the VNT might disconnect the latter creating several subsets of connected core DCs (connected components) whose contents will soon become outdated. Therefore, the VNT topology should be designed to cope with network failures. We assume the VNT is created as a tree topology and dynamic reconnection (similar to connection restoration) is used.

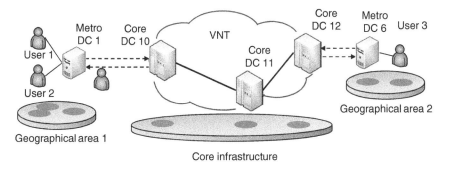

Figure 12.12 Hierarchical content distribution architecture.

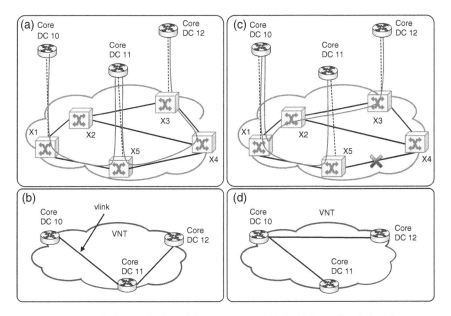

Figure 12.13 VNT before and after a failure in an optical link. (a) Optical and virtual networks before and (b) after a failure in an optical link.

Figure 12.13 presents an example with three core DCs (labeled as 10, 11, and 12); we assume that a Layer 2 (L2) switch connects each DC to the optical transport network. Core DCs are connected among them by a set of vlinks supported by optical connections established over the optical network (Figure 12.13a) forming a VNT (Figure 12.13b). When a metro DC needs to synchronize its local contents with those in the VNT, it establishes an optical anycast connection to any of the core DCs in the VNT (see Chapter 7).

In the event of a failure, vlinks can be affected; in Figure 12.13c a link failure has torn down connection 11–12, thus breaking the original VNT into two connected graph components ({10, 11} and {12}). Therefore, the VNT needs to be urgently reconnected into a single connected topology. In Figure 12.13c the VNT has been reconnected by creating the new vlink 10–12 and Figure 12.13d shows the new VNT topology after the reconnection.

Hence, two problems have been identified: VNT creation and reconnection. The next section formally defines each problem and proposes mathematical models and algorithms for solving them.

12.3.1 Mathematical Formulations and Algorithms

12.3.1.1 VNT_CREATE Problem

The VNT_CREATE problem consists of finding a tree connecting all the selected core DCs. Our approach consists of computing a Minimum Spanning Tree (MST) [Pe02] over an *auxiliary full-mesh* graph connecting the involved core DCs. The links in the resulting tree are the lightpaths to set up.

The problem can be stated as follows:

Given:

- the optical network topology represented by a graph $G(N_O, L)$, where N_O is the set of optical nodes and L the set of optical links.
- the number of available slices η_l in each optical link $l \in L$.
- the number of slices χ required to set up each lightpath supporting a vlink.

Output: the route for each bidirectional lightpath.

Objective: minimize the number of optical resources required.

We formulate the VNT_CREATE problem using a node-arc formulation (see Chapter 2) and assuming an opaque optical core network, so contiguity and continuity constraints are relaxed. To improve the availability of the VNT, each optical link is used just once on each VNT; therefore, a single link failure would disconnect VNTs into two connected graph components at the most.

The sets and parameters for this problem are:

Topology

N	Set of all nodes in the network, that is, OXCs in the optical network and Ethernet switches in the DCs, index n.
N_O	Subset of N containing the OXCs.
N_C	Subset of N containing the switches in the core DCs.
L	Set of optical links connecting two nodes, index l.
L_O	Subset of L containing the optical links connecting two optical nodes.
N_C	Subset of N_C containing the core DCs.
E	Set of candidate virtual links connecting two switches in N_C, index e.
$OD(e)$	Subset of N_C containing the end points of candidate virtual link e.

Other Parameters

δ_{nl} Equal to 1 if link l is incident to node n; 0 otherwise.

χ Number of slices required for each optical connection supporting a virtual link.

η_l Number of available spectrum slices in link l.

Decision Variables

x_{el} Binary. 1 if optical link l is used to support virtual link e; 0 otherwise.

y_e Binary. 1 if virtual link e is selected; 0 otherwise.

Then, the formulation for the VNT_CREATE problem is as follows:

$$\left(\text{VNT\_CREATE}\right) \quad \min \quad \sum_{e \in E} \sum_{l \in L} x_{el} \tag{12.25}$$

subject to:

$$\sum_{e \in E} y_e = |N_C| - 1 \tag{12.26}$$

$$\sum_{e \in (S,\bar{S})} y_e \geq 1 \quad \forall S \subset N_C \tag{12.27}$$

$$\sum_{l \in L} \delta_{nl} \cdot x_{el} = y_e \quad \forall e \in E, n \in OD(e) \tag{12.28}$$

$$\sum_{l \in L} \delta_{nl} \cdot x_{el} \leq 2 \quad \forall e \in E, n \in N \backslash OD(e) \tag{12.29}$$

$$\sum_{l' \in L, l' \neq l} \delta_{nl'} \cdot x_{el'} \geq \delta_{nl} \cdot x_{el} \quad \forall e \in E, n \in N \backslash OD(e), l \in L \tag{12.30}$$

$$\sum_{e \in E} \chi \cdot x_{el} \leq \eta_l \quad \forall l \in L \tag{12.31}$$

$$\sum_{e \in E} x_{el} \leq 1 \quad \forall l \in L_O \tag{12.32}$$

The objective function (12.25) minimizes the total number of optical links used to deploy the VNT.

Constraints (12.26) and (12.27) define the tree topology by restricting the VNT to have $|N_C| - 1$ vlinks and ensuring connectivity of the nodes belonging to the VNT, respectively. Note that constraint (12.27) is applied to any subset S of N_C but not to the whole set N_C.

Constraints (12.28)–(12.30) route the lightpaths for the vlinks through the optical network. Constraint (12.28) ensures that one lightpath for each selected

Table 12.5 VNT_CREATE heuristic algorithm.

INPUT: $G(N, L)$, N_C, χ, *mode*
OUTPUT: *LSP*

```
 1: G(N_C, E) ← Create_Full_Mesh_Undirected(N_C)
 2: for each e = (a, b) ∈ E do
 3:    e.metric ← |Shortest_Path(G(N_C, E), a, b)|
 4: while TRUE do
 5:    L_R ← ∅; LSP ← ∅
 6:    E_T ← MST(G(N_C, E));
 7:    if E_T = ∅ then return ∅
 8:    while E_T ≠ ∅ do
 9:       e = (a, b) ← argmax_{e.metric}(E_T)
10:       p ← Shortest_Path(G(N, L \ L_R), a, b, χ)
11:       if p = ∅ then
12:          E ← E \ {e}
13:          break
14:       LSP ← LSP ∪ {p}
15:       if mode = CREATE then L_R ← L_R ∪ p
16:       E_T ← E_T \ {e}
17:    if E_T = ∅ then return LSP
```

---

vlink is created with end nodes equal to the end points of the vlink. Constraint (12.29) and (12.30) guarantee that the route of each lightpath is connected and loopless. Constraint (12.31) ensures that selected optical links have enough optical resources, while constraint (12.32) guarantees that each optical link is used once at the most.

The VNT_CREATE problem requires a quadratic number of candidate vlinks to be considered with respect to the number of core DCs, that is, $|E| \sim |N_C|^2$. Thus, the number of variables is $O(|N_C|^2 \cdot |L|)$ and the number of constraints is $O(2^{|NC|} + |N_C|^2 \cdot |N| \cdot |L|)$. For instance, the number of variables and constraints for the 24-node 43-link US topology (presented in Chapter 7) is approximately 10^3 and 10^5, respectively. The size of the problem can be reduced and solved to optimality using large-scale optimization techniques, such as column generation [Ru13.2]. In this chapter, however, a heuristic algorithm is proposed.

The algorithm in Table 12.5 will be reused for the VNT_RECONNECT problem, so the *mode* parameter allows to specify the desired behavior; for the VNT_CREATE algorithm a value of *CREATE* is expected. The algorithm starts creating a *full-mesh* graph $G(N_C, E)$ and assigning a metric to each vlink in E as a function of the estimated number of optical hops for the lightpath supporting that vlink (lines 1–3). Then, an MST is computed over $G(N_C, E)$ (line 6). The vlinks in the tree are iteratively routed over the optical network (lines 9–16), and optical links already used (those in set L_R) are not considered for routing new lightpaths (line 10). If no path is found for a vlink, it is removed from the *full-mesh* graph, and the routing is restarted (lines 11–13); otherwise, the path is added to the *LSP* set and sets are updated (lines 14–16). The solution in *LSP* is eventually returned (line 17).

12.3.1.2 VNT_RECONNECT Problem

The VNT_RECONNECT problem aims at reconnecting the VNT in the event that it was affected by a single or multiple network failure.

As for the anycast problem presented in Chapter 7, our approach for solving this problem consists of creating an augmented graph where *dummy* nodes and links are added. In this case, one dummy node for each connected component is added and connected to every core DC in that connected component. Figure 12.14a illustrates a VNT topology where vlinks 10–11 and 11–12 have failed. Three dummy nodes have been added (labeled as *D1*, *D2*, and *D3* in Figure 12.14b). Then, finding a tree topology connecting these dummy nodes will reconnect the connected components into a connected VNT topology. We adapt the approach proposed for the VNT creation problem to compute the set of vlinks for the tree and to find feasible lightpaths to support the vlinks (Figure 12.4c), eventually reconnecting the VNT (Figure 12.14d).

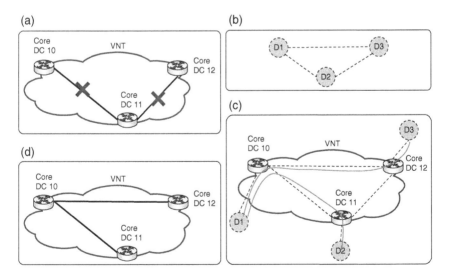

Figure 12.14 VNT reconnection: (a) VNT topology with failed vlinks; (b) dummy nodes for each connected component; (c) new vlinks to reconnect the virtual topology; and (d) final virtual topology.

The VNT_RECONNECT problem can be stated as follows:

Given:

- the optical network topology represented by a graph $G(N_O, L)$ where the failed elements (optical links or nodes) have been removed,
- the number of available slices η_l in each link $l \in L$,
- the disconnected VNT represented by a graph $G(N_C, E)$, where E is the set of operational vlinks after removing those affected by the failure. It is worth noting that even in the case of a single optical link failure, several virtual links might be affected,
- the number of slices χ required to set up each lightpath supporting a vlink.

Output: the route for the bidirectional LSPs used to reconnect the VNT.

Objective: minimize the optical resources used to reconnect the VNT.

The ILP formulation for the VNT_RECONNECT problem is similar to that of the VNT_CREATE one, and it is left to the interested reader to redefine such formulation by introducing the appropriate modifications. Since the VNT_RECONNECT problem needs to be solved in real time, for example, less than 1 s, we focus on proposing a heuristic algorithm to solve the problem. Table 12.6 presents the pseudo-code of the heuristic algorithm that makes use of the VNT_CREATE algorithm to reconnect VNTs by using one dummy node for each connected component of the VNT. The algorithm begins finding the set of connected components (line 2 in Table 12.6), adding the set of dummy nodes, and precomputing the links connecting them to the core DCs (lines 3–7). Next, the VNT_CREATE algorithm is run using the RECONNECT mode to connect the components among them (line 8), and the solution is returned.

Table 12.6 VNT_RECONNECT heuristic algorithm.

```
INPUT:  G(N, L), G(N_C, E), χ
OUTPUT: LSP
```

```
1: N_D ← ∅; E_D ← ∅
2: CC = {<N_P, E_P>} ← Find_Connected_Components(G(N_C, E))
3: for each cc ∈ CC do
4:    v ← Create_Dummy_Node()
5:    N_D ← N_D ∪ {v}
6:    for each n ∈ N_P do
7:       E_D ← E_D ∪ {(n, v)}
8: return VNT_CREATE(G(N ∪ N_D, L ∪ E_D), N_D, χ, RECONNECT)
```

12.3.2 Workflows and Protocol Extensions

In this section, we focus on implementing VNT creation and reconnection, that is, deploying the algorithms for the identified problems on the control and management planes, based on the ABNO architecture. We assume that a back-end PCE (bPCE) is responsible for dealing with computationally intensive algorithms, such as VNT creation and reconnection algorithms.

Next, workflows for the identified problems as well as related PCEP issues are analyzed, and extensions are proposed.

12.3.2.1 VNT_CREATE Workflow

The VNT_CREATE workflow (Figure 12.15) is triggered by the operator through the NMS. Its goal is twofold: first, to add the set of core DCs belonging to the VNT to the Application Service Orchestrator (ASO) service database; and, second, to set up the VNT itself. For the latter, the ASO issues a VNT creation request message (labeled as 1 in Figure 12.15) to the controller with the set of core DCs to be connected. The controller issues a PCReq message to the front-end PCE (fPCE) (2), which delegates the VNT topology computation to the bPCE (3). The bPCE runs the VNT_CREATE algorithm and sends the solution in a PCRep message (4). The fPCE issues a PCInit message to the provisioning manager (PM) so as to implement the related LSPs and waits for the PCRpt messages confirming their implementation (5). When the VNT topology has been created, the fPCE sends back a PCRep message to the controller (6) with the lightpaths implemented, which replies to the ASO (7) that updates its services database and informs the NMS.

PCReq messages (2) and (3) include one Request Parameters (RP) object for each candidate vlink connecting pairs of core DCs (specified by the ASO) together with its corresponding P2P END-POINT and BANDWIDTH objects.

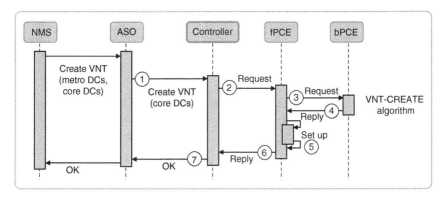

Figure 12.15 VNT_CREATE workflow.

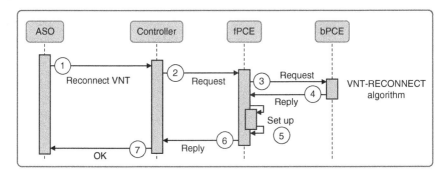

Figure 12.16 VNT_RECONNECT workflow.

To group all the candidate vlinks belonging to the VNT, a SVEC object is also included. The topology type, a tree in our case, is defined by adding an OF object. ERO and BANDWIDTH objects are used for the candidate vlinks selected to be part of the VNT, whereas NO-PATH objects are used for those vlinks not in the VNT. If no feasible solution was found, all requests are replied with NO-PATH objects.

12.3.2.2 VNT_RECONNECT Workflow

The VNT_RECONNECT workflow, presented in Figure 12.16, is initiated by the cloud resource manager when the VNT is affected by a failure. The ASO then issues a VNT reconnection request to the controller (labeled as 1 in Figure 12.16), which sends a PCReq message to the fPCE (2). VNT reconnection is delegated to the bPCE, so the fPCE forwards the message to the bPCE (3) that runs the VNT_RECONNECT algorithm. Since some vlinks could remain operational, both candidate and operational vlinks have to be included in the PCReq messages. To that end, a Symbolic-Path-Name type-length-value (TLV) is included in each RP object for the bPCE to know whether the related LSP already exists. Once a solution has been found, the bPCE sends a PCRep message (4) to the fPCE that issues a request to the PM to implement the lightpaths needed for recovery (5). The PM sends a PCRpt message after each lightpath has been set up and when all the PCRpt messages related to that VNT have been received, the fPCE informs the controller (6), which in turn informs the ASO (7). The ASO updates its services database and notifies the cloud resource manager.

12.3.3 Experimental Assessment

Let us now focus on the experimental validation of the VNT creation and reconnection workflows. ASO issues XML-encoded messages to the controller through its HTTP REST API. The network shown in Figure 12.13 was assumed.

```
No.    Source          Destination    Protocol  Length  Info
①  509 172.16.103.1   172.16.103.2   HTTP/XML   507 POST /abno/CPVNT_CREATE HTTP/1.1
②  511 172.16.103.2   172.16.103.3   PCEP       220 Path Computation Request
③  513 172.16.103.3   172.16.103.4   PCEP       196 Path Computation Request
④  514 172.16.103.4   172.16.103.3   PCEP       408 Path Computation Reply
┌  516 172.16.103.3   172.16.103.5   PCEP       444 Initiate
⑤  518 172.16.103.3   172.16.103.5   PCEP       240 Path Computation LSP State Report (PCRpt)
└  520 172.16.103.5   172.16.103.3   PCEP       276 Path Computation LSP State Report (PCRpt)
⑥  522 172.16.103.3   172.16.103.2   PCEP       408 Path Computation Reply
⑦  523 172.16.103.2   172.16.103.1   HTTP/XML  1277 HTTP/1.1 200 OK
```

Figure 12.17 VNT_CREATE message list.

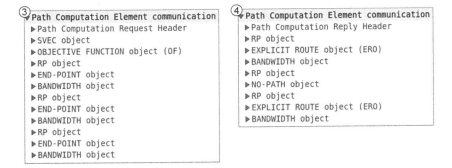

```
③ Path Computation Element communication      ④ Path Computation Element communication
▶ Path Computation Request Header               ▶ Path Computation Reply Header
▶ SVEC object                                   ▶ RP object
▶ OBJECTIVE FUNCTION object (OF)                ▶ EXPLICIT ROUTE object (ERO)
▶ RP object                                     ▶ BANDWIDTH object
▶ END-POINT object                              ▶ RP object
▶ BANDWIDTH object                              ▶ NO-PATH object
▶ RP object                                     ▶ RP object
▶ END-POINT object                              ▶ EXPLICIT ROUTE object (ERO)
▶ BANDWIDTH object                              ▶ BANDWIDTH object
▶ RP object
▶ END-POINT object
▶ BANDWIDTH object
```

Figure 12.18 VNT_CREATE PCReq and PCRep messages.

Control and management modules run in the IP subnetwork 172.16.103.X. Specifically, ASO runs in .1, the controller in .2, fPCE in .3, bPCE in .4, and PM in .5. An emulated data plane was deployed for the experiments.

Figure 12.17 shows the relevant messages for the VNT_CREATE workflow. The messages are identified with the same sequence numbers used in the workflow description. Figure 12.18 shows PCReq message (3) and PCRep message (4) details. The PCReq message contains a set of RP, END-POINT, and BANDWIDTH objects for each candidate link; a SVEC object groups the requests for a joint computation. The PCRep message indicates that two vlinks were selected since ERO and BANDWIDTH objects are included, whereas the other vlink is discarded. Note in Figure 12.17 that one single PCInit message was used for the two LSPs to be created, whereas two individual PCRpt reported each individual setup.

The control plane contribution to the VNT_CREATE workflow processing time was less than 9 ms, including messages exchange and the VNT_CREATE algorithm in the bPCE.

Figure 12.19 shows the relevant messages for the VNT_RECONNECT workflow that are similar to those of the VNT_CREATE workflow. The control plane contribution to the VNT_RECONNECT workflow processing time was less than 9 ms, including messages exchange and algorithms.

```
No.   Source          Destination    Protocol  Length  Info
① 1358 172.16.103.1   172.16.103.2   HTTP/XML   533 POST /abno/CPVNT_RECONNECT HTTP/1.1
② 1360 172.16.103.2   172.16.103.3   PCEP       228 Path Computation Request
③ 1362 172.16.103.3   172.16.103.4   PCEP       228 Path Computation Request
④ 1363 172.16.103.4   172.16.103.3   PCEP       288 Path Computation Reply
⑤ 1365 172.16.103.3   172.16.103.5   PCEP       276 Initiate
  1366 172.16.103.5   172.16.103.3   PCEP       276 Path Computation LSP State Report (PCRpt)
⑥ 1368 172.16.103.3   172.16.103.2   PCEP       288 Path Computation Reply
⑦ 1370 172.16.103.2   172.16.103.1   HTTP/XML   938 HTTP/1.1 200 OK
```

Figure 12.19 VNT_RECONNECT message list.

12.4 Conclusions

In the first part of this chapter, the problem of dynamic restoration in multilayer networks using a centralized PCE for restoration paths computation was studied. In such a scenario, an optical link failure may disrupt multiple MPLS paths, generating a large set of path computation requests to be served by the PCE. Applying a regular sequential PCE approach, the attained restorability was rather poor due to resource contention when signaling MPLS connections, as well as the scarce exploitation of grooming. To improve such restorability metric, reducing resource contention and leveraging grooming objectives, a PCE GCO architecture was designed and implemented to group all the requests and perform bulk path computation.

The DYnamic restorAtion in Multilayer MPLS-over-EON (DYNAMO) problem was formally stated and modeled using an MP formulation. Due to the stringent requirements in terms of restoration time, a randomized heuristic algorithm was conceived. Both approaches (sequential versus bulk) were experimentally validated and compared by means of three figures of merit: blocking probability, restorability, and restoration time.

Next, the problem of interconnecting a set of DCs by means of a survivable VNT was tackled; in particular, the creation and the reconnection of the VNT after a network failure were studied. These problems were formally stated, mathematically formulated, and then heuristic algorithms were eventually proposed for solving them in real time. Aiming at experimentally assessing the proposed algorithms, an ABNO-based control plane architecture was considered. As a result of the complexity of problems to be solved and the stringent time in which they need to be solved, a bPCE was used. Workflows were developed for the identified problems, and the PCEP feasibility for those workflows was studied.

Part V

Future Trends

13

High Capacity Optical Networks Based on Space Division Multiplexing

Behnam Shariati[1,2], Jaume Comellas[2], Dimitrios Klonidis[1],
Luis Velasco[2] and Ioannis Tomkos[1]

[1] *Athens Information Technology, Athens, Greece*
[2] *Universitat Politècnica de Catalunya, Barcelona, Spain*

Many physical attributes of light, including intensity, phase, and polarization, have been exploited so far to increase the capacity of optical transport systems and keep pace with the exponential traffic growth. However, we are rapidly approaching the fundamental spectral efficiency limit within the gain bandwidth of C-band erbium-doped fiber amplifiers (EDFAs). Current trends show that capacity upgrades may occur with the introduction of multiband systems as a midterm option, due to the emergence of sophisticated amplification systems supporting L- and S-bands. Nonetheless, the ultimate solution to address the capacity crunch in a cost-effective way relies on the introduction of some forms of space division multiplexing (SDM) techniques in optical transport networks.

In this chapter, we review the recent progress in the development of SDM-based optical networks. We describe the enabling technologies, including fibers and switches, for the efficient realization of SDM networks. It is unquestionable that the efficient use of space domain requires some form of spatial integration of network elements. Spatial integration decreases the architectural complexity, and, thus, the implementation cost of components. Nevertheless, it introduces extra crosstalk and additional constraints for planning of SDM networks. Therefore, for a given use case, a choice should be made of the most appropriate technology in terms of performance and implementation cost.

This chapter particularly describes three switching paradigms identified for SDM networks, which are highly correlated with three fiber categories proposed for SDM networks. It is then discussed that fiber types and switching paradigms are the two technology areas limiting the channel allocation in SDM networks. The capabilities/limitations brought about by different fiber types

Provisioning, Recovery, and In-operation Planning in Elastic Optical Networks,
First Edition. Luis Velasco and Marc Ruiz.
© 2017 John Wiley & Sons, Inc. Published 2017 by John Wiley & Sons, Inc.

and switching paradigms for resource allocation options in SDM networks are thoroughly investigated and several resource allocation schemes are proposed and their benefits/drawbacks are detailed. Ultimately, comparative performance evaluations are presented revealing benefits of SDM networks based on different switching paradigms.

Table of Contents and Tracks' Itineraries

13.1 Introduction

The traffic carried by core optical networks as well as the per-channel interface rates required by IP routers are growing at a remarkable pace year over year. Optical transmission and switching advancements have so far satisfied this huge traffic growth by delivering the content over the network infrastructure in a cost- and energy-efficient manner utilizing to the maximum extent the capabilities of optoelectronic and photonic subsystems and the available bandwidth of deployed optical fibers. However, we are rapidly approaching the fundamental spectral efficiency limits of single-mode fibers (SMFs) and the scientific and industrial telecommunications community have foreseen that the growth capabilities of conventional wavelength division multiplexing (WDM) networks operating on a fixed frequency grid are quite limited.

To address such limitations, over the last couple of years a large number of significant innovations able to offer, in practice, a capacity increase by a factor of around 10–20 (compared to legacy WDM systems at 10 Gb/s on a 50 GHz grid) have emerged. Initial efforts targeted innovative modulation/coding techniques [Wi06], novel switching subsystems [Gr10], and routing algorithms supporting flexible frequency allocations [Ch11], in an effort to increase the spectral density/utilization in the optical network. This eventually led to the definition of spectrally flexible/elastic optical networks (EONs) utilizing optical super-channels (Sp-Chs) together with spectrally flexible multiplexing schemes (e.g., Nyquist WDM) and advanced modulation formats, thus enabling the dynamic and adaptive allocation of end-to-end demands with variable connection characteristics (e.g., requested data rates) [Ji09]. However, while the spectrally flexible optical networking approaches can optimize network resources through increased spectral utilization compared to conventional fixed-grid networks, they have limited growth potential due to the nonlinear Shannon limit, which imposes an upper bound on the transport capacity of an SMF within the limited gain bandwidth of C-band Erbium-doped fiber amplifiers (EDFAs) [Es10]. A possible solution to increase the fiber capacity is to extend the amplification bandwidth. EDFA amplifiers working on the C+L band or Raman-based amplification systems with different designs—for example, all-Raman or hybrid Raman with EDFA amplifiers—can increase network capacity by amplifying broader spectral bands compared to conventional C-band EDFA systems. Even though current trends show that, as a midterm option, capacity upgrades may occur with the introduction of multiband systems [Sh16.1], the only evident long-term solution to extend the capacity of optical communication systems relies on the use of some form of SDM [Wi12]. The simplest way to achieve spatial multiplexing is to deploy multiple systems in parallel. However, by simply increasing the number of systems, the cost and power consumption also increases linearly. In order to limit the increase in cost and power consumption, component sharing and integration have to be introduced [Ri16].

Thus far, SDM research has focused on the development and performance evaluations of SDM fibers—for example, few-mode fiber (FMF), multicore fiber (MCF), or few-mode multicore fiber (FM-MCF)—transmission technologies, and integrated optical switches. Nonetheless, from the operational perspective, the additional degree of freedom introduced by SDM requires that, for a given use case, a choice be made of the most appropriate technology, including transmission media (fiber type), switching architectures, and transceiver (TRx), which ultimately introduce additional physical layer constraints. The extra physical impairments introduced by SDM media (compared to the case of SMF transmission) have been thoroughly investigated [Nk15], for example, MCF is mostly affected by intercore crosstalk (XT), while FMF is strongly impacted by mode coupling and/or differential mode group delay. For the purpose of our discussions in this chapter, SDM media can be categorized into three groups, according to whether they have (i) uncoupled/weakly coupled spatial dimensions (cores, modes, or parallel fibers), (ii) strongly coupled spatial dimensions, or (iii) subgroups of strongly coupled spatial dimensions.

The choice of a transmission medium belonging to one or another of these three categories is a key factor in determining the required architecture and properties of the optical switches at the nodes of an SDM-based network [Mr15]. For instance, strongly coupled MCFs or FMFs necessitate that all cores/modes be switched together, whereas, at the opposite extreme, the use of bundles of SMFs (BuSMFs) permits both independent and joint switching of the signals on each of the SMFs. A middle-ground solution, for example, FM-MCF with negligible coupling between cores, would allow independent switching of the mode groups on each of the cores (with a small reach penalty due to the existing, though small, intercore XT) or joint switching of all modes/cores. Therefore, switching strategies can be divided into several paradigms which strongly correlate with the SDM fiber categories defined earlier [Mr15]: (i) *independent switching* (Ind-Sw), whereby all spectral slices and spatial dimensions can be independently directed to any output port; (ii) *joint switching* (J-Sw), in which all S spatial dimensions (S is the number of core/modes per SDM fiber) are treated as a single entity, while spectral slices can still be independently switched; and (iii) *fractional joint switching* (FrJ-Sw), a hybrid approach in which a number of subgroups of G spatial dimensions out of S, as well as all spectral slices, can be independently switched to all output ports. The last two paradigms are categorized as *spatial group switching* (SG-Sw) solutions since the spatial resources are switched in groups rather than independently, as in the case of Ind-Sw. Note that several *spatial switching granularities* result from different levels of grouping of the spatial dimensions, which is determined by G. Evidently, G equal to 1 assumes individual fibers corresponding to the Ind-Sw case, thus offering the finest spatial granularity, while the value of G equal to S considers all spatial dimensions as one group and corresponds to the J-Sw case, which has the coarsest spatial granularity.

SDM will also have an impact on the TRx implementation [Ry12], [Wi12]. Historically, the TRx evolution toward more flexible and higher capacity solutions started with the single-carrier (SC) single-line-rate (SLR) TRx for 2.5 Gb/s data rate transmission. Later on, technology development led to higher transmission capacities to support SC-SLR systems at 10G, 40G, 100G and 200G. By employing multiple modulation formats, the SC multi-line-rate (SC-MLR) TRx was introduced to transmit traffic demands in a more flexible and spectral efficient way than before. Subsequently, multi-channel (MC) TRx architectures were proposed leveraging on the capability to assign variable spectrum, data rate, and modulation to connections. Sp-Ch is an alternative term referring to MC TRx. A Sp-Ch consists of a group of subchannels that are routed together across the network to connect two end nodes. In single-fiber systems, Sp-Chs (hereinafter called spectral Sp-Chs) can improve the spectral efficiency by decreasing the guard bands, but they reduce the routing flexibility since the number of channels that can be routed independently is reduced. A key question for SDM networks is how to best distribute demands over spectral and spatial resources. In that regard, two Sp-Ch allocation policies were explored [Kh16]: *spectral* and *spatial* Sp-Ch allocation policies. A spectral Sp-Ch, similar to the one in single-fiber systems, is the result of aggregating signals modulated on adjacent optical carriers in a single spatial dimension. A spatial Sp-Ch, on the other hand, results from the aggregation of signals modulated on a certain optical carrier across a number or all of the spatial dimensions of an SDM transmission medium. In the case of strong coupling among spatial channels, the only option is spatial Sp-Chs. Spatial Sp-Ch allows decreasing the optical switching complexity, since channels are switched in groups rather than independently—at the price, of course, of a potential reduction in routing flexibility—[Sh16.2]. SDM networks based on spatial Sp-Ch allocation can benefit from additional cost reduction due to the possibility of sharing network elements among different spatial dimensions [Ri16] (e.g., a number of Sp-Ch constituents can share lasers and digital signal processing (DSP) modules [Fe12], which can lead to cost and power consumption savings of integrated TRxs in SDM networks). A third possible policy would emerge from the combination of the aforementioned spectral and spatial Sp-Ch allocation options.

In Section 13.2, three types of SDM fibers are described in detail and their limitations imposed to the channel allocation problem are discussed. Section 13.3 is devoted to the SDM switching paradigms, where their architectures and their impact on the routing of spectral and spatial resources are explained. Built upon these two sections, in Section 13.4, resource allocation problem in SDM-based optical networks is thoroughly investigated. Extensive quantitative results provided in Sections 13.5 and 13.6 present the benefits and drawbacks of SDM networks utilizing various switching paradigms. The evaluations are done assuming different traffic profile. Ultimately, we summarize this chapter in the conclusion section.

13.2 SDM Fibers

The fundamental concept of SDM relies on placing numerous spatial channels in a single-fiber structure aiming at manufacturing an integrated fiber carrying manifold channels in a denser, lighter, and more cost-effective cable. The type of channels depends on the way SDM is exploited: multiple SMFs in a fiber bundle, many cores in a single cladding, multiplexed linearly polarized (LiPo) modes, or several cores each supporting few multiplexed LiPo modes [Nk15]. Even though many SDM fiber alternatives with different number of cores/ modes have been developed for future SDM networks, for the purpose of our

(a)   (b)

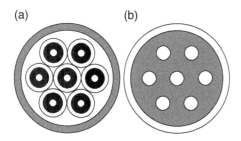

Figure 13.1 (a) SMFs arranged in a BuSMF, (b) weakly coupled MCF.

discussions in this chapter, SDM media can be categorized into three groups, according to whether they have (i) uncoupled/weakly coupled spatial dimensions (cores, modes, or parallel fibers), (ii) strongly coupled spatial dimensions, or (iii) subgroups of strongly coupled spatial dimensions [Mr15]. In the next couple of paragraphs, three fiber classes are discussed in more detail.

13.2.1 Uncoupled/Weakly Coupled Spatial Dimensions

Spatial channels of this type of fibers remain distinct in fiber propagation (i.e., there is no/negligible XT between spatial channels) as would be experienced in BuSMFs or weakly coupled MCF. Therefore, individual spatial channels can be freely switched from one link to another, or add/drop operations can be performed independently for any wavelength on any spatial channel. When these fibers are in place, it is also possible to route a wavelength channel from one spatial channel to another—the so-called lane changes (LCs) method of functioning of switching infrastructure [Mr15]. This type of fibers are designed to avoid any complex multi-input–multi-output (MIMO) processing at the receiver side; however, they require more complex and costly switching infrastructure as we will see in the next section. Figure 13.1 shows a simple illustration of a cross section of a bundle of seven SMFs and a seven-core weakly coupled MCF.

13.2.2 Strongly Coupled Spatial Dimensions

Spatial channels of this type of fibers strongly mix in fiber transmission, as occurs in FMFs and coupled MCFs. Coupled MCFs and FMFs are affected by intermodal impairments, for example, MCFs are affected by intercore XT and FMFs are impacted by mode coupling, differential group delay, mode dependent loss, and intermodal nonlinearity. As a consequence, MIMO processing is required in order to compensate for these impairments [Ry12]. Since the information is mixed across all spatial modes/cores, spatial channels must be switched together to other destinations and LC operation is not allowed. Even though this constraint sacrifices the routing flexibility, it significantly simplifies the switching infrastructure and results in huge cost savings [Ri16]. Figure 13.2 shows a cross section of a seven-core coupled MCF and a single-core FMF carrying several LP modes.

Figure 13.2 (a) Coupled MCF, (b) FMF carrying several LP modes.

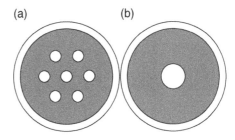

Figure 13.3 (a) FM-MCF, (b) SDM fiber having uncoupled subgroups of coupled cores.

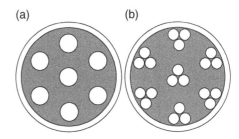

13.2.3 Subgroups of Strongly Coupled Spatial Dimensions

Spatial channels of this type of fibers mix only within the subgroups, as would be experienced in FM-MCF or SDM fibers with uncoupled subgroups of coupled cores. Subgroups are defined by the fiber design, for example, uncoupled cores of a FM-MCF, and the spatial channels belonging to a subgroup must be switched as a single entity. Subdividing the spatial channels into a smaller size eases the switching limitations with respect to the former type of SDM fibers, while not reaching the full flexibility of the first type (i.e., fibers with uncoupled spatial dimensions). Even though routing a wavelength channel from one spatial channel to another within a group is not allowed, routing a wavelength channel from one spatial group to another is applicable assuming that appropriate switching infrastructure is in place. Note that wherever LC is discussed in the rest of this chapter for this group of fibers, it indicates spatial group changes. Figure 13.3 shows a cross section of an FM-MCF and an SDM fiber having uncoupled subgroups of coupled cores.

Networks based on BuSMFs benefit from longer optical reach and lower power consumption in the receiver side (because a high-order MIMO processing is not necessary) [Di17]. Additionally, the utilization of already installed SMFs is likely to be the most cost-effective option for near-term realizations of SDM networks. However, in practice, the splicing of BuSMFs with a large number of fibers and their alignment with the TRx is a challenge since it makes the interfacing more difficult and costly compared to the FMF case. MCFs also pose the same problem. On the other hand, MCFs and FMFs offer reduced-size cables with higher density and lower weight.

Table 13.1 Classification of SDM switching paradigms and suitable fibers for each category.

Ind-Sw	FrJ-Sw	J-Sw
BuSMFs	BuSMFs	BuSMFs
Weakly coupled MCF	Weakly coupled MCF	Weakly coupled MCF
	Groups of MCFs	Groups of MCFs
	FM-MCFs	FM-MCFs
		Coupled MCF
		FMF

The choice of a fiber type belonging to one of these three classes has a considerable impact on the selection of the appropriate switching paradigm. Suitable fiber types for three SDM switching paradigms are listed in Table 13.1. SDM switching is the focus of the next section.

13.3 SDM Switching Paradigms

The implementation of switching solutions for the switching paradigms described in Section 13.1 requires the design of new SDM nodes [Fo15], [Mr15]. Ind-Sw can be realized by means of a node architecture such as the one shown in Figure 13.4a for a route-and-select reconfigurable optical add/drop multiplexer (ROADM) configuration. It is composed of a number of conventional wavelength selective switches (WSSs), one per spatial dimension, degree and ingress/egress port. Commercially available 1×5, 1×9 or 1×20 WSSs can be employed since the port count is not a limiting factor in this case. The selection of one or another WSS realization will depend on a number of factors, such as the required ROADM operation (e.g., whether "spatial LCs" are allowed) and the nodal degree—that is, the number of available directions—of the network nodes. Core switching is an alternative term referring to spatial LC when MCFs are discussed. Figure 13.4a depicts a node architecture enabling Ind-Sw without spatial LC for an SDM network with $S = 4$ spatial dimensions and a node with degree $D = 3$. A node based on Ind-Sw paradigm requires lots of WSSs of relatively low port count. Ind-Sw increases the complexity of the switching architecture, and hence the cost. On the other hand, it brings the highest level of flexibility for routing of spectral and spatial channels since it allows the allocation of demands over different spatial dimensions and spectral slices with variable widths.

Fr-Sw and J-Sw make necessary a redesign of the WSSs. They are configured to operate as $S \times (I \times O)$ WSSs, that is, they direct I input ports, each carrying S spatial modes/cores, toward O output ports using spatial diversity. This has the

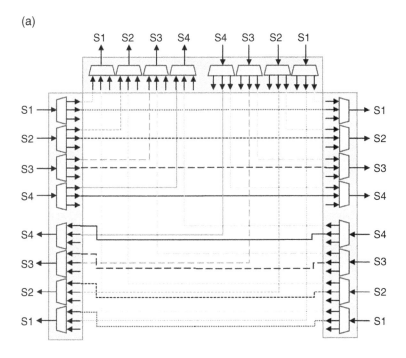

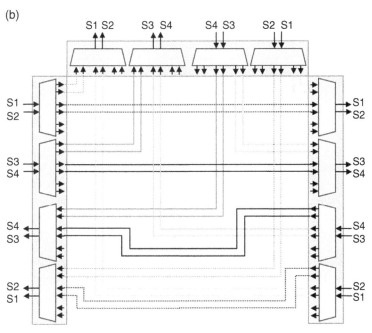

Figure 13.4 Node architectures, for $S=4$ and $D=3$, enabling (a) Ind-Sw, (b) FrJ-Sw with $G=2$, and (c) J-Sw. Ind-Sw, FrJ-Sw with $G=2$, and J-Sw require eight 1×3, four 2×6, and two 4×12 WSSs per degree, respectively.

(c)

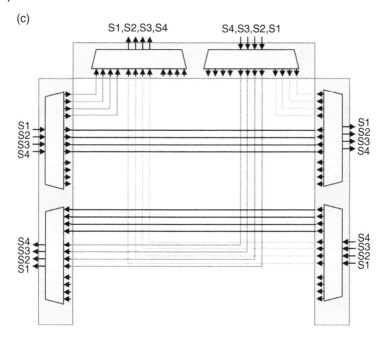

Figure 13.4 (Continued)

implication that for large *S*, WSSs with very high port count (HPC) are required [Sh16.2]. The required number of WSSs and their port count is shown in Table 13.2 for different switching paradigms with/without LC. By making use of spatial group WSSs, the FrJ-Sw and J-Sw paradigms enable reducing the number of necessary WSSs to $2 \cdot \lceil S/G \rceil$ and 2, respectively, per degree, as illustrated in Figure 13.4b and c, but the required port count increases by a factor of *G* and *S*, respectively, as provided in Table 13.2. In contrast to Ind-Sw, which has the highest flexibility in routing, in J-Sw the spectrum assignment is limited to a single connection across all spatial dimensions within a spectral slice, with the consequent drawback that the unused spatial dimensions at a certain spectral slot cannot be allocated to other demands. While the choice of switching technology can restrict the routing flexibility (given that the coarse granularity of J-Sw and FrJ-Sw penalizes the network spectral occupancy), it can also boost the economic feasibility of SDM solutions [Sh16.2]. For instance, J-Sw and FrJ-Sw allow the use of joint DSP at different degrees [Fe12], which can lead to cost and power consumption savings of integrated receivers in SDM networks [Ri16].

Node architectures with LC support for Ind-Sw and FrJ-Sw are not shown in Figure 13.4. In order to support LC operation, as provided in Table 13.1, WSSs with higher port count are required and internal connectivity of the nodes must be modified. Note that J-Sw cannot support any level of LC since all

Table 13.2 Number of WSSs required and the corresponding port count for SDM switching strategies.

Switching paradigm	Port count per WSS	Number of WSS per degree
Ind-Sw without LC	$1 \times (1 \times D)$	$2 \cdot S$
Ind-Sw with LC	$1 \times (1 \times [S \cdot (D-1)+1])$	
FrJ-Sw without LC	$G \times (1 \times D)$	$2 \lceil S/G \rceil$
FrJ-Sw with LC	$G \times (1 \times [(S/G) \cdot (D-1)+1])$	
J-Sw	$S \times (1 \times D)$	2

spatial channels are treated as a single entity and are directed to the corresponding spatial dimensions of each ROADM direction. For more details on switching technologies for SDM networks, see [Mr15] and [Mr17]. Add/drop modules (not shown in Figure 13.4), depending on the chosen add/drop module/transponder technology, can allow for several degrees of operational flexibility: they can be colorless, directionless and/or contentionless. Colorless, directionless, contentionless (CDC) ROADM architectures are based on multicast switches (MCSs) or $M \times N$ WSSs. For more details on CD(C) ROADM architectures for SDM networks, see [Ri17].

J-Sw is compatible with any type of SDM fibers, while FrJ-Sw and Ind-Sw are fiber-dependent. As provided in Table 13.1, alternatively, BuSMFs and weakly coupled MCFs are compatible with all switching paradigms, while coupled MCFs and FMFs require a particular type of switching paradigm (J-Sw).

13.4 Resource Allocation in SDM Networks

Resource allocation is a well-known problem in optical networks and has been widely investigated for WDM networks and EONs. Even though alternative algorithms/solutions have been proposed so far addressing this problem, the core idea in all of them is to find an available wavelength (in a WDM network) or a contiguous portion of available spectrum (in an EON) on a set of consecutive links, connecting two end nodes, to establish a lightpath between them. However, there are some technological/strategic constraints which must be taken into account while proposing a resource allocation option. For instance, the same wavelength or the same portion of contiguous spectrum must be available for all the links connecting two end nodes in order to avoid placing costly wavelength converters in the intermediate nodes where optical bypass operations take place. This constraint is known as *wavelength/spectrum continuity* which is the only one applying to routing and wavelength assignment

(RWA), the conventional resource allocation problem in WDM networks. By incorporating an extra degree of freedom introduced in EONs (i.e., the choice of variable channel spacing), an extra constraint has been added to the RWA problem. In order to address the routing and spectrum assignment (RSA), the popular resource allocation problem in EONs, in addition to spectrum continuity, *spectrum contiguity* must be considered. Spectrum contiguity means that if a certain connection requires more than one frequency slots, these frequency slots must be adjacent on the optical spectrum. Modulation format is another degree of freedom exploited in EONs due to the capability provided by bandwidth-variable TRxs. Thus, the RSA problem has been extended to consider modulation format and it becomes a routing, modulation format, and spectrum assignment (RMSA) problem. RMSA problems are typically optimized with the aim of minimizing spectrum utilization, energy consumption, or operational expenditure (OPEX).

With the introduction of space dimension, RMSA problems must be upgraded to take into account the space domain. Therefore, the RMSA problem becomes routing, space, modulation format, and spectrum assignment (RSMSA) in the context of SDM-based optical networks. In this section, we try to identify the potential capabilities/restrictions brought about by the space dimension. In particular, we discuss the extra constraints imposed by enabling technologies of SDM (i.e., fibers and switches) which affect the RSMSA problem. Ultimately, several channel allocation options will be presented and their benefits/drawbacks will be discussed.

The two main technology areas limiting channel allocation and routing options in an SDM-based optical network are *fiber type* and *switching paradigms*. SDM fibers affect the resource allocation options by imposing various physical impairments to spectrally/spatially multiplexed channels (i.e., Sp-Chs), while switching paradigms do so by determining the routing properties of the multiplexed channel at the SDM nodes. Figure 13.5 shows the routing options in a 4-degree SDM node utilizing alternative SDM switching paradigms assuming that an SDM fiber with six spatial channels is in place. In order to illustrate how three SDM switching paradigms determine the routing operation, some exemplary occupied spectral and spatial slots (each one of those dark, gray, and white rectangle-like shapes denotes one spectral–spatial slot) are considered on degree 1 of the depicted node of Figure 13.5. Note, even though the same spectrum status is assumed on degree 1 of all options, they are distributed among different number of demands for different switching paradigms. The spectral–spatial slots bounded by dashed-dotted lines, dashed lines, and dotted lines are routed together from degree 1 to degrees 2, 3, and 4, respectively. Each of these rectangles denotes an independent demand required to be routed separately. As is shown in Figure 13.5a, a certain portion of spectrum on all spatial dimensions must be switched together when J-Sw is used. Therefore, if a demand is small and it only spreads over some of the spatial

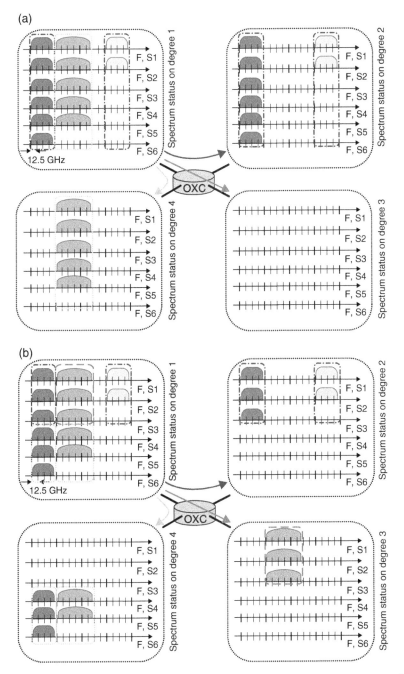

Figure 13.5 Routing of some exemplary demands from degree 1 to other degrees of an SDM node utilizing different switching paradigms. The areas bound by dashed-dotted lines, dashed lines, and dotted lines show the spectral–spatial slots to be routed from degree 1 to degrees 2, 3, and 4, respectively. Filled bounded slots show the spectral–spatial slots which undergo LC operation. (a) J-Sw, (b) FrJ-Sw (G = 3) without LC, (c) FrJ-Sw (G = 3) with group LC, (d) Ind-Sw without LC, and (e) Ind-Sw with LC.

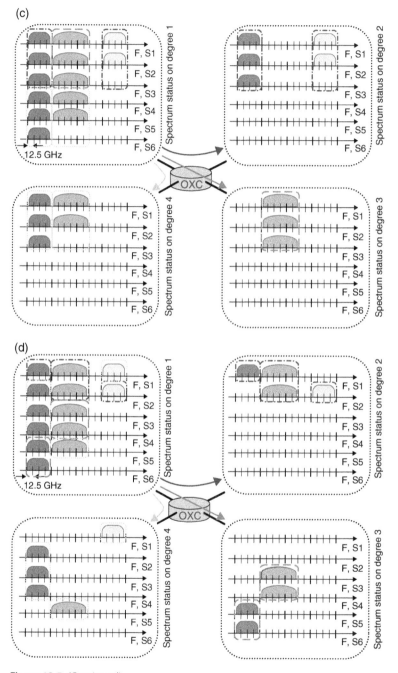

Figure 13.5 (Continued)

(e)

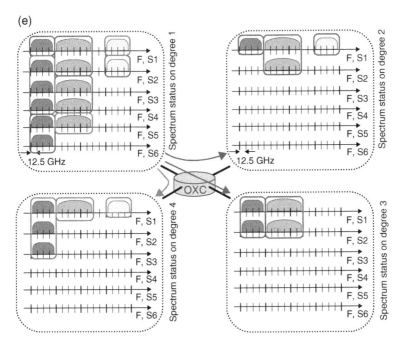

Figure 13.5 (Continued)

dimensions, the available spectrum on the rest of the spatial dimensions cannot be allocated to other demands and remains unutilized due to the coarse spatial granularity of J-Sw. Utilizing switching paradigms with finer spatial granularity relaxes this constraint if we are not limited by the physical impairment of the fibers. Routing of spectral–spatial slots assuming FrJ-Sw with $G = 3$ is presented in Figure 13.5b. The value of G equal to 3 means that the six spatial dimensions are divided into two groups of size 3 and each group can be freely switched from/to any degree of the ROADM. This feature allows the allocation of distinctive demands on individual groups. However, if a single demand is big enough to be spread over all spatial dimensions, FrJ-Sw routes all dimensions together as a single entity, similar to the case of J-Sw.

In Figure 13.5a and b, the spectral–spatial slots are kept on the same portion of optical spectrum and the same spatial dimensions while routing to another degree. It means that, in addition to *spectrum continuity* constraint, *space continuity* is considered. Space continuity is a constraint which must be ensured while devising RSMSA algorithms for SDM networks based on coupled transmission media. The space continuity constraint is granted in the architecture of J-Sw and FrJ-Sw shown in Figure 13.5. This constraint is relaxed for fibers having uncoupled spatial groups (e.g., FM-MCFs) or uncoupled individual spatial dimensions (e.g., weakly coupled MCFs), thus allowing LC operation at switching nodes. It is worth noting that while LC operation is allowed between groups for fibers with uncoupled spatial groups, space continuity must still be considered inside each group. FrJ-Sw with LC allows the routing of spatial groups independently of any group of spatial dimensions on any direction of the ROADM as shown in Figure 13.5c. Filled bounded slots show the spectral–spatial slots which undergo LC operation. The two filled bounded slots illustrated by dotted lines in Figure 13.5c are routed from one spatial group on degree 1 to another spatial group on degree 4. This property allows the implementation of more flexible/advanced RSMSA algorithms. Finally, routing of spectral–spatial slots exploiting Ind-Sw is presented in Figure 13.5d and e. In Figure 13.5d spectral–spatial slots can be freely routed to any direction, but the space continuity constraint is applied. Figure 13.5e shows the most flexible routing scheme in which the space continuity constraint is relaxed and any spectral–spatial slot can be independently directed to any spatial dimension on any degree of the ROADM. Ind-Sw with LC can be exploited for SDM networks based on weakly coupled MCF or BuSMFs and it must be avoided for SDM networks based on other fiber types (e.g., FM-MCF, FMF). The capability of performing LC increases the complexity and thus the implementation cost of the SDM node (remember that Figure 13.4a shows only the architecture of an SDM node based on Ind-Sw without LC). Note that wavelength conversion is not allowed in any of these options and therefore spectrum continuity must always be ensured.

As previously discussed, RSMSA algorithms depend on the enabling technologies, mainly the fiber type and the switching paradigm, on which an SDM network is deployed. Ind-Sw with LC provides the highest level of flexibility for RSMSA algorithms; however, it requires the most complex and costly node architecture. At the opposite extreme, J-Sw offers the lowest level of flexibility for RSMSA algorithms and the simplest and most cost-effective node architecture. Even though some switching paradigms suffer from lower routing flexibility, they perform well in some specific cases. We discuss this issue further in the next two sections where extensive comparative simulation results are presented evaluating SDM switching paradigm in various cases.

Even though the fiber types and switching paradigms are two main technology areas determining the channel allocation properties, spectral/spatial multiplexing techniques, forming end-to-end optical transport channels, are defined by the TRx technology. Three channel allocation options, which are followed by three approaches that a TRx can form an optical transport channel, are presented in Figure 13.6. Figure 13.6a shows the case in which demands are transported in the form of spectral Sp-Chs, that is, demands spread over a contiguous portion of the spectrum, enough to accommodate them, in a single spatial dimension. Spectral Sp-Chs can improve the spectral efficiency by decreasing the guard bands between subchannels. Nonetheless, SDM networks based on spectral Sp-Chs require Ind-Sw-based switching nodes which are the most

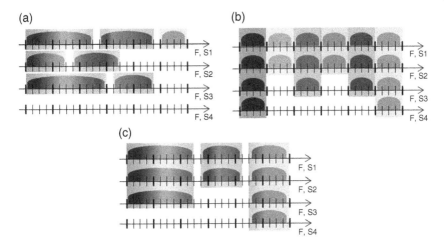

Figure 13.6 Three resource allocation policies identified for SDM networks based on different transceiver technologies. (a) Spectral Sp-Ch, (b) spatial Sp-Ch, and (c) spectral–spatial Sp-Ch.

expensive switching elements [Ri16]. Practically speaking, spectral Sp-Ch allocation policy can be used only for SDM networks based on uncoupled/weakly coupled transmission media. On the other hand, Figure 13.6b illustrates the case in which demands are transported in the form of spatial Sp-Chs, that is, demands spread across a number or all of the spatial dimensions over a given spectral slice (e.g., one optical carrier). Spatial Sp-Ch allocation policy can be utilized with any type of transmission medium but is necessary for SDM networks based on coupled transmission media. Spatial Sp-Ch allocation policy allows (i) decreasing the optical switching complexity, since spatial channels are switched in groups (i.e., Sp-Chs) rather than independently—at the price, of course, of a potential reduction in routing flexibility—[Fe15]; and (ii) SDM networks based on spatial Sp-Ch allocation can benefit from additional cost reduction due to the possibility of sharing network elements among different spatial dimensions (e.g., a number of Sp-Ch constituents can share lasers and DSP modules, which can lead to cost and power consumption savings of integrated TRxs in SDM networks [Ri16]). A third possible policy is spectral–spatial Sp-Ch, shown in Figure 13.6c, would emerge from the combination of the aforementioned spectral and spatial Sp-Ch allocation policies, that is, demands spread across a number or all of the spatial dimensions over a contiguous portion of the spectrum. One can say that spatial Sp-Ch is a particular case of spectral–spatial Sp-Ch when the spectral width of spectral–spatial Sp-Chs is as narrow as a single optical carrier.

The space dimension provides a huge capacity and flexibility to allocate spectral and spatial channels. However, in order to propose an RSMSA algorithm that efficiently utilizes spectral and spatial resources, we must take into account the physical characteristics of transmission media, the routing constraints of switching paradigms, and the capabilities of the TRxs. For more details on resource allocation policies for SDM networks, see [Kl15], [Sh16.3], and [To17].

13.5 Impact of Traffic Profile on the Performance of Spatial Sp-Ch Switching in SDM Networks

The additional degree of freedom introduced by SDM requires that, for a given scenario, a choice be made of the most appropriate technology. As discussed in previous sections, one of the key technologies in determining the overall performance of an SDM-based optical network is the switching technology. In this section, we present extensive simulation results revealing the benefits and drawbacks of different switching paradigms under various traffic profiles.

Large aggregated demands are the most typical type of traffic in core networks today, as we have traffic aggregation at the edge of the network. However, due to the introduction of application-centric services and the increasing dependency of traffic on (i) the type of network (access/metro/core), (ii) the services offered by the network (e.g., 4G/5G connectivity, FTTX, TV on demand), and (iii) the type of offered applications (e.g., file sharing, video conferencing), traffic deaggregation will be a possible networking policy. Depending on the use case, traffic increase may not be caused by the increase in the size of the demands, but by the growth of the number of demands [Sh16.4]. Therefore, in order to capture the effect of various traffic dimensioning approaches on the performance of SDM switching paradigms, different cases in which lightpaths might be established based on (i) a large number of small demands (typically seen in regional part of networks), (ii) a small number of large demands (applicable for example in inter-datacenter communications), and (iii) any combinations of the last two options (found, e.g., in national-scale networks serving heterogeneous type of traffic demands) should be investigated. We performed extensive simulations with the aim of revealing benefits/drawbacks of different switching paradigms under various traffic profiles. In order to perform simulations, we use the Spanish national network of Telefonica. It comprises 30 nodes (average nodal degree 3.7, maximum 5), 14 of which build in add/drop capability (A/D), as well as 56 links with an average length of 148 km. In order to have a fair comparison among the three SDM switching paradigms, regardless of any transmission medium–related performance constraints, bundles of 12 SMFs were considered for all links in the network. Moreover, based on the network characteristics and the related performance evaluation studies [Kh16], DP-8QAM at 32 Gbaud was chosen as the modulation format offering the best compromise between transmission reach and spectral efficiency. A 50 GHz channel spacing is used.

According to the discussion earlier and considering an available spectrum per fiber equal to 4.8 THz (C-band) on the ITU-T 12.5 GHz grid, discrete event simulation studies were carried out for the purpose of performance evaluation. More specifically, the routing, space, and spectrum allocation (RSSA) problem is solved with a k-shortest path ($k = 3$) and spatial Sp-Ch allocation algorithm that follows a first-fit strategy, starting with the shortest computed path. The load generation followed a Poisson distribution process. Traffic demands for each source–destination pair were generated randomly following a normal distribution with mean μ and standard deviation σ over the range of the study, namely, 50 Gb/s to 2.25 Tb/s. Blocking Probability (BP) was used as a quantitative performance measure.

13.5.1 Illustrative Results

In the first part of the study, we consider three traffic profiles corresponding to the three previously mentioned cases: small, large, and medium-sized demands. The distribution of demands for the three different mean demand values and fixed σ is shown in Figure 13.7a, while the effect of the deviated demands over a fixed mean value is shown in Figure 13.7b. The lower and upper mean values were chosen according to the following: (a) for $\mu = 700$ Gb/s, 98% of demands requires less than half of the 12 spatial dimensions (i.e., BuSMFs in our case study) to be allocated, (b) for $\mu = 1600$ Gb/s, we have large aggregated demands that result in more than 98% of them requiring more than half of the 12 spatial dimensions to be allocated. The three traffic profiles of Figure 13.7a are used to obtain the results shown in Figure 13.8. It is noted that in all cases traffic dimensioning is realized by varying the number of live connections per A/D node.

For high mean traffic demands (Figure 13.8a), the three switching paradigms show the same performance. Since most demands require more than half of the spatial resources, the unutilized resources of FrJ-Sw and Ind-Sw cases cannot be allocated, thus leading to the same results as the J-Sw case. For a traffic profile with diverse and relatively medium-sized demands (Figure 13.8b), FrJ-Sw and Ind-Sw start performing better than J-Sw in terms of BP, since now part of the incoming demands that require less than six spatial dimensions to be allocated can fit within the free spatial resources that FrJ-Sw and Ind-Sw enable them to use. For small mean traffic demands (Figure 13.8c) the performance difference between J-Sw and FrJ-/Ind-Sw is more pronounced since the allocation options in space dimension are increased and small demands can fit in available spatial slots. The previous discussion on Figure 13.8 strongly suggests that the optimal switching paradigm for an SDM network, in fact, depends on the nature of its traffic, specifically whether there is a prevalence of relatively small or large demands. However, since most of the demands in the core networks are aggregated traffics, J-Sw would be a suitable choice, considering its cost–benefit [Ri16].

In order to see the impact of traffic diversity on the performance of SDM switching paradigms, a complementary set of simulations is carried out, where

(a)

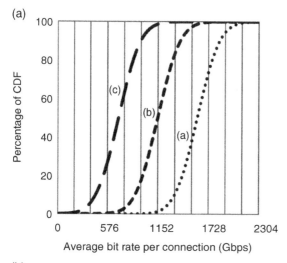

(b)

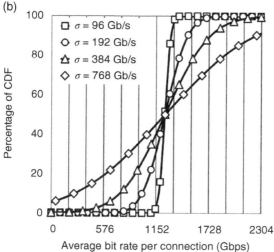

Figure 13.7 CDFs of the assumed traffic profiles with fixed $\sigma=200$ Gb/s and $\mu=700$, 1150, and 1600 Gb/s are plotted with dotted line, dashed line, and dotted-dashed line, respectively, in (a) and with fixed $\mu=1248$ Gb/s, and $\sigma=96$, 192, 384, 768 Gb/s in (b).

the traffic dimensioning is done by varying σ and keeping μ and the number of live connections fixed (Figure 13.7b). Note, in this study, the total offered load to the network during the whole range of the simulation is fixed to $14\times112\times1248$ Gb/s = ~1.95 Pb/s.

Results plotted in Figure 13.9 show that at the beginning the performance of three SDM switching is the same (similar to Figure 13.8a), and by increasing σ,

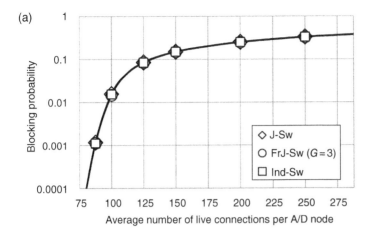

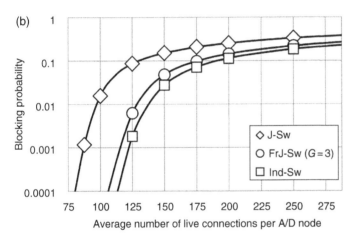

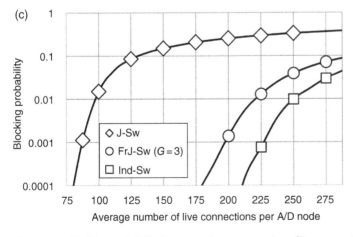

Figure 13.8 Blocking probability in terms of average number of live connections per A/D node for three profiles of traffic forming of small, large, and medium-sized demands, the distributions of which are plotted in Figure 13.5: (a) $\mu = 1600$, (b) $\mu = 1150$, and (c) $\mu = 700\,\text{Gb/s}$.

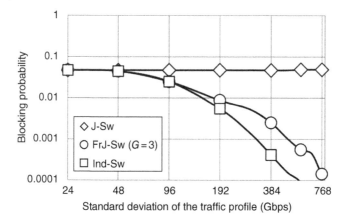

Figure 13.9 Blocking probability in terms of the standard deviation when $\mu = 1248$ Gb/s and average number connection per A/D node is 112.

which is equivalent to the diversity level of the traffic profile, Ind-Sw and FrJ-Sw show a remarkable improvement, which justifies their suitability for networks with high level of diversity in their traffic profile. In conclusion, the performance of the three SDM switching paradigms is highly dependent on the traffic profile. While J-Sw shows similar performance as Ind-Sw for large demands, it presents a reduced performance when the network is fed by a large number of small demands. J-Sw shows better performance for large demands because when the load increases the capacity of a spatial Sp-Ch becomes comparable to the demand that has to be served. Therefore, if we can form spatial Sp-Chs with lower capacity spreading across all spatial dimensions, smaller demands can fill most of the Sp-Ch container and, thus, reduce the unutilized spectral and spatial resources compared to the spatial Sp-Chs with higher capacity. One of the required technologies to realize narrower spatial Sp-Chs is the WSSs with finer spatial granularity capable of switching narrower spectral bands. In Section 13.6, we investigate the impact of spatial and spectral granularity on the performance of SDM networks based on spatial Sp-Ch switching.

13.6 Impact of Spatial and Spectral Granularity on the Performance of SDM Networks Based on Spatial Sp-Ch Switching

For SDM switching paradigms, the spatial granularity is related to the grouping of the spatial resources, which is determined by G. Evidently, G equal to 1 assumes individual fibers corresponding to the Ind-Sw case, thus offering the finest spatial granularity, while the value of G equal to S considers all spatial

Table 13.3 Values of selected ChBWs with the amount of corresponding spectral contents supported by two WSS technologies.

ChBW (GHz)	50	43.75	37.5	31.25	25
Current WSS technology (Gbaud)	32	25.75	19.5	13.25	7
Improved resolution WSS technology (Gbaud)	41	34.75	28.5	22.25	16

dimensions as one group and corresponds to the J-Sw case, which has the coarsest spatial granularity. On the other hand, the spectral granularity depends on the channel baud rate and the spectral resolution supported by WSSs. In the previous section, the performance of SDM switching paradigms are examined only with a fixed spectral channel width (ChBW)—defined as the spectrum over each of the spatial dimensions used to allocate the spatial Sp-Ch constituents—equal to 50 GHz. Since the current WSS technology allows occupying 32 GHz on a 50 GHz grid or similarly—by operating the WSS on the 6.25 GHz grid—25.75 GHz on 43.75 GHz, 19.5 GHz on 37.5 GHz, 13.25 GHz on 31.25 GHz, or 7 GHz on 25 GHz, spectral and spatial resources can be allocated to spatial Sp-Chs with narrower spectral widths (i.e., smaller ChBW), which can enhance the performance of SDM networks based on FrJ-/J-Sw cases. Furthermore, to quantify the spectrum utilization improvement for the FrJ-/J-Sw cases utilizing an enhanced resolution WSS, we consider a WSS technology with a factor 2 resolution improvement (i.e., requiring 9 GHz for guard band instead of the 18 GHz considered earlier). A summary of the ChBW values selected for this study and the corresponding amount of spectral contents that can be switched by current and future (factor 2 resolution improvement) WSS realizations is provided in Table 13.3.

In contrast to the previous section, where the results are obtained using a dynamic SDM network simulator, the results presented in this section are obtained using an offline planning tool which requires a static traffic matrix. Therefore, we use the Spanish national network of Telefonica and its traffic matrix. The network is characterized by a significantly heterogeneous traffic matrix with 84 demands between two subsets of seven transit nodes. The heterogeneous nature of the traffic matrix is the consequence of connection requests being mostly exchanged between highly populated transit areas and Madrid/Barcelona (more than 70% of the total traffic flows are from/to Madrid), where the Internet exchange points are located. This leads to the existence of ~30% of *hot links* (i.e., links with more than twice the average spectrum utilization per link in the network), ~20% of *underutilized links* (i.e., links with <1/3 the average spectrum utilization per link in the network), and ~50% of *moderately utilized links*.

We implemented an RSMSA algorithm consisting of a diverse routing computation element (a *k*-fixed alternate shortest path with maximal disjoint links for each source–destination pair) and a resource allocation module in which spatial and spectral resources are assigned to demands in the form of spatial Sp-Ch, following a first-fit strategy, starting from the shortest path and the lowest indexed spatial/spectral resource. In order to alleviate the problem posed by the *hot links*, as described earlier, we implemented a load-balancing engine including a *request-breakdown* element, which breaks up connections larger than the capacity of one spatial Sp-Ch, that is, the number of SMFs in the bundle. The load-balancing engine distributes big connection requests proportionally over *underutilized* and *moderately utilized* links when the shortest paths between end nodes become very congested. Ultimately, the selection of the best path and the most adequate spatial/spectral resources to establish a connection is carried out by a simulated annealing meta-heuristic optimization tool equipped with a multi-starting-point generator to avoid local minima, thus yielding a nearly optimal global spectrum utilization.

13.6.1 Illustrative Results

In this section, we compare the performance of spatial Sp-Ch switching paradigms under different spatial and spectral granularities in a network planning scenario for the Telefonica Spain national network. We assume bundles of 12 SMFs with 4.8 THz available spectrum per fiber across all links as a near-term SDM solution. The performance evaluation is done in terms of load and its growth. Since there are a fixed number of demands in the traffic matrix (84 demands), traffic growth is achieved by increasing the size of the demands. Therefore, small values of load correspond to smaller demands and large values of load to larger ones. For the spatial switching granularity, we consider groups (G) of 1, 2, 3, 4, 6, and 12 fibers out of 12 fibers in the bundle of SMFs, where $G = 1$ and 12 correspond to the cases of Ind-Sw and J-Sw, respectively, which offer the finest (Ind-Sw) and the coarsest (J-Sw) spatial granularities. Intermediate values represent FrJ-Sw with spatial groups formed of two to six SMFs.

Figure 13.10 shows the results considering present-day WSS technology requiring 18 GHz guard band. Figure 13.10a presents the results for the case of fixed-grid 50 GHz WDM ChBW, in which 32 Gbaud is selected for the contents of each ChBW as it is the maximum baud rate supported by the WSS resolution. The average spectrum utilization per link per fiber is used as a quantitative network performance metric. Thus, for example, the coarsest granularity in our studies corresponds to $G = 12$ and ChBW = 50 GHz, which, for the Dual Polarization Binary Phase Shift Keying (DP-BPSK) format, results in a minimum spatial–spectral bandwidth slot of 768 Gb/s. Any demand smaller than that will simply occupy the whole spatial–spectral slot, resulting in a significant amount of unutilized bandwidth. At the finest granularity of $G = 1$ and ChBW = 50 GHz, the equivalent minimum spatial–spectral bandwidth slot amounts to 64 Gb/s.

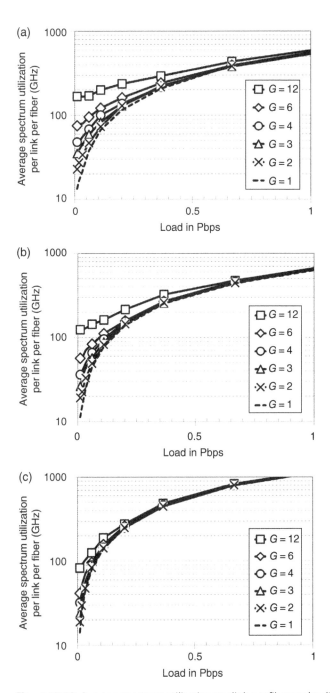

Figure 13.10 Average spectrum utilization per link per fiber under different spatial Sp-Ch switching paradigms considering current WSS technology for (a) 50 GHz, (b) 37.5 GHz, and (c) 25 GHz ChBWs at a baud rate of 32, 19.5, and 7 Gbaud, respectively, in terms of load.

In Figure 13.10a, Ind-Sw shows the best performance for all loads, that is, lowest utilization for the given traffic load, since it offers the finest granularity. We use it as the benchmark to estimate the unutilized bandwidth due to grouping of spatial dimensions. The performance of the rest of spatial Sp-Ch switching paradigms is seen to converge to that of Ind-Sw as load increases. This is due to the fact that when the load increases the spectral–spatial slot becomes comparable to the demand that has to be served and therefore the amount of unutilized bandwidth due to SG-Sw reduces. Additionally, we observe that, independently of the load, but more noticeably for smaller loads, the curves for the SG-Sw cases with lower values of G (i.e., finer spatial granularity) are closer to the Ind-Sw curve. This is a consequence of the higher flexibility that SG-Sw with low G offers to allocate smaller demands in the space dimension. However, as we will see in the next section, finer spatial granularity results in higher switching infrastructure cost. Current WSS technology with 6.25 GHz resolution enables the switching of smaller ChBWs, which allows us to evaluate the impact of spectral switching granularity on the performance of spatial switching paradigms.

To carry out this investigation, we repeated the simulations discussed earlier for 37.5 GHz ChBW at a baud rate of 19.5 Gbaud (Figure 13.10b), and 25 GHz ChBW at 7 Gbaud (Figure 13.10c). As observed in Figure 13.10b and c, for small values of load, all curves show improved performance compared to the 50 GHz ChBW case. It is noteworthy that the performance of J-Sw (switching paradigm with the coarsest spatial granularity) converges to that of Ind-Sw for smaller load values, as the ChBW decreases, compared to the case of 50 GHz ChBW. Therefore, we can conclude that, for small load values, the utilization of WSSs with finer spectral switching granularity can compensate for the spatial granularity rigidity of SG-Sws. For larger load values, on the other hand, the performance of all switching paradigms is degraded (i.e., the average spectrum utilization increases) as ChBW is decreased. This is due to a less efficient utilization of the spectrum arising from a lower amount of occupied spectrum containing actual traffic compared to the required guard band for the WSSs (i.e., 177% ≈ 32/18 versus 38% ≈ 7/18 for ChBW equal to 50 and 25 GHz, respectively).

In order to evaluate the improvement of the SG-Sw performance resulting from the utilization of WSSs with improved resolution, we repeat the studies discussed earlier for a WSS technology with a factor 2 resolution improvement (i.e., requiring a 9 GHz guard band instead of the 18 GHz considered previously).

Figure 13.11 presents the results of a detailed performance evaluation of SG-Sw paradigms for the values of ChBWs indicated in Table 13.3 and for two WSS technologies with coarser and finer resolutions. For the sake of clarity, the results are shown only for J-Sw. Figure 13.11a shows the average spectrum utilization with the current WSS technology. For small loads (<80 Tb/s), as shown

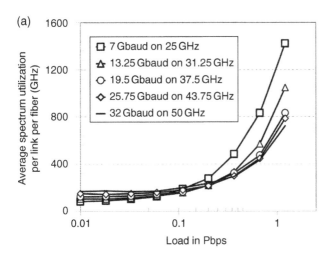

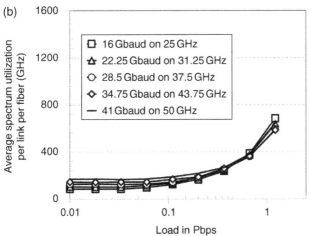

Figure 13.11 Average spectrum utilization per link per fiber for J-Sw considering (a) the current WSS technology which requires 18 GHz for guard band and (b) an improved resolution WSS which requires 9 GHz for the guard band. Five values of ChBWs are assumed for the simulations.

previously, smaller ChBW values lead to better J-Sw performance. However, as traffic increases, smaller ChBW values result in significant performance degradation. Figure 13.11b shows the results when the improved resolution WSS is used. Due to the more efficient utilization of the spectrum, smaller ChBW values lead to better performance for loads lower than 800 Tb/s. Even if the performance of J-Sw with smaller values of ChBW reduces for loads larger than 800 Tb/s, this is remarkably better than in the case of WSSs with lower resolution. Another important finding is that, for small and large loads, the best J-Sw performance is obtained for the lowest and highest values of ChBWs, respectively. Consequently, ChBWs must be adaptable to the load level in order to achieve a globally optimum spectrum utilization in an SDM network. This highlights the importance of utilizing flexgrid transmission enabled by spectrally flexible ROADMs and bandwidth variable TRxs when SG-Sw paradigms are considered. Understanding the benefits/drawbacks of spectrally flexgrid technologies require further investigation.

13.7 Conclusions

SDM has been proposed as the ultimate solution to address the capacity crunch with a cost-per-bit reduction delivered to the end-users. Cost reduction is anticipated to be realized by introducing some level of component sharing and integration. However, component integration introduces extra XT. The denser the integration of the spatial channels becomes, the more significant is the XT interactions among them which necessitates the use of complex Multiple Input Multiple Output-Digital Signal Processing (MIMO-DSP) to compensate for and thus results in high implementation cost and power consumption. Therefore, in order to find the most appropriate technology for a given use case, network-wide performance evaluations are required to reveal the benefits/drawbacks of alternative solutions.

The fundamental concept of SDM relies on placing numerous spatial channels in a single-fiber structure aiming at manufacturing an integrated fiber carrying manifold channels in a denser, lighter, and more cost-effective cable. SDM fiber types can be categorized according to whether they have (i) uncoupled spatial dimensions (e.g., BuSMFs), (ii) coupled spatial dimensions (e.g., FMFs), and (iii) uncoupled groups of coupled spatial dimensions (e.g., FM-MCFs). The choice of a fiber belonging to one or another of these three categories is a key factor in determining the required architecture and properties of the optical switches at the nodes of an SDM-based network. Three SDM switching paradigms have been identified for SDM networks, which highly correlate with three fiber classes—Ind-Sw, FrJ-Sw, and J-Sw—in order of decreasing flexibility and increasing hardware efficiency. Various levels of flexibility offered by different switching paradigms are directly related to their implementation's complexity and, hence, their cost. The complexity of an SDM node is determined by the number of required WSSs and their corresponding port counts. Ind-Sw provides the highest level of flexibility for resource allocation; however, it requires lots of WSSs of relatively low port count. On the other hand, J-Sw sacrifices flexibility for architectural simplicity. It requires just two WSSs per degree, but of very high port count. Finally, FrJ-Sw offers intermediate flexibility and requires intermediate complexity in comparison with Ind-Sw and J-Sw.

The two main technology areas limiting channel allocation and routing options in an SDM-based optical network are *fiber type* and *switching paradigms*. SDM fibers affect the resource allocation options by imposing various physical impairments to spectrally/spatially multiplexed channels (i.e., Sp-Chs) while switching paradigms do so by determining the routing properties of the multiplexed channel at the SDM nodes. In addition, spectral/spatial multiplexing techniques, forming end-to-end optical transport channels (i.e., Sp-Chs), are a key player in determining the resource allocation options in SDM networks. Spectral and spatial channels can be multiplexed to form (i) spectral Sp-Ch, (ii) spatial Sp-Ch, or (ii) spectral–spatial Sp-Ch.

Spectral Sp-Chs can improve the spectral efficiency by decreasing the guard bands between subchannels. Nonetheless, SDM networks based on spectral Sp-Ch require Ind-Sw-based switching nodes which are the most expensive switching elements. Practically speaking, spectral Sp-Ch allocation option can be used only for SDM networks based on uncoupled/weakly coupled transmission media. Spatial Sp-Ch allocation option can be utilized with any type of transmission medium but is necessary for SDM networks based on coupled transmission media. Spatial Sp-Ch allocation policy allows (i) decreasing the optical switching complexity, at the price, of course, of a potential reduction in routing flexibility; and (ii) SDM networks based on spatial Sp-Ch allocation can benefit from additional cost reduction due to the possibility of sharing network elements among different spatial dimensions (e.g., a number of Sp-Ch constituents can share lasers and DSP modules, which can lead to cost and power consumption savings of integrated TRxs in SDM networks).

In order to reveal the benefits/drawbacks of SDM networks based on alternative SDM switching paradigms for various use cases, we compared their performance considering different spatial and spectral granularities. The spatial granularity is related to the grouping of the spatial resources leading also to different switching paradigms. The spectral granularity is related to the supported spectral resolution, which can be switched by Liquid Crystal on Silicon (LCoS)-based WSSs. We showed that the performance of all switching paradigms converges as traffic increases. However, the use of switches with finer spatial granularity or even better with finer spectral resolution can lead to significant performance improvements for small values of traffic.

Additionally, we evaluated the impact of various traffic profiles and dimensioning approaches on the performance of SDM switching paradigms. We showed that the performance of different SDM switching paradigms is highly traffic dependent. More specifically, for the case of BuSMFs, Ind-Sw and FrJ-Sw perform well for networks with a high level of traffic diversity, while J-Sw is a favorable option for networks with large demands, considering also its cost benefits. However, J-Sw can perform significantly better in high diverse traffic scenarios, if spatial Sp-Chs occupy smaller spectral width which can be switched by WSSs with finer granularity.

14

Dynamic Connectivity Services in Support of Future Mobile Networks

Adrián Asensio[1], Luis Miguel Contreras[2], Marc Ruiz[1] and Luis Velasco[1]

[1] *Universitat Politècnica de Catalunya, Barcelona, Spain*
[2] *Telefonica Investigación y Desarrollo, Madrid, Spain*

Aiming at satisfying in a cost-effective manner the forecast traffic growth that future mobile networks will need to support, traditional distributed Radio Access Networks (Rans) are evolving toward centralized architectures. Specifically, the Cloud-RAN (C-RAN) architecture has shown that it can alleviate to some extent the ever increasing Total Cost of Ownership (TCO) in mobile networks. The current trend in C-RAN is to separate Remote Radio Heads (Rrhs) with radio frequency (RF) functions and Baseband Units (Bbus) gathering baseband processing. This functional split allows keeping RF modules close to the antennas while placing BBUs at centralized locations, so they can be shared among different sites and even be virtualized. However, some issues still need to be addressed in future mobile networks, especially due to the dynamicity of services and the strict constraints imposed by their interfaces. In fact, connectivity reconfiguration for X2 and S1 backhaul interfaces needs to be provided as an *all-or-nothing* request to enable mobile resources reconfiguration in a geographical area. In view of that, we propose dynamic Customer Virtual Network (CVN) reconfiguration to be supported in both metro and core network segments. Such CVN requests must include Quality of Service (QoS) constraints to ensure specific delay requirements, as well as bitrate guarantees to avoid service interruption. A mathematical formulation and a heuristic algorithm are presented for the CVN reconfiguration problem, and exhaustive simulation results study its performance in realistic scenarios.

Provisioning, Recovery, and In-operation Planning in Elastic Optical Networks,
First Edition. Luis Velasco and Marc Ruiz.
© 2017 John Wiley & Sons, Inc. Published 2017 by John Wiley & Sons, Inc.

Table of Contents and Tracks' Itineraries

14.1 Introduction

Forecast of mobile data traffic shows that it will reach up to 367 Exabytes per year by 2020 [CISCO16]. Such huge demand growth needs to be satisfied in a cost-effective manner, which imposes challenging requirements to future mobile networks. In consequence, next generation mobile networks are currently a hot topic within the research community.

From the mobile side, huge efforts have been carried out to study architectures and techniques to increase capacity offered to users, for example, Heterogeneous Networks (HetNets) [Hw13] and Multiple-input–Multiple-output (MIMO) [Ge07]. However, those solutions may result in higher interference and costs increments. To alleviate to some extent the ever increasing Total Cost of Ownership (TCO) in mobile networks (including both Capital Expenditures (CAPEX) and Operational Expenditures (OPEX)) while satisfying the expected cells' demand growth, research work has focused on centralized Radio Access Network (RAN) architectures, including the Cloud-RAN (C-RAN) concept [CMRI]. In centralized RAN architectures, baseband processing is gathered in Baseband Units (BBUs) and placed in centralized locations (e.g., central offices (COs)) separated from Remote Radio Heads (RRHs) with radio frequency (RF) functions, so they can be shared among different sites and even be virtualized to run in the form of virtual machines. A fronthaul network connects RRHs and BBUs and entails stringent requirements to support protocols such as the Common Public Radio Interface (CPRI) [CPRI]. To connect BBUs to core COs (hosting the Mobility Management Entity (MME) or the Serving Gateway (S-GW)), connections also need to be established over the so-called backhaul network.

Benefits from shared BBU pools have been studied lately. To support this, authors in [Ve15] proposed the telecom cloud architecture focused on orchestrating cloud with heterogeneous access and core networks to maximize the performance in terms of throughput and latency while minimizing TCO. In addition, noticeable OPEX reduction can be obtained by dynamically reconfiguring the C-RAN mobile network to adapt it to the actual load [As16.1].

Several works can be found in the literature where optical networks are proposed to support C-RAN architectures. In [Po13], the authors proposed using the Wavelength Division Multiplexing (WDM) technology in the access/aggregation network. Focusing on the same network segment, the authors in [Cp14] proposed an energy-efficient WDM aggregation network and formally defined the BBU placement optimization problem as a Mixed Integer Linear Programming (MILP) model, aiming at optimizing the aggregation network in terms of power consumption. The authors in [Mu16] recently proposed an MILP model for optimal BBU hotel placement over WDM networks in centralized RAN.

Nonetheless, some issues need to be addressed to support the expected dynamicity in future mobile networks, specifically related to the wide range of services that those networks are expected to support. As a result, communication interfaces required in future RAN impose very restrictive constraints, mainly in terms of capacity and delay, which transport networks need to satisfy. In fact, service-specific parameters in Service Level Agreements (SLAs) become crucial to guarantee both the required Quality of Service (QoS) and a minimum bitrate guaranteed aiming at avoiding service interruption.

To provide service-specific network services, network virtualization techniques on Multiprotocol Label Switching (MPLS) or Optical Transport Network (OTN) multilayer networks can be considered. Network virtualization allows network operators to optimize their infrastructure and resources utilization while offering Customer Virtual Networks (CVNs) to, among others, mobile operators providing C-RAN services. The latter can manage their mobile infrastructure and resources using their own Software Defined Network (SDN) controller and request reconfiguring their virtual network to add or release capacity on demand subject to some constraints, such as delay and bitrate guarantees. Note that the network operator controls the network infrastructure, is aware of resource availability, and is able to collect performance monitoring data, such as effective throughput and delay, and correlate them into QoS indicators.

To facilitate network resources virtualization, the Internet Engineering Task Force (IETF) is working on the Abstraction and Control of Transport Networks (ACTN) framework [ACTN]. Moreover, ACTN is supported by the standardized Application-Based Network Operations (ABNO) architecture [ABNO] (see Chapter 4), which can accept on-demand connection requests via an ABNO controller. In addition, ABNO can be complemented with an Application Service Orchestrator (ASO) on top implementing a northbound interface aiming at facilitating applications' requests using their own semantic (see Chapter 6).

Several works can be found in the literature studying related topics (see, e.g., [Tz13], [Yo14]). Focused on scenarios requiring both virtual topology reconfiguration and additional service-related requirements, authors in [Tz13] studied reconfiguration in scenarios requiring service resilience and security based on periodic preplanning, whilst authors in [Yo14] proposed to control virtual networks that adapt to traffic changes.

Since the requirements for C-RAN fronthaul and backhaul networks are different, differentiated approaches need to be considered. An important requirement to support C-RAN backhauling is that connectivity should be survivable by offering bitrate guarantees in case a failure affects the underlying network. Such bitrate guarantees can be based on recovery techniques such as $1:1$ (or $1+1$) protection, where two Shared Risk Link Group (SRLG)-disjoint paths are established [Gr04]. Another option is diversity, where two SRLG-disjoint paths are set up with a combined capacity to satisfy that requested and being the minimum path capacity equal to that to be guaranteed. It is worth noting that diversity, being supported by two paths, entails less capacity to be reserved compared to protection. Authors in [Gu07] proposed using SRLG for protection in WDM networks with differentiated reliable requirements.

An important feature of C-RAN is the ability to reconfigure the resources in a geographical area to adapt the mobile network to the current load. In that regard, when a reconfiguration is decided, connectivity reconfiguration needs to be provided as an *all-or-nothing* request. In this chapter, we focus on the backhaul network and propose dynamic CVN provisioning with quality constraints and bitrate guarantees to support such all-or-nothing connectivity requests.

14.2 C-RAN Requirements and CVN Support

In this section, we first present the C-RAN scenario considered in this chapter and summarize its requirements for the backhaul links. Finally, our proposal to provide CVN with QoS constraints and bitrate guarantees is described.

14.2.1 C-RAN Architecture Model

Recently, the functional splitting of RF and baseband processing into RRHs and BBUs, respectively, allows keeping the RF modules close to the antennas while placing BBUs at certain distant locations. RRH and BBU are usually connected using an optical fronthaul network to support CPRI, which requires huge capacity and strict delay constraints. At the backhaul, two interfaces are needed [TS23.401]:

- X2 interfaces are required between base stations of neighboring cells to assist handover and to support rapid coordination schemes;
- S1 interfaces between base stations and mobile core entities, such as MME or S-GW, are also required.

Finally, C-RAN assumes that virtualized BBU pools are hosted in different central locations and can be flexibly configured to serve RRHs. This reconfiguration requires thus from connectivity services provided by the underlying network supporting C-RAN backhauling.

Without loss of generality, in this chapter we consider a reference scenario for future mobile networks based on C-RAN, where a set of geographically distributed RRHs cover certain regions and virtualized BBU pools are hosted in COs in the metro segment; mobile core entities, MME and S-GW, are placed in the core network and their functions virtualized in a core CO. Aiming at satisfying the required capacity in different areas, RRHs corresponding to Macro Base Stations (MBSs) and RRHs corresponding to small cells that can be activated or deactivated for capacity management according to the traffic demand fluctuation at different hours are considered. Let us assume that activation/deactivation of those RRHs can be done through the corresponding entity in charge of the control and management of the C-RAN. In addition, we assume that a given RRH can be served from different virtualized BBU pools along the time.

Moreover, handover and tight coordination schemes among active and neighboring RRHs need to be considered; thus, X2 interfaces between virtualized BBUs in BBU pools are required. It is worth noting that, due to strict delay limitations required in X2 interfaces, not all BBUs in BBU pools hosted in distant COs might be accessible among them. Finally, connections to support the S1 interface toward the core CO (hosting MME and S-GW entities) also need to be established over the backhaul network.

14.2.2 Backhaul Requirements in C-RAN

S1 and X2 interfaces present the following requirements and limitations that backhaul networks need to satisfy:

1) S1 and X2 required capacity depends not only on the site configuration (technology adopted, MIMO, and spectrum width, among others) but also on the load density, which varies with time. Moreover, although configurations

such as 8×8 MIMO and 100 MHz spectrum width require about 3 Gb/s [COMBO] according to the S1 peak rate, higher capacities will be required in future mobile networks, targeting at cell throughputs between 10 and 100 Gb/s.

2) From the C-RAN operator side, the required network topology in the backhaul can also vary with time: during periods of low dense demand, when few RRHs are active to serve a given load, certain BBU pools may remain unused, and no connections are required to support neither X2 nor S1 interfaces. On the contrary, during peak hours, more BBUs need to be allocated, and connectivity increased to serve the load. As a result, when a reconfiguration is decided in a geographical area, connectivity reconfiguration for all the affected X2 and S1 backhaul interfaces needs to be provided as an all-or-nothing request. Figure 14.1 shows an example of the virtual topology required from a

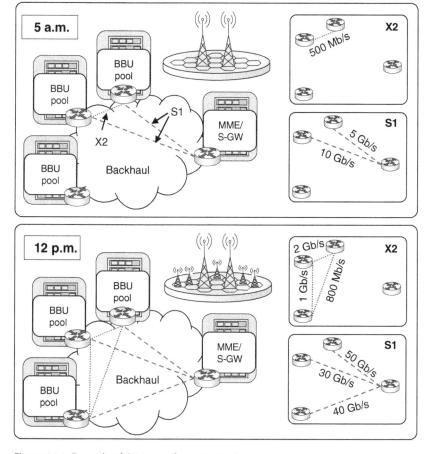

Figure 14.1 Example of CVN reconfiguration in C-RAN scenarios.

C-RAN operator to satisfy backhaul connectivity requirements for X2 and S1 at two different hours of the day (5 a.m. and 12 p.m.).

3) Delay constraints need to be considered in both interfaces, although they are especially restrictive in the X2 interface. Currently, delays in the order of tens of milliseconds are allowed in S1 interfaces, whereas tight coordination schemes between base stations lead to maximum delays allowed in the order of hundreds of microseconds for the X2 interface (round-trip delay time under 1 ms).

4) Failures in links supporting connections in the backhaul would impact several sites simultaneously and interrupt service for a wide range of users. To avoid such service interruption, some kind of bitrate guarantees is needed for both X2 and S1 interfaces.

Generally speaking, C-RAN and the transport network are independently operated. C-RAN operators can compute their RAN architecture reconfiguration dynamically (e.g., according to the load density), which in turn translates into a reconfiguration on the backhaul connections, in terms of both connectivity and capacity required. However, C-RAN operators are not aware of transport network resources. Therefore, once the C-RAN operator has computed its solution to reconfigure its RAN, it requests the new topology (and capacities) required in the backhaul to implement the computed solution effectively.

Finally, although both X2 and S1 interfaces belong to the backhaul network (as seen from the C-RAN operator), from the point of view of the transport operator, X2 connectivity will be supported on metro/regional networks, whereas S1 will be supported on core national networks.

14.2.3 CVN Reconfiguration

In view that connectivity reconfiguration needs to be provided as an all-or-nothing request, we propose dynamic CVN reconfiguration based on the ACTN framework to be supported in metro and core networks. In brief, three key entities can be differentiated in the ACTN framework: customers, service providers, and network providers. Customers can request on-demand connectivity between their endpoints (EPs) to reconfigure their virtual topology. In the case of C-RAN, EPs are COs and mobile core entities. We assume that CVNs are provided on top of a multilayer MPLS network supported by an Elastic Optical Networks (EONs).

Figure 14.2 represents such a three-layered network. The following layers can be identified from top to bottom:

- the customer layer with CVNs connecting customer's EPs. Every CVN's link is supported by one or more MPLS paths;

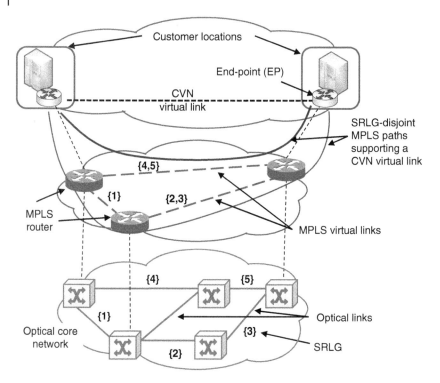

Figure 14.2 MPLS-over-optical layered network to support CVNs.

- the network operator's MPLS network layer consisting of a number of MPLS routers connected through virtual links (*vlink*) supported by optical connections;
- the network operator's EON consisting of a number of optical nodes and optical links.

Although this architecture can be deployed for single or multiple network operators providing connectivity services to a set of service operators, for example, C-RAN operators, in this chapter we restrict ourselves to the single network operator scenario for the sake of clarity. In such case, the network operator owns and controls the transport network infrastructure, thus being aware of resource availability and able to collect performance monitoring data, such as effective throughput and delay, and correlate them with QoS indicators. C-RAN operators own and control their mobile infrastructure and require virtual network services to connect EPs in geographically disperse locations; a C-RAN service orchestrator requiring on-demand connectivity between separated BBU pools (X2) or between BBU pools and the mobile core entities (S1) is assumed.

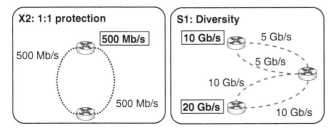

Figure 14.3 Proposed bitrate guarantees for X2 and S1 interfaces.

According to the described architecture and assumptions, C-RAN operators can request CVN reconfiguration to the network operator, including service-specific parameters for each CVN's link such as the maximum end-to-end (e2e) delay between EPs or bitrate guarantees, in addition to the required capacity.

Regarding bitrate guarantees, the requested bitrate for X2 interfaces needs to be guaranteed in case of failures, so protection is strictly required. In contrast, a reduction in the available bitrate in S1 interfaces would entail a reduction in users' data rate that can be acceptable for the C-RAN operator. Therefore, diversity can be considered for the S1 interface, where a proportion (e.g., 50%) of the requested bitrate is guaranteed in case of a failure. This results in a more cost-effective solution compared to protection, also considering the difference in the required bitrate and the network segment that will support each of the interfaces. Figure 14.3 illustrates the selected bitrate guarantees options. Note that both bitrate guarantees options are supported by the assumed MPLS-over-optical network. In this regard, note that the CVN link shown in Figure 14.2 is supported by two SRLG-disjoint MPLS paths, where one of them uses SRLGs {4, 5} and the other uses SRLGs {1, 2, 3}.

To control such a layered network, the architecture in Figure 14.4 is assumed, with C-RAN SDN controllers on top controlling each CVN and requesting their reconfiguration to the ASO. The ASO module maintains service-related databases (DBs); the Customer Virtual Networks-Data Bade (CVN-DB) describes the current state of each CVN in terms of EPs and CVN links, whereas the service DB includes, among others, SLA-related data. In contrast, ABNO maintains network-related DBs, that is, the Traffic Engineering Database (TED) and Label-Switched Path-Database (LSP-DB), and it is in charge of the control of the transport network.

The next section formally states the CVN reconfiguration with QoS constraints and the bitrate guarantees (CUVINET) problem and presents an MILP formulation to model it. In view of its complexity and the short time in which it must be solved, a heuristic algorithm is eventually proposed.

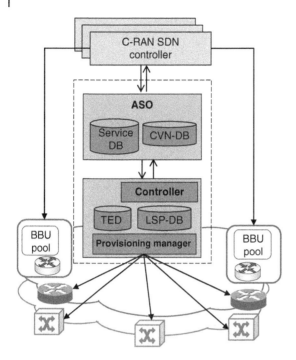

Figure 14.4 Control and management architecture.

14.3   The CUVINET Problem

14.3.1   Problem Statement

The CUVINET problem can be formally stated as follows:

Given:

- an optical network represented by the graph $G^O(N, L)$, N being the set of optical nodes and L the set optical of links;
- an MPLS network represented by a graph $G^V(V, E)$, V being the set of MPLS nodes and E the set of vlinks, where each vlink can be supported by several lightpaths.
- a set of customers C; each $c \in C$ manages its own CVN service, the topology of which is represented by a fully meshed graph $G_c(V_c, E_c)$, V_c being the set of EPs and E_c the set of CVN vlinks of c.
- a CVN service reconfiguration request coming from customer c that is represented by the tuple $\{B^r, Q^r, W^r\}$; where B^r is the capacity matrix of CVN vlinks between EPs, Q^r is the QoS matrix, and W^r is the matrix with CVN links capacity to be guaranteed. We assume that the number of

SRLG-disjoint MPLS paths is restricted to two and thus the maximum guaranteed bitrate cannot exceed 50% of the total capacity in the case of diversity ($w^r \leq 0.5 \times b^r$).

Output: the set of MPLS paths over G^V and lightpaths over G^O to be established to serve the CVN reconfiguration request.

Objective: minimize the cost of the used resources in both the optical and the MPLS layer.

14.3.2 MILP Formulation

As previously stated, we assume that the optical network is based on EON. In that regard, we solve the Routing and Spectrum Allocation (RSA) following a link-path formulation (see Chapter 3). Aiming at allowing distinct bitrate guarantees options, a parameter α is used to define whether diversity or protection is selected to support every CVN link.

The following sets and parameters have been defined.

Topology and Spectrum

N	Set of optical nodes, index n.
L	Set of optical links, index l.
P	Set of optical routes, index p.
S	Set of frequency slices, index s.
K	Set of frequency slots, index k.
A	Set of SRLG identifiers, index a.
β	Capacity of lightpaths in Gb/s.
δ_{pl}	Equal to 1 if route p uses optical link l; 0 otherwise.
δ_{ks}	Equal to 1 if slot k uses slice s; 0 otherwise.
δ_{ls}	Equal to 1 if slice s in link l is available; 0 otherwise.
δ_{la}	Equal to 1 if optical link l is supported by SRLG a; 0 otherwise.
q_p	Delay introduced by route p.

MPLS Network

V	Set of MPLS routers, index v.
E^+	Augmented set of vlinks to connect every pair of MPLS routers, index e.
$E^+(v)$	Subset of E^+ with the vlinks incident to node v.
M	Set of MPLS paths, index m.
$P(e)$	Subset of P with optical routes for MPLS vlink e.
δ_{me}	Equal to 1 if MPLS path m uses MPLS vlink e; 0 otherwise.
δ_{ea}	Equal to 1 if MPLS vlink e is supported by SRLG a; 0 otherwise.
φ_e	Available capacity in vlink e; in Gb/s.
t_v	Number of available transponders in MPLS node v.

q_V	Delay introduced by MPLS routers.
q_e	Delay introduced by MPLS vlink e; 0 for those vlinks not supported by any lightpath.

CVN Reconfiguration

V_c	Set of CVN c nodes.
$E_c = D$	Set of CVN vlinks, index d.
$M(d)$	Subset of M with the MPLS paths for CVN link d.
b_d	Requested capacity for CVN link d.
q_d	Requested delay for CVN link d.
w_d	Requested capacity to be guaranteed for CVN link d.

Finally, the additional parameters have been defined:

κ_e	Cost per Mb/s from using MPLS vlink e.
κ_l	Cost per Mb/s from using optical link l supporting an MPLS vlink.
α	1 if diversity is considered; 0 for protection.

The decision variables are:

h_e	Continuous, with the maximum delay of vlink e.
h_m	Continuous, with the maximum delay of MPLS path m.
u_{de}	Continuous, with the amount of flow in MPLS vlink e from CVN link d.
x_{ma}	Binary. Equal to 1 if MPLS path m is supported by SRLG a; 0 otherwise.
y_{epk}	Binary. Equal to 1 if MPLS vlink e uses optical route p and slot k; 0 otherwise.
z_{dm}	Binary. Equal to 1 if CVN link d is supported by MPLS path m as primary path; 0 otherwise.
z_{dma}	Binary. Equal to 1 if CVN link d is supported by primary path MPLS path m and SRLG a; 0 otherwise.
r_{dm}	Binary. 1 if MPLS path m is selected as the SRLG-disjoint MPLS path to guarantee w_d for CVN link d; 0 otherwise.
r_{dma}	Binary. Equal to 1 if CVN link d is supported by SRLG-disjoint MPLS path m and SRLG a; 0 otherwise.

The MILP model for the CUVINET problem is as follows:

$$\text{Minimize} \quad \sum_{e \in E^+} \sum_{d \in D} \kappa_e \cdot u_{de} + \sum_{l \in L} \sum_{e \in E^+} \sum_{p \in P(e)} \sum_{k \in K} \kappa_l \cdot \delta_{pl} \cdot y_{epk} \qquad (14.1)$$

subject to:

$$\sum_{m \in M(d)} z_{dm} = 1 \quad \forall d \in D \tag{14.2}$$

$$\sum_{m \in M(d)} r_{dm} = 1 \quad \forall d \in D \tag{14.3}$$

$$u_{de} = \sum_{m \in M(d)} \delta_{me} \cdot \left[\left(b_d - \alpha \cdot w_d \right) \cdot z_{dm} + \left(b_d - \alpha \cdot \left(b_d - w_d \right) \right) \cdot r_{dm} \right] \tag{14.4}$$
$$\forall e \in E^+, d \in D$$

$$\sum_{d \in D} u_{de} \leq \varphi_e + \beta \cdot \sum_{p \in P(e)} \sum_{k \in K} y_{epk} \quad \forall e \in E^+ \tag{14.5}$$

$$\sum_{e \in E^+} \sum_{p \in P(e)} \sum_{k \in K} \delta_{ks} \cdot \delta_{pl} \cdot y_{epk} \leq \delta_{ls} \quad \forall l \in L, s \in S \tag{14.6}$$

$$\sum_{e \in E^+(v)} \sum_{p \in P(e)} \sum_{k \in K} y_{epk} \leq t_v \quad \forall v \in V \tag{14.7}$$

$$h_e \geq q_e \quad \forall e \in E^+ \tag{14.8}$$

$$h_e \geq \sum_{k \in K} q_p \cdot y_{epk} \quad \forall e \in E^+, p \in P(e) \tag{14.9}$$

$$h_m = \sum_{e \in E^+} \delta_{me} \cdot \left(q_V + h_e \right) + q_V \quad \forall m \in M \tag{14.10}$$

$$q_d + \left(1 - z_{dm} \right) \cdot bigM \geq h_m \quad \forall d \in D, m \in M(d) \tag{14.11}$$

$$q_d + \left(1 - r_{dm} \right) \cdot bigM \geq h_m \quad \forall d \in D, m \in M(d) \tag{14.12}$$

$$\sum_{e \in E^+} \delta_{me} \cdot \left(\delta_{ea} + \sum_{p \in P(e)} \sum_{l \in L} \sum_{k \in K} \delta_{pl} \cdot \delta_{la} \cdot y_{epk} \right) \leq bigM \cdot x_{ma} \quad \forall m \in M, a \in A \tag{14.13}$$

$$z_{dma} + 1 \geq z_{dm} + x_{ma} \quad \forall d \in D, m \in M(d), a \in A \tag{14.14}$$

$$z_{dma} \leq z_{dm} \quad \forall d \in D, m \in M(d), a \in A \tag{14.15}$$

$$z_{dma} \leq x_{ma} \quad \forall d \in D, m \in M(d), a \in A \tag{14.16}$$

$$r_{dma} + 1 \geq r_{dm} + x_{ma} \quad \forall d \in D, m \in M(d), a \in A \tag{14.17}$$

$$r_{dma} \leq r_{dm} \quad \forall d \in D, m \in M(d), a \in A \tag{14.18}$$

$$r_{dma} \leq x_{ma} \quad \forall d \in D, m \in M(d), a \in A \tag{14.19}$$

$$\sum_{m \in M(d)} z_{dma} + \sum_{m \in M(d)} r_{dma} \leq 1 \quad \forall d \in D, a \in A \tag{14.20}$$

The objective function (14.1) minimizes the cost of using resources in MPLS vlinks and that of the optical resources to support new MPLS vlinks.

Constraints (14.2)–(14.7) deal with MPLS paths and lightpaths serving demands. Constraints (14.2) and (14.3) ensure that exactly two MPLS paths support each CVN link d disregarding whether diversity or protection is selected. Constraint (14.4) computes the amount of bitrate from a given demand that is conveyed through each vlink, which depends on the selected bitrate guarantees option. When diversity is considered ($\alpha = 1$), the disjoint path must meet the guaranteed bitrate, whereas in the case of protection ($\alpha = 0$) the requested capacity must be ensured in both primary and disjoint paths. Constraint (14.5) ensures that every vlink has enough aggregated capacity and forces a new lightpath to be set up if necessary. Constraint (14.6) guarantees that every frequency slice is used by one new lightpath at the most, provided it was unused. Constraint (14.7) limits the number of new lightpaths that are set up to the number of available transponders.

Constraints (14.8)–(14.12) ensure the requested QoS. Constraints (14.8) and (14.9) compute the maximum delay of every vlink considering its current delay and the delay of every new lightpath supporting such link, respectively. Constraint (14.10) computes the delay of every MPLS path as the sum of the delay in every vlink and the MPLS router it traverses. Constraints (14.11) and (14.12) guarantee that the delay of both primary and SRLG-disjoint paths does not exceed the requested delay for the CVN link they support; otherwise, the problem becomes unfeasible.

Finally, constraints (14.13)–(14.20) deal with SRLGs. Constraint (14.13) computes SRLG support for every MPLS path. Constraints (14.14)–(14.16) guarantee that a CVN vlink using a primary MPLS path is assigned to the SRLG supporting that path. Specifically, these constraints compute z_{dma} as the product between z_{dm} and x_{ma} variables by means of linear equations. Similarly, constraints (14.17)–(14.19) compute that product for the SRLG-disjoint path of every CVN vlink. Constraint (14.20) ensures that the primary path and the SRLG-disjoint path are not supported by any common SRLG for every demand.

The CUVINET problem is *NP* (nondeterministically polynomial)-*hard* since it is based on the RSA problem that has been proved to be *NP-hard*. Regarding problem size, the number of variables is $O(|D| \cdot |M| \cdot |A| + |D| \cdot |V|^2 + |V|^2 \cdot |P| \cdot |K|)$ and the number of constraints is $O(|D| \cdot |M| \cdot |A| + |V|^2 \cdot (|D| + |P|) + |L| \cdot |S|)$. Considering a realistic scenario on a network like those in the next section, the number of variables is in the order of 10^6 and the number of constraints in the order of 10^5, which makes the MILP model discussed unsolvable within the times required for serving on-demand CVN reconfiguration requests, even using state-of-the-art computer hardware and the latest commercially available solvers, as will be shown in Section 14.4. As a result, a heuristic algorithm is needed aiming at providing near-optimal solutions for realistic scenarios.

14.3.3 Heuristic Algorithm

To solve the CUVINET problem, we propose the randomized algorithm in Table 14.1. The algorithm runs a number of iterations (line 2 in Table 14.1), where at every iteration the set of requested CVN links (demands) is randomly sorted and served sequentially in the resulting order (lines 3–14). At the end of each iteration, the obtained solution is compared against the best solution obtained so far and, in case the latter is improved, it is updated (lines 15–16). The best solution is eventually returned (line 17).

The algorithm first updates those CVN links with unchanged or decreased requirements to release resources that can be reused afterward (lines 4–6). The rest of the CVN links are deallocated from G_c and added to the set D (lines 7–9). The set D is randomly sorted (line 10) and every CVN link is then set up (lines 11–14) using the *set upCVNLink* algorithm described in Table 14.2. The

Table 14.1 Algorithm for the CUVINET problem.

INPUT: $G^o(N, L)$, $G^V(V, E)$, $G_c(V_c, E_c)$, B^r, Q^r, W^r, α, maxIter
OUTPUT: Ω

```
 1: BestSol ← Ø
 2: for 1..maxIter do
 3:   D ← Ø, Sol ← Ø
 4:   for each e ∈ E_c do
 5:     if B^r(e) ≤ B(e) and Q^r(e) ≤ Q(e) and W^r(e) ≤ W(e) then
 6:       update (e, B^r(e), Q^r(e), W^r(e))
 7:     else
 8:       dealloc (e, G_c)
 9:       D ← D ∪ {e}
10:   shuffle (D)
11:   for each d ∈ D do
12:     Ω ← set upCVNLink (d, B^r(d), Q^r(d), W^r(d), G^o, G^V, G_c, α)
13:     if Ω = Ø then break
14:     Sol ← Sol ∪ { Ω }
15:   if BestSol = Ø OR Sol.fitness > BestSol.fitness then
16:     BestSol ← Sol
17: return BestSol
```

Table 14.2 *Set up CVNLink* algorithm.

INPUT: d, b, q, w, G^o, G^V, G_c, α
OUTPUT: Ω

```
 1: Ω ← ∅
 2: if α = 0 then bₚ = b; bₛ = b
 3: else bₚ = b - w; bₛ = w
 4: R ← findPath(d, bₚ, q, Gⱽ)
 5: if R = ∅ then
 6:     Ω← Ω U set upMPLSLinks (d, bₚ, q, Gᵒ, Gⱽ)
 7:     R ← findPath(d, bₚ, q, Gⱽ)
 8:     if R = ∅ then return ∅
 9: Ω ← Ω U allocate(d, R, bₚ, Gᵒ, Gⱽ)
10: S ← findDisjointPath (d, q, R, bₛ, Gⱽ)
11: if S = ∅ then
12:     Ω ← Ω U set upDisjointMPLSLinks (d, bₛ, q, R, Gᵒ, Gⱽ)
13:     S ← findDisjointPath (d, q, R, bₛ, Gⱽ)
14:     if S = ∅ then
15:        deallocate(Ω, Gᵒ, Gⱽ)
16:        return ∅
17: Ω ← Ω U allocate(d, S, bₛ, Gᵒ, Gⱽ)
18: return Ω
```

---

set upCVNLink algorithm returns the set Ω with the MPLS paths and light-paths to be established. Note that, if one of the CVN links cannot be updated, then the complete CVN reconfiguration request is blocked (line 13).

The *set upCVNLink* algorithm (Table 14.2) starts computing the capacity to be assigned to each MPLS path supporting the CVN link depending on the considered bitrate guarantees option. In the case that protection is selected, b is the bitrate to be carried by both the primary and the disjoint path; otherwise, the required capacity for CVN links is split (lines 2–3). Then, function *findPath* finds an MPLS path R with capacity b_p in G^V for the primary path to guarantee the requested QoS q (line 4). In the case that no path is found, the algorithm tries to increase capacity in G^V, so as to serve d by adding new MPLS vlinks, and another try is performed. If no path is found, CVN link d cannot be served (lines 5–8); otherwise, the primary path is allocated in G^V (line 9). Next, the algorithm calls function *findDisjointPath* to find an MPLS path S guaranteeing SRLG disjointness with R, as well as meeting capacity and delay requirements (line 10). In case no disjoint path is found in G^V, a similar procedure as the one described for the primary path is followed to increase its capacity G^V (lines 11–16). The set Ω with the MPLS paths and lightpaths to be established is eventually returned (line 18).

14.4 Illustrative Numerical Results

In this section, we first evaluate the proposed heuristic algorithm and then we use it to study CVNs from both the network operator's and the C-RAN operator's viewpoint. From the network operator's viewpoint, we evaluate two alternative MPLS network approaches to support CVN services for different kinds of customers, which include but are not limited to C-RAN operators. From the C-RAN operator's viewpoint, we evaluate the performance of CVNs to support C-RAN backhauling.

14.4.1 Network Scenario

We use the realistic network topologies depicted in Figure 14.5, representing Telefonica's Spanish network topology (TEL) and Orange's French network topology (FR).

The management architecture including customer (e.g., C-RAN) controllers, ASO (running the heuristic algorithm described in Section 14.3.3), and ABNO were developed in C++ and integrated into an ad hoc event-driven simulator based on OMNET++. A set of customers (e.g., C-RAN operators) require CVN reconfiguration and send CVN reconfiguration requests to ASO according to an exponential distribution with mean 1 h following the traffic profiles in [CMRI]. A CVN reconfiguration request is accepted provided that all the requested CVN link's capacity, QoS constraints, and bitrate to be guaranteed is served; otherwise, the request is blocked.

Regarding the optical layer, fiber links with spectrum width equal to 2 THz are considered. An MPLS router is colocated with every optical node and

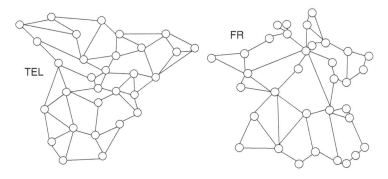

Figure 14.5 Considered network topologies.

connected to it through 100 Gb/s bandwidth-variable transponders (BVTs) using 37.5 GHz. Note that MPLS vlinks are supported by lightpaths on the optical layer, and CUVINET creates and releases them to serve the actual load.

14.4.2 Heuristic Algorithm Validation

In order to validate the proposed heuristic algorithm, we compared its performance against solving the MILP formulation in terms of quality of obtained solutions and solving time. Several instances of an X2 connectivity service reconfiguration among a number of COs distributed over the TEL topology were generated and solved with both MILP and heuristic methods using an Intel(R) Core(TM) i7-4790K CPU @4.00GHz machine with a 4GB RAM running Linux.

The results showed that solving the MILP formulation took up to 40 min for instances of the size considered in the following studies (i.e., five COs). As for the proposed heuristic algorithm, it reached the optimum in less than 500 ms, which validates the proposed heuristic algorithm as a scalable method for solving the CUVINET problem. In consequence, we use the proposed heuristic for the studies in this section.

14.4.3 Approaches to Support CVNs

The performance from the network operator's viewpoint is studied considering two incremental approaches: (i) a *static* MPLS network approach in which the MPLS virtual network is preplanned beforehand, and (ii) a *dynamic* MPLS approach allowing MPLS network reconfiguration.

Regarding customers, two differentiated services have been considered: one requiring regional five-EP CVN topologies during office hours and 50% bandwidth guaranteed; the other requiring nation-wide CVN topologies during off-peak periods. EPs are connected to the closest MPLS router. The maximum delay allowed between EPs is 10 ms. CVN reconfiguration requests arrive during service-defined periods, where the capacity requested between EPs is randomly chosen in the range [1–10] Gb/s. The offered load is related to the number of customers being served.

For evaluation purposes, an MPLS virtual network topology was preplanned targeted at providing the same performance as that of the MPLS dynamic approach at 1% of service blocking probability (normalized load 0.85). Each point in the results is the average of 10 runs with more than 10,000 CVN reconfiguration requests entailing capacity increment on one CVN link at least.

Figure 14.6a focuses on QoS for the TEL network topology. As shown, the MPLS preplanning approach provides the best QoS for both average and maximum delay. As for the dynamic approach, it provides QoS that increases with the load but always under the specified maximum; on average delay, however,

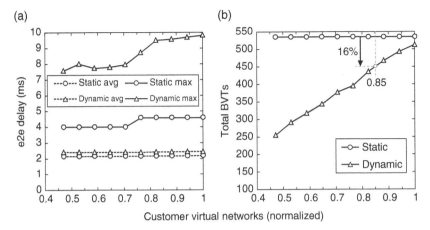

Figure 14.6 End-to-end delay (a) and total number of BVTs (b) versus number of CVNs (normalized).

it is comparable to that of the static approach. Next, Figure 14.6b concentrates on the total number of BVTs to equip on the TEL network topology. As observed, the required number of BVTs increases with the load in the dynamic approach, in contrast to the static one. CAPEX savings as high as 16% for 1% service blocking probability are shown. Similar results were obtained for the FR network topology.

This clearly shows that dynamic MPLS-over-optical networks provide noticeable benefits to the network operator to support CVNs.

14.4.4 Performance Evaluation

Let us now evaluate the performance of CVNs to support C-RAN backhauling. Because of their different capacities, QoS, and bitrate guarantees requirements, we assume that specific CVNs are requested for X2 and S1 connectivity. The performance of CVNs supporting the X2 interface is studied in terms of delay and the affected connections in the event of a failure in an optical link, comparing the case that no bitrate guarantees are requested (referred to as *no guarantees*) against with bitrate guarantees considering 1:1 protection (referred to as *protection*). Regarding CVNs supporting the S1 interface, performance in terms of service blocking probability and affected connections in the event of a failure is compared for no bitrate guarantees and 50% bitrate guarantees considering diversity (referred to as *diversity*).

The network topologies in Figure 14.5 are used here for both metro and core segments. However, to capture real distances (and delays), the network diameter is scaled down to about 50 km when those topologies are used for the metro segment.

To evaluate the performance in failure scenarios, the network was loaded, and the traffic generation was stopped at noon peak hour (12 p.m.). Then, we simulated that a failure occurs on each optical link in the route of MPLS paths supporting CVN vlinks and computed the CVN vlinks affected.

Finally, each point in the results is the average of 10 runs with more than 10,000 CVN reconfiguration requests entailing capacity increment on one CVN link at least.

14.4.4.1 C-RAN CVN Provisioning to Support X2

To represent X2 CVN scenarios, we consider a set of five COs geographically distributed, hosting virtualized BBU pools and connected to the network through packet nodes. The number of X2 CVNs ranges from 10 to 90 for TEL and from 55 to 75 for FR topologies, resulting the maximum values of CVNs in 1% of service blocking probability when protection is considered. The capacity of each link is in the range [200 Mb/s–5 Gb/s] according to a uniform distribution. X2 CVNs are full mesh topologies, where the probability that an X2 connection between two specific COs is required depends on the hour of the day, being higher during daytime hours and lower during night hours, thus representing the flexible assignment between RRHs and BBUs along the day to a different number of COs. Regarding delay constraints, we limited the maximum e2e delay to 0.5 ms for X2 interfaces; note, however, that per-CVN link values can be specified in the CVN reconfiguration request, for example, to satisfy more restrictive delay constraints.

Figure 14.7 illustrates the results obtained for the two topologies, each with its own COs placement. Figure 14.7a and b show the average and maximum delay values for TEL and FR, respectively. Delay limit is never exceeded, the maximum values being close to 0.5 ms. Interestingly, if we focus on the average values, the protection approach results in slightly lower times than when no bitrate guarantees are considered. The reasoning behind this is that a large number of MPLS vlinks are established to support SRLG-disjoint MPLS paths aiming at guaranteeing whole capacity. In contrast, fewer MPLS vlinks need to be established under the no guarantees approach, thus reducing the number of routes.

Although using protection resulted in slightly lower delays, the cost in terms of the number of transponders needed increases, on average, by 81% for the TEL and 110% for the FR topologies, respectively, compared to the no guarantees approach. Regarding failures, it is worth noting that when a failure affects the working path of a connection supported by 1:1 protection, the delay of the protected path is considered.

Table 14.3 summarizes representative values for the length of MPLS paths. Note that when protection is applied, only values for the working path are considered. As can be seen in Figure 14.7c and d, the affected connections when protection is required (about 10% for TEL and 12% for FR) is lower

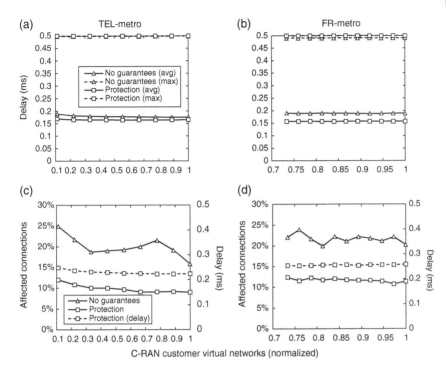

Figure 14.7 Delay and affected connections versus the number of X2 CVNs. Average and maximum delay values for (a) TEL and (b) FR networks. Affected connections and delay for (c) TEL and (d) FR networks.

Table 14.3 Average length (km) of MPLS paths supporting X2.

	TEL-metro (# CVN normalized)				
	0.2	**0.4**	**0.6**	**0.8**	**1.0**
No guarantees	36.2	35.3	35.2	35.0	34.9
Working path	33.0	32.8	32.7	32.7	32.9
Protected path	47.7	45.9	45.1	44.7	44.8
	FR-metro (# CVN normalized)				
	0.8	**0.85**	**0.90**	**0.95**	**1.0**
No guarantees	38.0	37.9	37.8	37.8	37.9
Working path	31.4	31.4	31.4	31.4	31.4
Protected path	51.2	51.3	50.8	51.3	51.7

than when no guarantees are requested (20% for TEL and 22% for FR). When protection is used, larger routes will be considered when a failure affects a connection, according to values shown in Table 14.3, thus increasing average delay (without exceeding the 0.5 ms limit) as shown in Figure 14.7c and d. In contrast, when no guarantees are requested, service supported by the affected connections will be disrupted, thus impacting several cells simultaneously.

14.4.4.2 C-RAN CVN Provisioning to Support S1

Compared to X2, S1 connections require higher capacities but less restrictive delay constraints. To reproduce S1 scenarios as described in Section 14.2, we consider regions, each containing one or more sets of COs (each set of COs with three to five COs) to create star topologies. A set of C-RAN operators manage their own CVNs and require connectivity to support S1 interfaces between COs and a centralized CO hosting mobile core entities. To represent the asymmetric traffic related to S1, the capacity required from COs to the central location is about one-third of the capacity required from the central location to COs, which varies following a uniform distribution between 5 and 50 Gb/s. Moreover, a given CO may or may not require connectivity with a certain probability, depending on the hour of the day; fewer connections are required during night hours than during working hours. In addition, we consider that the maximum allowed e2e delay is 10 ms.

Aimed at providing solid conclusions for a wide range of scenarios a lower number of C-RAN operators but CVNs requiring a large number of connections were considered for the TEL network, whereas the opposite situation was explored in FR (a large number of C-RAN operators with smaller CVNs).

Plots in Figure 14.8a and b show the service blocking probability for the TEL and FR topologies, respectively. In the evaluated scenarios, the performance of the diversity approach slightly improves that of the approach without bitrate guarantees. The rationale behind that is that under the diversity approach two disjoint paths with 50% capacity (50% of bitrate to be guaranteed has been considered) are used, in contrast to the single path used when no guarantees are requested.

When failures are considered (Figure 14.8c and d), the percentage of affected connections is higher when bitrate guarantees are required, since two MPLS paths are considered to support every single CVN vlink. Notwithstanding the reduction in the available capacity in affected connections, service will not be interrupted when diversity is implemented. This is in contrast to service disruption in the case that no bitrate guarantees are requested. Additionally, Figure 14.8c and d illustrate connections' average delay when diversity is considered.

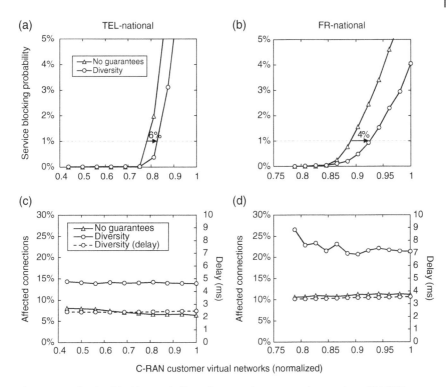

Figure 14.8 Service blocking and affected connections versus the number of S1 CVNs. Service blocking probability for (a) TEL and (b) FR networks. Affected connections and delay for (c) TEL and (d) FR networks.

14.5 Conclusions

Dynamic customer virtual topology services have been proposed to support C-RAN backhauling (X2 and S1 interfaces). Customers (i.e., C-RAN operators) request their CVNs to be reconfigured while taking into account delay constraints imposed by the interfaces and guaranteeing bitrate. To that end, an MILP formulation was presented for the CVN reconfiguration problem, named as CUVINET. In view of the complexity and the size of the problem, a heuristic algorithm was proposed providing a better trade-off between complexity and optimality.

The performance of the proposed CUVINET heuristic to adapt the underlying MPLS virtual network by creating and releasing MPLS vlinks has been demonstrated through simulation; delay was kept under the limit while the amount of BVTs to be equipped in the network was noticeably reduced when the MPLS network was dynamically managed with respect to the static MPLS alternative.

Next, exhaustive simulations were carried out in scenarios focusing on the X2 interface, where bitrate guarantees are implemented using protection 1:1, and on the S1 interface, where guarantees are implemented considering diversity. Results showed that using 1:1 protection for X2 backhauling to avoid service interruption slightly increases the average delay in the connections in scenarios with failures. Regarding S1 backhauling, diversity was considered to guarantee 50% of the requested bitrate in the case of a single link failure, thus avoiding service disruption. In addition, benefits from using more than one path to support a CVN link were observed in the scenarios evaluated, which resulted in the fact that the CVNs supported was slightly increased for a given blocking probability.

15

Toward Cognitive In-operation Planning

Alba Pérez Vela[1], Filippo Cugini[2], Marc Ruiz[1] and Luis Velasco[1]

[1] *Universitat Politècnica de Catalunya, Barcelona, Spain*
[2] *CNIT, Pisa, Italy*

Data analytics is an interesting and challenging task to find patterns in heterogeneous data coming from different sources. In networking, data comes from monitoring the data plane, for example, received power, impairments, and errors in the optical layer or service traffic in the Multi-Protocol Label Switching (MPLS) layer. The output of data analytics can be used for automating network operation and reconfiguration (e.g., see Chapter 11), detecting traffic anomalies, or transmission degradation. This way of doing network management is collectively known as the observe–analyze–act (OAA) loop since it links together monitoring, data analytics, and operation (with or without reoptimization). In this chapter, the OAA loop is applied to a variety of use cases that result into a proposed architecture based on the Application-Based Network Operation (ABNO) one to support the OAA loop in Elastic Optical Networks (EONs).

Table of Contents and Tracks' Itineraries

Provisioning, Recovery, and In-operation Planning in Elastic Optical Networks,
First Edition. Luis Velasco and Marc Ruiz.
© 2017 John Wiley & Sons, Inc. Published 2017 by John Wiley & Sons, Inc.

15.1 Introduction

Traffic monitoring is an essential task for network operators since it allows evaluation of network performance. Monitoring data can be collected in a repository for further analysis, for example, to localize failures and identify their cause or to create predicted traffic matrices for the near future. According to the International Telecommunication Union-Telecommunication Standardization Sector (ITU-T) [M2120], performance events are counted second by second over every 15 min period. At the end of a period, they are stored in historical registers usually in the network manager.

In multilayer networks, virtual network topologies (VNTs) are created by connecting Multi-Protocol Label Switching (MPLS) routers through virtual links (vlinks) supported by lightpaths in the optical layer; traffic flows are conveyed through MPLS paths along vlinks on the VNT. In such networks, a failure affecting the optical layer might affect a large number of service connections. In case the failure produces a cut of a lightpath, recovery mechanisms can be applied both at the optical and at the MPLS layer. However, it is more difficult to localize and identify the cause of a failure when it produces transmission errors. Therefore, among the different use cases of data analytics for networking, that of identifying failures is undoubtedly of interest for many network operators, since affected services can be reconfigured to improve the committed service level agreements (SLAs).

Another interesting use case of data analytics for networking is that of VNT reconfiguration, where transponders installed in routers can be used to increase the capacity of existing vlinks or even support new vlinks, so as to follow traffic variations as was shown in Chapter 11. However, VNT reconfiguration is not restricted to such use case; another use case is to detect traffic anomalies, which entails analyzing monitoring data to

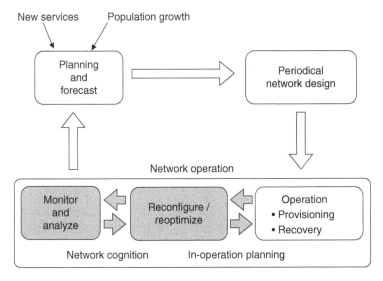

New services   Population growth

Figure 15.1 Extended network life cycle to support the OOA loop.

anticipate traffic congestion, since after the anomaly is detected lightpath provisioning can be triggered.

These examples illustrate the observe–analyze–act (OAA) loop, that is, how monitoring-based data analytics together with in-operation planning derives in adding cognition to the network. In fact, we are extending the classical network life cycle presented in Chapter 2 (see Figure 15.1), so new methodologies can be devised to improve network resources and thus to reduce costs.

The next few sections provide insights on the proposed OAA loop by means of some relevant use cases where monitoring and analysis of data from both optical and MPLS layers enhance in-operation planning capabilities.

15.2 Data Analytics for Failure Localization

Service layer connections are usually set up on a VNT, where vlinks are supported by lightpaths in the optical layer. Thus, errors in lightpaths translate to errors in those connections that might cause packet losses and retransmissions and lead to unacceptable Quality of Service (QoS). In this section, we assume that the service layer is the one detecting degradation.

Physical layer monitoring is key to verifying the fulfillment of SLAs and, in the case of faults or degradations, to localizing the failed elements [Ta15], [Ch16] and to taking actions for preserving the services. Information retrieved by commonly used power monitors can be combined with monitoring data accessible through emerging transponders based on coherent detection [Na15]. In particular, such transponders offer the possibility to monitor several parameters associated with connections or to the traversed links, for example, pre-forward error correction bit error rate (pre-FEC BER) or linear dispersion. In this section, we propose an algorithm that analyzes monitoring time series of collected Quality of Transmission (QoT) monitoring parameters that include received power (P_{Rx}) and pre-FEC BER. Based on the expected patterns of the considered failure causes, the algorithm localizes and identifies the most probable cause of failure at the optical layer affecting a given service. For the algorithm, the effects on QoT monitoring parameters of several failures on the optical layer need to be studied, specifically those identified as *tight filtering* and *interchannel interference*.

In the following, we present an algorithm for failure identification/localization based on Bayesian Networks (BNs) [Bi06], as well as the experimental results obtained. Some generic modules that should be available in the architecture to support failure identification/localization are eventually reviewed.

15.2.1 Algorithm for Failure Identification/Localization

For illustrative purposes, Figure 15.2 presents monitoring data series for the two considered causes of failure affecting a given optical connection.

Tight filtering happens when a very narrow filter configuration distorts the signal. Such effect may become even more relevant when the signal drifts (e.g., due to a laser drift) toward the rising edge of the filter. As will be shown in the experiments in Section 15.2.2, in the case of laser drift, P_{Rx} decreases because of the filtered power and BER is degraded; in fact, periods with degraded BER are followed by others with normal BER, which makes it difficult to localize the failure cause. In fact, BER degradation is not always caused by P_{Rx} decrease, as shown in the interchannel interference example, where the allocation of a neighboring lightpath results in a sudden increment observed in the target lightpath. Hence, failure localization entails deep analysis of monitoring data from several lightpaths.

With this in mind, we propose a probabilistic failure localization algorithm based on BN. A BN is a directed acyclic graph where nodes represent features and edges represent the conditional dependency between a pair of features. Each node is associated with a probability function that takes values from the parent nodes and returns the probability of the feature represented by the node.

The proposed BN is trained to locate different causes of failures and return the probability. Before training, several experimental tests for each of the possible causes of failure, as well as for the no failure case, need to be carried out to obtain monitoring data series similar to the ones in Figure 15.2 (see Section 15.2.2). Then, those data series are transformed into relevant *descriptive features* collecting the main characteristics of the data series, such

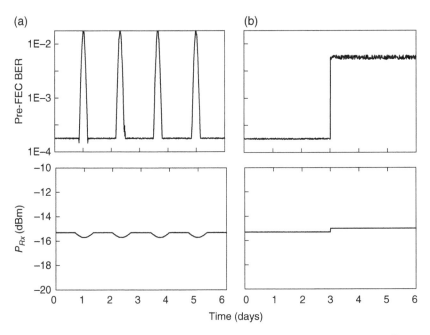

Figure 15.2 BER and P_{Rx} monitoring time series for the considered failures. (a) Tight filtering and (b) interchannel interference.

as minimum, maximum, average, trend, stepped change presence and size. Since BNs require categorical features (i.e., with a finite range of levels), continuous features can be easily discretized by applying a clustering algorithm to find the number and ranges of each of the levels [Ag13]. The type of failure is also added as *response feature*.

Figure 15.3 shows an example of the proposed failure localization algorithm. Let us imagine that a service using a connection between R1 and R3 has detected and notified service degradation to the service provider. The monitoring data of the two lightpaths supporting the service connection are analyzed. As observed, *p1* monitoring data series shows an almost constant value for both power and BER, whereas *p2* BER suffered a steeped increase at some point in the past. From this available data, the failure localization algorithm returns no failure with probability 95% for *p1* and identifies interference with 70% and tight filtering with 25% for *p2*.

According to the probabilities discussed, the scope of network reconfiguration is first focused on *p2* and those lightpaths sharing an optical link in the route of *p2*. A deeper analysis identifies that *p2* BER increment is correlated with the lightpath *p* set up. Therefore, by slightly shifting *p* in the spectrum away from *p2*, its BER should be improved and the detected service

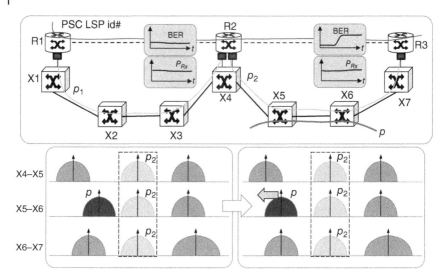

Figure 15.3 Example of failure localization caused by interchannel interference.

Table 15.1 Failure localization algorithm.

INPUT: s, BN

OUTPUT: A

```
1: P ← getLightpaths(s)
2: A ← ∅
3: for each p in P do
4:     H ← getMonitorDataSeries(p)
5:     F ← BN.computeFeatures(H)
6:     F' ← BN.discretize(F)
7:     D ← BN.predict(F')
8:     A.p ← sortProblemList(D)
9: return A
```

---

degradation eventually reduced. However, a monitoring period after reconfiguration is needed to verify that BER degradation has been completely solved. In the case that *p2* BER has not reduced to normal values after *p2* shifting, the second action in the list is taken, which might include making filters wider to overcome the probable tight filtering failure. With this second reconfiguration step, service degradation should be solved finally.

The proposed algorithm that integrates the BN is presented in Table 15.1; it receives as input the affected service connection *s* and the previously trained BN.

After retrieving the set P of lightpaths supporting s from the operational database (DB) (line 1 in Table 15.1), every single lightpath is sequentially processed as follows: first, the available BER and P_{Rx} monitoring data series are retrieved from the monitoring repository in the form of variable-length time series of continuous monitoring samples (line 4). Then, continuous features are computed and transformed into categorical values (lines 5–6). The prediction D returns, for each of the failures, the probability that it actually occurs (line 7). Such probabilities allow sorting the list of failures for every single lightpath, which is eventually returned (lines 8–9).

15.2.2 Experiments and Results

We demonstrated the effectiveness of the proposed algorithm by exploiting monitoring data collected from an experimental test bed. Two experiments were carried out: first, filtering effects are assumed upon laser drift in one lightpath; second, interchannel interference is induced by reducing the channel spacing between two lightpaths.

Figure 15.4 reports BER and differential P_{Rx} measurements for both experiments (i.e., the difference between the actual received power and the received power in normal conditions). It is worth noting the differences between these plots and those in Figure 15.2, where historical time series are plotted. Considering interchannel interference, BER increases when the channel spacing between two lightpaths decreases, while P_{Rx} increases when channel spacing decreases. This is due to the fact that part of one of the optical signals enters in the bandwidth of the other.

Assuming tight filtering, a similar behavior on the BER is experienced. Indeed, BER increases with laser drift since the impact of filtering effects becomes more relevant. The behavior of the power with respect to the interchannel interference case is different. Indeed, received power decreases while the laser drift increases since part of the power is cut by the filter. The different behaviors of P_{Rx} may drive the decision to discern between interchannel interference and excessive filtering effects because of laser drift. This is considered by the proposed decision algorithm.

Synthetic monitoring time series for the normal signal and the considered failure cases were generated according to the experimental values in Figure 15.4. Aiming at illustrating how to use experimental values to generate synthetic monitoring time series, let us assume that we need to generate a time series corresponding to interchannel interference failure. First, a number of measures for BER and P_{Rx} are randomly generated following the probability distribution of the normal signal case; in our experiments, we observed normal fluctuations of BER and P_{Rx} around average values 2E-4 and −15.3 dBm, respectively, with a narrow Gaussian random noise. Then, a channel spacing value in the range of Figure 15.4 is randomly chosen, for example, 29.5 GHz;

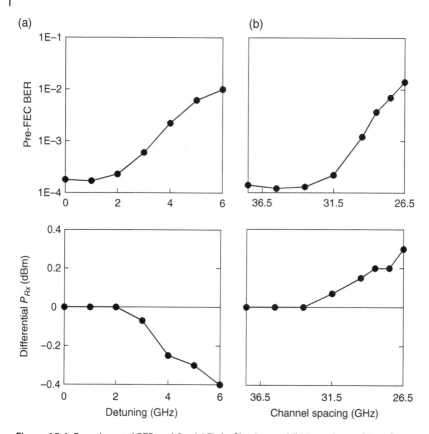

Figure 15.4 Experimental BER and P_{Rx}. (a) Tight filtering and (b) interchannel interference.

and, finally, a number of measures affected by the failure are generated and concatenated to previously generated normal ones. Note that, in the case of interchannel interference, random Gaussian fluctuations still remain in the range of the normal signal, whereas the average is shifted according to the values in Figure 15.4 for the selected channel spacing that is, $-15.15\,\mathrm{dBm}$ for P_{Rx} and 1.2E-3 for BER.

A set of 5000 randomly generated time series were used to first train the BN and next 500 additional ones were used for testing. Table 15.2 reports the obtained goodness-of-fit computed as the probability that the BN predicts the actual failure cause as the first option. Note that only 0.8% error was observed in some tests where a normal signal was predicted instead of a tight filtering failure. In such cases, the second most probable cause of failure was tight filtering failure. This demonstrates the validity of the proposed procedure and BN to localize and identify failures in the optical layer.

Table 15.2 BN goodness-of-fit.

		Real		
		Normal	Filtering	Interference
Prediction	Normal	**99.2%**	0.8%	0%
	Filtering	0%	**100%**	0%
	Interference	0%	0%	**100%**

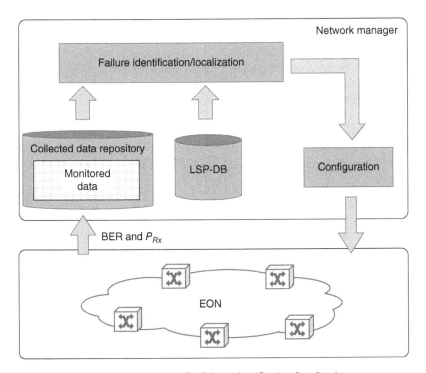

Figure 15.5 Applying the OAA loop for failure identification/localization.

15.2.3 Generic Modules to Implement the OAA Loop

To automate failure identification/localization, BER and P_{Rx} need to be monitored in the transponders and stored in a DB in the network manager.

Therefore, optical nodes need to collect the measured values for every installed transponder and send those monitored data to the network manager that stores them in a *collected data repository* (Figure 15.5). When the algorithm for failure identification/localization runs, it needs to access the monitoring data, as well as the DB with the established lightpaths (Label Switched Path (LSP)). Once a failure has been identified, the network manager is in charge of reconfiguring the optical network, for example, by shifting a lightpath in the spectrum.

15.3 Data Analytics to Model Origin–Destination Traffic

The VENTURE problem was proposed in Chapter 11 as an approach for VNT reconfiguration based on real traffic measurements. That approach regularly reconfigures the VNT for the next period to adapt its topology to the expected traffic volume and direction for the next period. However, the method behind the computation of the expected traffic was not detailed in that chapter. In this section, we concentrate on proposing a machine learning procedure that can be used to generate traffic models, which can be used afterward in a variety of use cases not restricted to VNT reconfiguration, for example, traffic anomaly detection.

The next subsections propose some generic modules that should be available in the architecture to support traffic modeling and a machine learning procedure based on artificial neural network (ANN) [Em08] for traffic estimation.

15.3.1 Generic Modules for VNT Reconfiguration Based on Traffic Modeling

To automate traffic modeling, traffic needs to be monitored in the packet nodes and counters need to be accessible in the network manager.

We assume that traffic monitoring data is collected at the edge MPLS routers at regular intervals, for example, every 15 min. Every edge router collects a set of samples for the traffic to every other destination router, which, as in the failure identification/localization case, is stored in a *collected data repository* (Figure 15.6). Note that since we focus on origin–destination (OD) traffic monitoring, $|N|\cdot(|N|-1)$ traffic samples need to be stored at every monitoring interval, where $|N|$ is the number of routers.

In contrast to the failure identification/localization case, some additional modules are defined for traffic modeling. Following a predefined time period, for example, every hour, a time series from the collected data repository is retrieved for each OD pair and preprocessed applying data stream mining *sketches* to conveniently summarize collected data, thus producing modeled data representing the OD pair that is stored in a *modeled data repository*. Modeled data includes for every OD, among others, the *minimum, maximum, average,* and *last* collected bitrate within the hour.

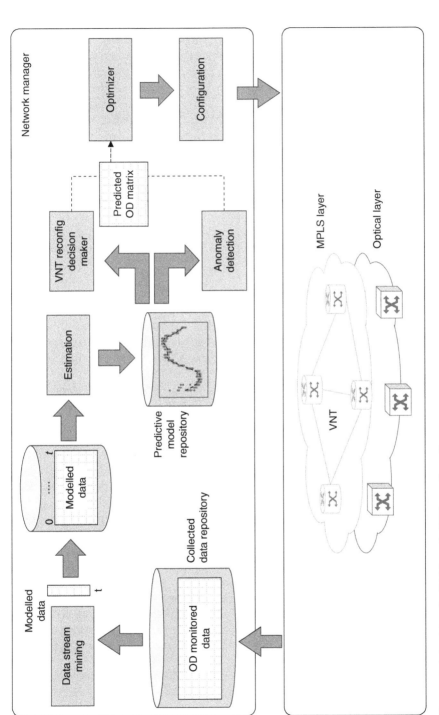

Figure 15.6 Applying the OAA loop for VNT reconfiguration.

The set of modeled variables for the current period t is stored in a repository together with variables belonging to previous periods. An *estimation module* based on machine learning techniques can generate the OD traffic model that can be used to predict traffic for any given period; models are stored in a *model repository*.

Such prediction can be used by a *decision maker* module to decide, for instance, whether the current VNT needs to be reconfigured (as in the VNT reconfiguration case in Chapter 11). Another use of the generated traffic model is to check whether the monitored OD traffic actually follows the predicted traffic or, on the contrary, an OD traffic anomaly is detected and the VNT needs to be reconfigured to cope with that unexpected traffic change. In case a VNT reconfiguration needs to be performed, the predicted OD traffic matrices are provided to the *VNT optimizer* to adapt the VNT. Once the algorithm finds a solution, the network manager is responsible for implementing the changes in the data plane.

15.3.2 Machine Learning Procedure for Traffic Estimation

Let us now propose an estimation module that consists of ANN-based models, selected because of its inherent capability of adapting to traffic changes in a nonsupervised manner. We consider different ANNs to estimate the traffic of each OD pair separately. Each ANN returns the expected bitrate at time t of the corresponding OD pair taking as inputs a number p of modeled data values obtained before t. Here, the *average* as well as the *maximum* traffic can be considered depending on the target traffic model. For instance, for the VNT reconfiguration use case, considering *maximum* instead of *average* bitrate allows adapting the VNT to the maximum expected traffic, hence ensuring a better grade of service. Once a modeled variable has been selected, two models for the same target variable can be obtained: one for predicting the expected value (i.e., the mean) and one for predicting the variance (or alternatively, the standard deviation). Note that the combination of both models opens the opportunity to provide centered estimations within a confidence interval.

Since the size of an ANN depends on the number of inputs, hidden layers, and neurons, we consider ANN models with p inputs, s neurons in a single hidden layer, and one output. Consequently, $s \cdot (p + 1)$ coefficients need to be found to specify every ANN. Aiming at keeping the number of coefficients small, we designed the algorithm in Figure 15.7 that has to be triggered every time an ANN needs to be refitted. It consists of three phases: (i) input data preprocessing, (ii) selection of significant inputs, and (iii) dimensioning of the hidden layer.

In the first phase, a time series X with the target modeled variable (*maximum* in this case) for the selected OD pair is retrieved from the modeled data repository. The *auto-correlation function* (ACF) is applied to X, and a list of

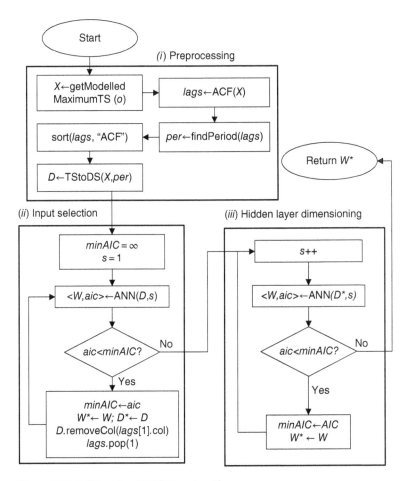

Figure 15.7 Self-learning ANN fitting algorithm.

lags is returned, where the *i*th lag contains the average correlation between every value in the time series and its *i*th previous value. Based on the lag analysis, a method is triggered to detect whether a periodic repetitive (seasonal) pattern is observable in X [Hm94]. The resulting period *per* defines the number of inputs of the ANN; in the case of nonseasonal data without observable periodical behavior, we assume *per* = 24 (i.e., one day) for convenience. Once *per* is obtained, X is transformed into a dataset D used for ANN fitting. Every row in D corresponds to a time t within the time series, and every column corresponds to a lag within *per*.

The second phase is an iterative procedure that finds the ANN with the best trade-off between accuracy and number of inputs. This trade-off is captured numerically by the *Akaike Information Criterion* (AIC) [Em08]. Starting with

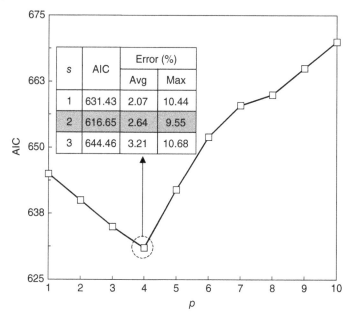

The table embedded in the figure:

s	AIC	Error (%)	
		Avg	Max
1	631.43	2.07	10.44
2	616.65	2.64	9.55
3	644.46	3.21	10.68

Figure 15.8 ANN goodness-of-fit.

$p = per$, the ANN routine fits an ANN from dataset D and returns the corresponding AIC value. While the AIC value obtained improves the lowest one obtained so far, the best ANN is stored, and p is decremented, effectively removing one input. Aiming at reducing the complexity of choosing the input to be removed, we select the lag with lowest ACF. When the minimum AIC is reached, the third phase is executed to increase even more the accuracy of the model by adding hidden neurons until the AIC does not improve. The best ANN is eventually returned.

The ANN models were trained applying the fitting algorithm in Figure 15.7 on a training dataset with modeled data belonging to the last week. Results in Figure 15.8 illustrate the average size and goodness-of-fit of ANN models for predicting the expected value of the maximum. Recall that during the input selection phase, the number of inputs p is decreased aiming at minimizing the AIC value. We observe that the minimum AIC is on average reached at $p = 4$, those inputs from t-1 to t-4 being mainly selected. Results from the hidden layer dimensioning phase are shown in the table embedded in Figure 15.8 for a number of hidden neurons ranging from 1 to 3. Note that the minimum AIC is obtained for $s = 2$, which results in an ANN model with 10 coefficients that accurately estimates the output variable with a good trade-off between average and maximum relative errors (2.64 and 9.55%, respectively).

Once the proposed methodology for traffic estimation has been validated, two use cases that take advantage of such OD traffic models are detailed next.

15.3.3 Use Case I: Anomaly Detection

Traffic anomalies are *short-living* events that do not follow expected patterns (see a survey in [Ch09]). Detecting anomalies is a difficult task because anomalous patterns need to be extracted and interpreted from large amounts of high-dimensional, noisy data.

It is clear that when analytics are applied to data collected every 15 min, as specified in [M2120], expected traffic anomaly detection times will be in that order of magnitude as well. Note that traffic anomalies can create network congestion and stress resource utilization in routers and hence its prompt detection becomes essential since it allows preparing the network, for example, by modifying routing tables or reconfiguring the VNT.

In this use case, we evaluate the performance of different OD traffic anomaly detection methods and propose monitoring strategies and architectural approaches to find the combination with the best trade-off between detection time and amount of collected data.

In our approach, OD traffic models are computed for the modeled variable *average*, which predicts two response variables: the mean ($\mu(t)$) and the standard deviation ($\sigma(t)$). Figure 15.9 illustrates the main steps of the proposed method. In Figure 15.9a, traffic samples for every OD pair are collected from the packet nodes at a given monitoring period. Besides, Figure 15.9b shows traffic samples and the upper and lower bounds, computed as $\mu \pm 3\sigma$, for a given OD pair and for a typical day. Although out-of-bound samples are considered as *atypical* (Figure 15.9c), their detection does not entail a traffic anomaly. In fact, the decision regarding whether an atypical sample can be considered a traffic anomaly cannot be based on just one single sample, but in observing some previous samples. This is depicted in Figure 15.9d, where an anomaly is detected after receiving two out-of-bound samples and considering some other previous within-bound samples.

To efficiently implement OD-based traffic anomaly detection methods, we use the modules depicted in Figure 15.6. Specifically, the anomaly detection module is first assumed to be placed in the network manager. In such architecture, OD traffic models predict response variables for the average OD traffic (i.e., $\mu(t)$ and $\sigma(t)$). The traffic anomaly detection module is in charge of detecting traffic anomalies; it first verifies whether a just arrived OD traffic sample is out of bounds and only in such case a machine learning algorithm is run to detect traffic anomalies.

15.3.3.1 Methods for Anomaly Detection

Two different methods for anomaly detection are studied in this section: *threshold-based* and *probability-based*. Both methods have already been proposed in different contexts, as in traffic changes detection and anomaly detection based on hypothesis testing [Ku00]; here, we adapt them for OD traffic anomaly detection. The adapted *threshold-based* method involves

(a)

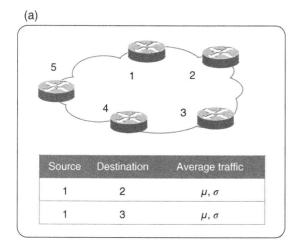

(b)   (c)

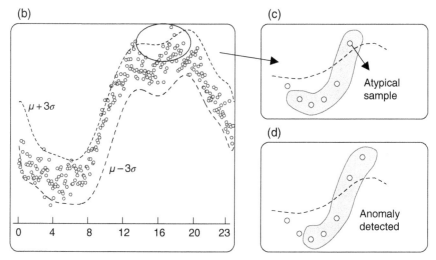

Figure 15.9 (a) Per OD statistical models. (b) Monitoring samples versus estimation. (c) Atypical and (d) anomaly detection.

detecting anomalies after receiving a number of out-of-bound traffic samples with respect to the $\mu \pm 3\sigma$ confidence interval. The *probability-based* method is a self-learning classifier with two labels for the response: normal and anomaly.

The probability-based algorithm (see Figure 15.10) is based on a multiresponse model to predict whether a sequence of consecutive traffic samples belongs to the normal class or, on the contrary, there is sufficient evidence to declare it as anomalous. The algorithm starts when a traffic sample $x(t)$ is received in time t; let $x'(t)$ be the normalized value of $x(t)$ with respect to the average model, that is, $x'(t) = (x(t) - \mu(t))/\sigma(t)$. Note that a normalized value equal to k means that $x(t) = \mu(t) + k\sigma(t)$.

After normalization, $x'(t)$ is stored in a fixed-size data series H, containing the last normalized traffic samples received (hereafter referred to as *features*). As a result of traffic normalization, every feature H_i follows, in normal conditions, the standard Gaussian distribution, that is, $H_i \sim N(0,1)$. Let us define a random variable $Z \sim N(0,1)$ and the probability $p_i = P(Z > |H_i|)$ that feature i takes a value above its current absolute value, that is, feature i actually belongs to the normal class. Therefore, it is likely to assume that smaller (i.e., less probable) p_i values will be observed in the case of an anomaly. The classifier basically consists of (i) a distance function $w(H)$ to compute how likely a features vector H belongs to the normal class, and (ii) a distance threshold w_{thr} that normal data series H do not practically exceed. The distance function $w(H)$ is defined as the probability that all the features in H belong to the normal class. As a result of traffic normalization, every feature H_i is distributed independently, therefore this probability being the product of the individual probabilities $p_i (w(H) = \prod_{1...|H|} p_i)$.

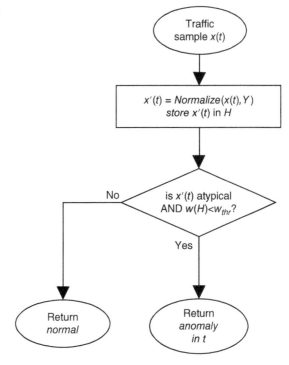

Figure 15.10 Probability-based algorithm for anomaly detection.

To decide whether an anomaly is detected, we simply compare $w(H)$ against the w_{thr} threshold, that is, when $w(H) < w_{thr}$. Under the assumption of feature independence and considering $p_i = 0.05$ for each individual feature as a commonly used limit for accepting the Gaussian null hypothesis, we fix $w_{thr} = 0.05^{|H|}$.

15.3.3.2 Proposed Monitoring Strategies

As discussed, anomaly detection time is directly related to the monitoring frequency, and thus it is a key parameter for study. However, increasing monitoring frequency entails increasing in the same proportion the amount of monitoring data to be conveyed to the repositories in the network manager. Aiming at keeping the latter amount under control, we propose studying the performance of the following monitoring strategies:

- the traditional *fixed monitoring* period strategy but reducing its period to accelerate anomaly detection;
- a *dynamic monitoring* strategy, where the monitoring period can be reprogrammed during the day;
- a *reactive monitoring* strategy $(c:f)$ that uses a coarse monitoring period (c) and reconfigures it to a finer period (f) after detecting the first out-of-bound traffic sample.

From the possible combination of methods and strategies, we focus on studying the four most relevant approaches:

- threshold-based with fixed monitoring (*threshold-fixed*),
- probability-based with fixed monitoring (*probability-fixed*),
- probability-based with dynamic monitoring (*dynamic*), and
- probability-based with reactive monitoring (*reactive*).

15.3.3.3 Illustrative Numerical Results

For evaluation purposes, we developed an ad hoc event-driven simulator in OMNeT++ that was used to generate normal traffic and traffic anomalies separately and subsequently combined (Figure 15.11).

Traffic (Figure 15.11a) is generated as the summation of two different functions, mean and noise, where the traffic mean represents a normal day with values varying along day hours, while the noise is a random function with mean zero and a given standard deviation.

Regarding anomalies, they are used as a multiplicative factor over traffic and are generated following a pulse function (Figure 15.11b), where the raising front consists of an exponential function (Figure 15.11c). Anomalies can be configured to be triggered at any specific time and with any specific duration and scaling factor. As an illustrative example, an anomaly can be generated to multiply traffic by ×1.5, last for 2 h, and characterized by reaching the 90% of its maximum value in the first 30 min.

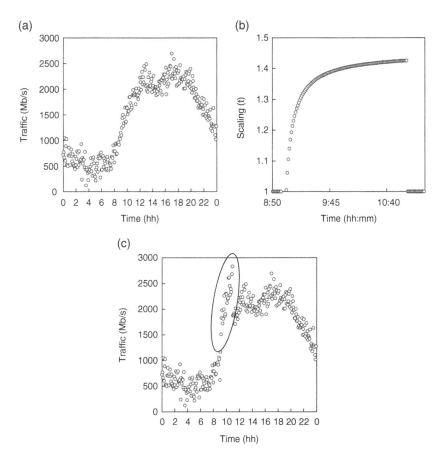

Figure 15.11 (a) OD traffic profile in a typical day. (b) Traffic anomaly at 9 a.m. (c) OD traffic with anomaly.

The traffic estimator module was implemented in C++ and integrated into the simulator, whereas the anomaly detection module implementing the proposed approaches was developed in R and kept as a stand-alone separate module.

Graphs in Figure 15.12 plot, for several hours of the day, the detection time for the threshold-fixed, probability-fixed, and reactive anomaly detection approaches, where the monitoring period is in the interval [1–5] min. We observe that although anomaly detection time varies for the different considered hours of the day, detection time increases remarkably with the threshold-fixed approach when the monitoring period increases. This is in contrast to the moderated increment achieved by the probability-fixed one. In case of the reactive approach, where we assume a ($c = 5 : f = 1$) min monitoring strategy, slightly lower detection times with respect to the previous approaches can be observed.

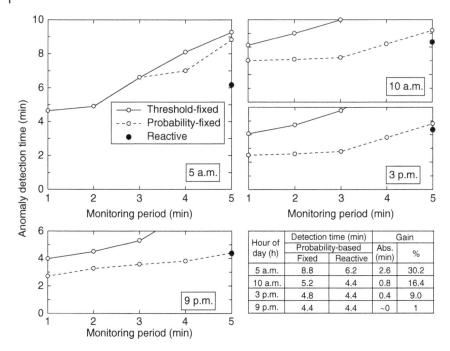

Figure 15.12 Monitoring period as a function of the target traffic anomaly detection time for different hours of a day.

The table in Figure 15.12 reports the gains in detection time for the studied hours of the day, where using a finer monitoring period after an out-of-bound traffic sample is detected provides gains between 1 and 30%, depending on the hour of the day.

Figure 15.13a focuses on studying in depth the potentials of the probability-based method and illustrates how anomaly detection depends on different factors, such as the changes in volume traffic among the different hours of the day for the same monitoring period. This opens the opportunity to dynamically adapt the monitoring period for different hours of the day and achieve the same anomaly detection times (dynamic monitoring). Note that this is positive since to achieve low detection times, a 1 min period should be fixed. Hence, by relaxing the monitoring period we are effectively reducing the amount of monitoring data to be collected.

Finally, Figure 15.13b shows the amount of monitored data to be collected along the day when reducing the monitoring period. For example, assuming a 5 min period, 288 monitoring samples per OD and day need to be collected achieving 5.14 and 4.62 min detection times for the probability-fixed and the reactive approaches, respectively.

(a)

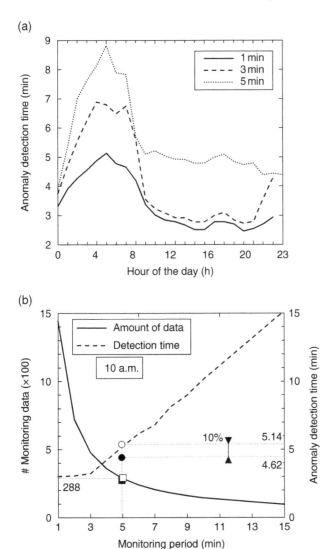

(b)

Figure 15.13 Anomaly detection time versus hours of the day for different monitoring periods (a). Amount of data and anomaly detection time versus monitoring period (b).

15.3.3.4 Concluding Discussion

The earlier discussion showed that 15 min monitoring could not provide the short anomaly detection times required to react against unexpected traffic changes. Consequently, we studied four different approaches mixing detection methods and monitoring strategies and showed that the shortest anomaly

Table 15.3 Comparative function placement.

	Centralized	Distributed
Network manager	• Monitoring period programming • Traffic estimation • OD anomaly detection	Traffic estimation
Node	Monitoring period reconfiguration	OD anomaly detection

detection times are achieved when monitoring every 1 min. Nevertheless, this comes at the cost of collecting and storing a large amount of data in the network manager.

In view of this, we propose to bring the proposed data analytics method for OD anomaly detection to the network nodes, thus relaxing data collection from the network manager to the traditional 15 min period that can still be used for traffic modeling and estimation purposes. Table 15.3 compares function placement in the centralized and distributed architectures.

15.3.4 Use Case II: VNT Reconfiguration Triggered by Anomaly Detection

This section presents a second use case of VNT reoptimization triggered by the detection of an anomalous amount of traffic in an OD pair.

To adapt the VNT to the current traffic, both the threshold-based and the VENTURE methods reviewed in Chapter 11 can be used. The former increases the capacity of an existing vlink by setting up a parallel lightpath when the volume of traffic through that vlink reaches some threshold (e.g., 90% of its capacity). The threshold value must be configured in a way that traffic increments can be conveyed with the remaining capacity while the new lightpath is being established (e.g., some minutes). The latter method works periodically and adapts the VNT for the traffic conditions that are estimated for the next period. However, when the traffic experiences an abrupt increment in a short period of time vlinks' capacity might be exceeded, and thus traffic losses appear until the new lightpath becomes available. Thus, a prompt detection of *traffic anomalies* becomes essential to trigger VNT reconfiguration, so as to anticipate the capacity for the traffic increment (see Section 15.3.3).

An example is shown in Figure 15.14, where OD traffic is monitored (Figure 15.14a) and a traffic anomaly in OD pair 1->3 is detected. The VNT is reconfigured by creating new vlink 1->3 and the traffic is rerouted (Figure 15.14b). As a result of the anticipated VNT reconfiguration, traffic losses might be greatly decreased.

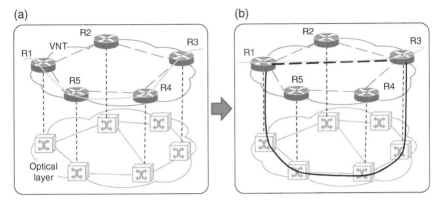

Figure 15.14 Initial VNT and OD 1->3 routing (a). Reconfigured VNT and OD pair 1->3 routing after traffic anomaly detection. (a) A traffic anomaly is detected. (b) The VNT is reconfigured.

Based on this, the OD traffic anomaly-triggered topology reconfiguration (ODEON) problem can be formally stated as:

Given:

- a graph $G(V, E)$, where V is the set of MPLS routers and E is the set of directed vlinks connecting two routers. The remaining capacity of each vlink e (b_e) is also known;
- a set $TP(v)$ of transponders installed in each router v. Some of these transponders can be currently used for the existing vlinks and some of them might be unused;
- the OD pair d affected by the anomaly; d is defined by the tuple $<o, t, b>$, where o and t are the source and target routers, respectively, and b is the maximum expected bitrate during anomaly lifetime.

Output: The capacity increments in existing vlinks and the new vlinks to be created, so as to serve d. In addition, the new path for d needs to be specified.

Objective: Minimize the use of resources to serve d, including transponders' utilization.

To tackle the ODEON problem, we propose the algorithm in Table 15.4. The resources currently allocated to pair d are first released (line 1 in Table 15.4), and the vlinks without enough capacity are removed from the graph (lines 2–3). Next, the graph G is augmented to $G'(V, E')$, where new vlinks ($E' \backslash E$) are created between routers with available transponders (line 4). A shortest path between end routers o and t is computed (line 5) assuming vlinks cost as follows:

$$c_e = \begin{cases} 1 & \text{if } e \in E \\ |E| + 1 & \text{otherwise (i.e., if } e \in E' \backslash E) \end{cases} \tag{15.1}$$

Table 15.4 ODEON algorithm.

```
INPUT: G(V, E), d = <o, t, b>
OUTPUT: <L, p>
```

```
 1: deallocate(G, d)
 2: for each e in E do
 3:   if availablecapacity(a) ≤ b then E ← E \ {e}
 4:   G'(V, E') ← augment(G)
 5:   p ← SP(G', <o, t>)
 6: if p = ∅ then return INFEASIBLE
 7: L ← ∅
 8: for each e in p do
 9:   if e = (v₁, v₂) ∈ E'\E then
10:     l ← RMSA(G', <v₁, v₂, b>)
11:     L ← L U {l}
12: return <L, p>
```

Table 15.5 Improvement using ODEON.

Max OD traffic (Gb/s)	Loss (Gb) without ODEON	Loss (Gb) with ODEON	Loss reduction (%)			
			Avg	1 a.m.	9 a.m.	5 p.m.
40	15	10	33	9	22	20
60	20	13	33	58	13	56
80	28	18	38	21	18	54
100	45	28	37	43	17	54
120	49	21	56	40	36	76
140	48	27	44	29	18	80

Finally, the routing, modulation, and spectrum allocation (RMSA) problem (see Chapter 5) is solved for every new vlink and for existing vlinks where the capacity needs to be increased.

For evaluation purposes, we developed an ad hoc event-driven simulator to evaluate the performance of the proposed ODEON algorithm that reconfigures the network in case of a sudden traffic anomaly. Lightpath setup time was set to 1 min. The gain in terms of traffic losses is analyzed for different values of maximum traffic expected per OD.

Table 15.5 shows the results of traffic losses with and without ODEON assuming that an anomaly can occur at any hour, as well as the reduction of traffic losses at three different times of the day. In view of the results, we can conclude that ODEON reduces traffic losses in more than 30% in average, reaching up to 80% of reduction for specific hours and loads.

15.4   Adding Cognition to the ABNO Architecture

To support the generic modules introduced earlier for both traffic modeling and failure localization/identification, we propose the architecture in Figure 15.15. Monitored traffic data is sent to the Application-Based Network Operation (ABNO) [ABNO]-based network manager, ready to be analyzed. A big data repository consisting of a distributed DB and a data collector is used to store collected data. Specifically for traffic modeling, ABNO's Operations, Administration and Maintenance (OAM) handler includes data stream mining sketches, the modeled data repository, the estimation module running the proposed ANN to anticipate next period traffic conditions, and the decision maker.

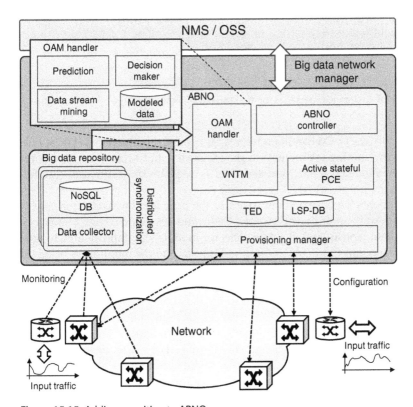

Figure 15.15 Adding cognition to ABNO.

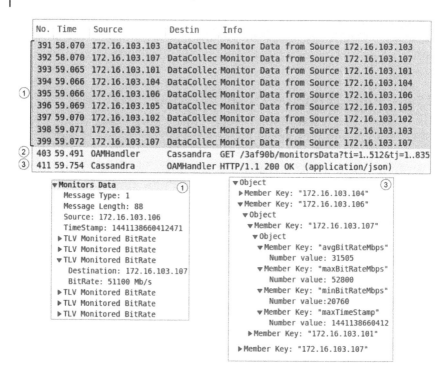

Figure 15.16 Exchanged messages for monitored traffic collection.

Periodically, the OAM handler retrieves aggregated monitored data and applies data stream mining techniques on the received data to transform monitored data into modeled data. Modeled data is used by the estimation module running the proposed ANN.

Experiments were carried out to assess the proposed big data extensions to ABNO for traffic monitoring. Apache Cassandra DB [CASSANDRA] was used as a big data repository and a data collector module was implemented to offer a User Datagram Protocol (UDP)-based interface to the monitors, storing the received data in Cassandra. Apache Spark [SPARK] was used to implement data stream mining and machine learning techniques.

Figure 15.16 illustrates monitored traffic data being periodically sent by the packet nodes to the data collector, as well as the request that the OAM handler issues to Cassandra's Representational State Transfer Application Programming Interface (REST API) to collect monitored data. Monitoring messages (messages 1) contain, among others, the source node and the timestamp of the sample, and, for each aggregated flow leaving the node to a destination, its destination node and bitrate. After selecting and aggregating monitored data between the selected times t_i and t_j, (2) Cassandra replies with a JSON-encoded matrix (3) specifying for each pair of source–destination the average, minimum, and maximum bitrate.

15.5 Conclusions

This final chapter has explored new trends in networking, where the use of data analytics together with optimization (OAA loop) showed interesting new applications and opened opportunities to bring cognition to the network facilitating its operation and maintenance.

Related to failure localization and identification, two different failure causes at the optical layer were experimentally characterized and the obtained measurements used to generate time series to train a BN. When a service detects excessive errors, an algorithm uses the trained BN to localize and identify the most probable cause of the errors at the optical layer by using monitoring data collated in a repository in the network manager. Results showed the effectiveness of the algorithm.

Another use of monitoring was related to collect OD traffic in the MPLS routers, where data analytics were applied to learn predictive models that can be used for traffic anomaly detection, as well as for VNT reconfiguration. In particular, an ANN for every OD pair was proposed as a predictive model along with an algorithm to obtain a highly accurate model using as few coefficients as possible.

Finally, extensions to the ABNO architecture were proposed to support the OAA loop; monitoring data is collected by an extended OAM handler with big data capabilities.

List of Acronyms

ABNO	Application-based Network Operations
AC_RECOVER	Anycast connection recovery
ACF	Auto-Correlation Function
AC-RSA	AnyCast RSA
ACTN	Abstraction and Control of Transport Networks
AFRO	After-Failure Repair Optimization
AIC	Akaike Information Criterion
ANN	Artificial Neural Network
AP	Arc-Path
API	Application Programming Interface
ASO	Application Service Orchestrator
BATIDO	Bitrate Squeezing and Multipath Restoration
BBU	Baseband Units
BER	Bit Error Rate
BGP	Border Gateway Protocol
BGP-LS	BGP link state
BGPUpd	BGP Update
BN	Bayesian Network
bPCE	Back-end Path Computation Element
BRKGA	Biased Random Key Genetic Algorithm
BT	British Telecom
BuSMF	Bundle of Single Mode Fibers
BV	Bandwidth-Variable
BVT	Bandwidth-Variable Transponder
CAPEX	Capital Expenditures
CA-RSA	Slot Assignment RSA
CD	Chromatic Dispersion
CF	Central Frequency
ChBW	Spectral Channel Width

Provisioning, Recovery, and In-operation Planning in Elastic Optical Networks,
First Edition. Luis Velasco and Marc Ruiz.
© 2017 John Wiley & Sons, Inc. Published 2017 by John Wiley & Sons, Inc.

CO	Central Office
CPRI	Common Public Radio Interface
C-RAN	Cloud-RAN
CVN	Customer Virtual Network
CVN-DB	Customer Virtual Networks-Data Bade
CWDM	Coarse Wavelength Division Multiplexing
DB	Database
DC	Datacenter
DC2DC	Datacenter-to-datacenter
DP	Dual Polarization
DP-BPSK	Dual Polarization Binary Phase Shift Keying
DPP	Dedicated Path Protection
DSP	Digital Signal Processing
DWDM	Dense Wavelength Division Multiplexing
DYNAMO	DYNamic restorAtion in Multilayer MPLS-over-EON
e2e	end-to-end
EB	Exabytes
EDFA	Erbium Doped Fiber Amplifier
EON	Elastic Optical Networks
ERO	Explicit Route Object
ERO	Explicit Route Object
ETC	Explicit Transponder Control
FA	Forwarding Adjacency
FEC	Forward Error Correction
FF	First Fit
FMF	Few-Mode Fiber
FM-MCF	Few-Mode Multi-Core Fiber
fPCE	Front-end Path Computation Element
FrJ-Sw	Fractional Joint Switching
FRR	Fast Reroute
GCO	Global Concurrent Optimization
GHz	GigaHertz
GMPLS	Generalized Multi-Protocol Label Switching
GRASP	Greedy Randomized Adaptive Search Procedure
HPC	High Port Count
HT	Holding Time
IA-RSA	Impairment-Aware Routing and Spectrum Allocation
IETF	Internet Engineering Task Force
ILP	Integer Linear Programming
Ind-Sw	Independent Switching
IP	Internet Protocol
ITU-T	International Telecommunication Union-Telecommunication Standardization Sector

J-Sw	Joint Switching
kSP	*k*-Shortest Paths
LC	Lane Change
LCoS	Liquid Crystal on Silicon
LiPo	Linearly Polarized
LP	Linear Programming
LSC	Lambda-Switch Capable
LSP	Label-Switched Path
LSP-DB	LSP Database
MBS	Macro Base Stations
MC	Multi-Carrier
MCF	Multi-Core Fiber
MF-Restoration	Modulation Format-Aware Restoration
MILP	Mixed Integer Linear Programming
MIMO	Multi-Input–Multi-Output
MIMO-DSP	Multiple Input Multiple Output Digital Signal Processing
MLR	Multi Line Rate
MME	Mobility Management Entity
MP	Mathematical Programming
MP-AFRO	Multipath AFRO
MPLS	Multi-Protocol Label Switching
MST	Minimum Spanning Tree
MTTF	Mean Time to Failure
MTTR	Mean Time to Repair
NA	Node-Arc
NetConf	Network Configuration Protocol
NMS	Network Management System
OAA	Observe–Analyze–Act
OADM	Optical Add/Drop Multiplexer
OAM	Operations, Administration, and Maintenance
OD	Origin–Destination
ODEON	OD Traffic Anomaly-Triggered Topology Reconfiguration
OEO	Optical–Electronic–Optical
OF	Objective Function
OFDM	Orthogonal Frequency-Division Multiplexing
OPEX	Operational Expenditures
OSNR	Optical Signal to Noise Ratio
OSPF-TE	Open Shortest Path First for Traffic Engineering
OSS	Operations Support System
OTN	Optical Transport Network
OXC	Optical Cross-Connect
P2MP	Point-to-multipoint
P2P	Point-to-point

Pb	Blocking probability
PCC	Path Computation Client
PCE	Path Computation Element
PCEP	PCE Communication Protocol
PCInit	Path Computation Initiate
PCNtf	Path Computation Notification
PCRep	Path Computation Reply
PCReq	Path Computation Request
PCRpt	Path Computation Report
PCUpd	Path Computation Update
PM	Provisioning Manager
PMD	Polarization Mode Dispersion
PR	Path Relinking
PSC	Packet Switch Capable
PSC	Packet-Switch Capable
QAM	Quadrature Amplitude Modulation
QoS	Quality of Service
QoT	Quality of Transmission
QPSK	Quadrature Phase Shift Keying
RAN	Radio Access Networks
RCL	Restricted Candidate List
REST	Representational State Transfer
RF	Radio Frequency
RMSA	Routing Modulation format, and Spectrum Assignment
RO	Route Object
ROADM	Reconfigurable Optical Add/Drop Multiplexer
RP	Request Parameters
RRH	Remote Radio Heads
RRO	Record Route Object
RSA	Routing and Spectrum Allocation
RSMSA	Routing, Space, Modulation Format, and Spectrum Assignment
RSSA	Routing and Scheduled Spectrum Allocation
sTED	Scheduled Traffic Engineering Database
RSVP-TE	Resource Reservation Protocol-Traffic Engineering
RWA	Routing and Wavelength Assignment
SA	Spectrum Allocation
SA-RSA	Slice Assignment RSA
SBVT	Sliceable Bandwidth Variable Transponder
SC	Single Carrier
SDM	Space Division Multiplexing
SDN	Software Defined Networking
SERO	Secondary Explicit Route Object
S-GW	Serving Gateway

SLA	Service Level Agreement
SLR	Single Line Rate
SMF	Single Mode Fiber
SP	Shortest Path
Sp-Ch	Super-channel
SPP	Shared Path Protection
SPRESSO	Spectrum Reallocation
SPRING	Spectrum Shifting
SPT	Shortest Path Tree
SRLG	Shared Risk Link Group
SSA-RSA	Starting Slice Assignment RSA
SVEC	Synchronization Vector
TCO	Total Cost of Ownership
TCP	Transmission Control Protocol
TE	Traffic Engineering
TED	TE Database
TEL	Telefonica
TLV	Type-length-value
TP	Traffic Profile
TRx	Transceiver
U2DC	User-to-Datacenter
UDP	User Datagram Protocol
USB	Universal Serial Bus
VENTURE	VNT Reconfiguration Approach Based on Traffic Measurements
vlink	Virtual Link
VM	Virtual Machine
VNT	Virtual Network Topology
VNTM	Virtual Network Topology Manager
WCC	Wavelength Continuity Constraint
WDM	Wavelength Division Multiplexing
WSS	Wavelength Selective Switch
XRO	eXclude Route Object
XT	Crosstalk

References

[ABNO] D. King and A. Farrel, "A PCE-based Architecture for Application-based Network Operations," IETF RFC 7491, 2015. https://www.ietf.org/ (accessed April 18, 2017).

[ACTN] D. Ceccarelli and Y. Lee, "Framework for Abstraction and Control of Traffic Engineered Networks," IETF draft-ietf-teas-actn-framework-04, work-in-progress, 2017. https://www.ietf.org/ (accessed April 18, 2017).

[Ag09] F. Agraz, L. Velasco, J. Perelló, M. Ruiz, S. Spadaro, G. Junyent, and J. Comellas, "Design and Implementation of a GMPLS-Controlled Grooming-Capable Optical Transport Network," IEEE/OSA Journal of Optical Communications and Networking (JOCN), vol. 1, pp. A258–A269, 2009.

[Ag13] C. Aggarwal and C. Reddy, *Data Clustering: Algorithms and Applications*, Chapman & Hall/CRC, Boca Raton, 2013.

[Ag15] A. Aguado, M. Davis, S. Peng, M. Álvarez, V. López, T. Szyrkowiec, A. Autenrieth, R. Vilalta, A. Mayoral, R. Muñoz, R. Casellas, R. Martínez, N. Yoshikane, T. Tsuritani, R. Nejabati, and D. Simeonidou, "Dynamic Virtual Network Reconfiguration over SDN Orchestrated Multi-Technology Optical Transport Domains," IEEE/OSA Journal of Lightwave Technology, vol. 34, pp. 1933–1938, 2016.

[Ah93] R.K. Ahuja, T.L. Magnanti, and J.B. Orlin, *Network Flows: Theory, Algorithms and Applications*, Prentice Hall, Upper Saddle River, 1993.

[Al16] Z. Ali, C. Filsfils, R. Varga, V. Lopez, O. Gonzalez de Dios, and X. Zhang, "Path Computation Element Communication Protocol (PCEP) Extensions for Remote-Initiated GMPLS LSP Set up," IETF draft-ietf-pce-remote-initiated-gmpls-lsp-03.txt, work in progress, 2016. https://www.ietf.org/ (accessed April 18, 2017).

[Am11] N. Amaya, M. Irfan, G. Zervas, K. Banias, M. Garrich, I. Henning, D. Simeonidou, Y. Zhou, A. Lord, K. Smith, V. Rancano, S. Liu, P. Petropoulos, and D. Richardson, "Gridless Optical Networking Field Trial: Flexible Spectrum Switching, Defragmentation and Transport of 10G/40G/100G/555G over 620-km Field Fiber," in Proceedings of the European Conference on Optical Communication (ECOC), 2011, Geneva, 18–22 September.

Provisioning, Recovery, and In-operation Planning in Elastic Optical Networks,
First Edition. Luis Velasco and Marc Ruiz.
© 2017 John Wiley & Sons, Inc. Published 2017 by John Wiley & Sons, Inc.

[Ar10] M. Armbrust, A. Fox, R. Griffith, A.D. Joseph, R.H. Katz, A. Konwinski, G. Lee, D.A. Patterson, A. Rabkin, I. Stoica, and M. Zaharia, "A View of Cloud Computing," Communications of the ACM, vol. 53, pp. 50–58, 2010.

[As13] A. Asensio, M. Klinkowski, M. Ruiz, V. López, A. Castro, L. Velasco, and J. Comellas, "Impact of Aggregation Level on the Performance of Dynamic Lightpath Adaptation under Time-Varying Traffic," in Proceedings of the IEEE International Conference on Optical Network Design and Modeling (ONDM), 2013, Brest, April 15–18.

[As14] A. Asensio and L. Velasco, "Managing Transfer-based Datacenter Connections," IEEE/OSA Journal of Optical Communications and Networking (JOCN), vol. 6, pp. 660–669, 2014.

[As16.1] A. Asensio, P. Saengudomlert, M. Ruiz, and L. Velasco, "Study of the Centralization Level of Optical Network-Supported Cloud RAN," in Proceedings of the IEEE International Conference on Optical Network Design and Modeling (ONDM), 2016, Cartagena, May 9–12.

[As16.2] A. Asensio, M. Ruiz, L.M. Contreras, and L. Velasco, "Dynamic Virtual Network Connectivity Services to Support C-RAN Backhauling," IEEE/OSA Journal of Optical Communications and Networking (JOCN), vol. 8, pp. B93–B103, 2016.

[ASON] ITU-T, "Architecture for the Automatically Switched Optical Network," Rec. G.8080, 2012. https://www.itu.int/rec/T-REC-G.8080/en (accessed April 18, 2017).

[Bh99] R. Bhandari, *Survivable Networks: Algorithms for Diverse Routing*, Kluwer Academic Publishers, Norwell, 1999.

[Bi06] Ch. Bishop, *Pattern Recognition and Machine Learning*, Springer-Verlag, New York, 2006.

[Ca12.1] A. Castro, L. Velasco, M. Ruiz, and J. Comellas, "Single-Path Provisioning with Multi-Path Recovery in Flexgrid Optical Networks," International Workshop on Reliable Networks Design and Modeling (RNDM), 2012, St. Petersburg, October 3–5.

[Ca12.2] A. Castro, L. Velasco, M. Ruiz, M. Klinkowski, J.P. Fernández-Palacios, and D. Careglio, "Dynamic Routing and Spectrum (Re)Allocation in Future Flexgrid Optical Networks," Elsevier Computers Networks, vol. 56, pp. 2869–2883, 2012.

[Ca12.3] A. Castro, R. Martínez, L. Velasco, R. Casellas, R. Muñoz, and J. Comellas, "Experimental Evaluation of a Dynamic PCE-based Regenerator-Efficient IA-RWA Algorithm in Translucent WSON," European Conference on Optical Communication (ECOC), 2012, Amsterdam, 16–20 September.

[Ca13] A. Castro, L. Velasco, J. Comellas, and G. Junyent, "Dynamic Restoration in Multi-Layer IP/MPLS-Over-Flexgrid Networks," in Proceedings of the IEEE Design of Reliable Communication Networks (DRCN), 2013, Budapest, March 4–7.

[Ca14] A. Castro, R. Martínez, R. Casellas, L. Velasco, R. Muñoz, R. Vilalta, and J. Comellas, "Experimental Assessment of Bulk Path Restoration in Multi-Layer

Networks Using PCE-based Global Concurrent Optimization," IEEE/OSA Journal of Lightwave Technology (JLT), vol. 32, pp. 81–90, 2014.

[CASSANDRA] Apache Cassandra, Available: http://cassandra.apache.org/ (accessed March 23, 2017).

[Ch08] X. Chu, H. Yin, and X. Li, "Lightpath Rerouting in Wavelength-Routed DWDM Networks," OSA Journal of Optical Networking (JON), vol. 7, pp. 721–735, 2008.

[Ch09] V. Chandola, A. Banerjee, and V. Kumar, "Anomaly Detection: A Survey," ACM Computing Surveys, vol. 41, pp. 1–72, 2009.

[Ch10] P. Cholda and A. Jajszczyk, "Recovery and Its Quality in Multilayer Networks," IEEE/OSA Journal of Lightwave Technology (JLT), vol. 28, pp. 372–389, 2010.

[Ch11] K. Christodoulopoulos, I. Tomkos, and E. Varvarigos, "Elastic Bandwidth Allocation in Flexible OFDM based Optical Networks," IEEE/OSA Journal of Lightwave Technology (JLT), vol. 29, pp. 1354–1366, 2011.

[Ch13] K. Christodoulopoulos, I. Tomkos, and E. Varvarigos, "Time-Varying Spectrum Allocation Policies and Blocking Analysis in Flexible Optical Networks," IEEE Journal on Selected Areas in Communications (JSAC), vol. 31, no. 1, pp. 13–25, 2013.

[Ch16] K. Christodoulopoulos, N. Sambo, and E. Varvarigos, "Exploiting Network Kriging for Fault Localization," in Proceedings of the Optical Fiber Communication Conference (OFC), 2016, Anaheim, 20–24 March.

[Ch83] V. Chvatal, *Linear Programming*, Freeman, 1983.

[Ch99] M. Charikar, Ch. Chekuri, T. Cheung, Z. Dai, A. Goel, S. Guha, and M. Li, "Approximation Algorithms for Directed Steiner Problems," Journal of Algorithms, vol. 33, pp. 73–91, 1999.

[CISCO12] CISCO Global Cloud Index, 2012. http://www.cisco.com/c/en_uk/ solutions/cloud/global-cloud-index-gci.html (accessed April 18, 2017).

[CISCO16] CISCO Visual Networking Index (VNI), 2016. http://www.cisco. com/c/en/us/solutions/service-provider/visual-networking-index-vni/ vni-infographic.html (accessed April 18, 2017).

[Cl02] M. Clouqueur and W. Grover, "Availability Analysis of Span-Restorable Mesh Networks," IEEE Journal of Selected Areas in Communications, vol. 20, pp. 810–821, 2002.

[CMRI] China Mobile Research Institute, "C-RAN the Road Towards Green RAN," 2011. http://labs.chinamobile.com/cran/wp-content/uploads/CRAN_ white_paper_v2_5_EN.pdf (accessed April 18, 2017).

[Co12] L.M. Contreras, V. López, O. González, and J.P. Fernández-Palacios, "Towards Cloud-Ready Transport Networks," IEEE Communications Magazine (ComMag), vol. 50, pp. 48–55, 2012.

[COMBO] COMBO EU project, "Deliverable D3.3—Analysis of Transport Network Architectures for Structural Convergence," 2015. http://www. ict-combo.eu/ (accessed April 18, 2017).

[Cp14] N. Carapellese, M. Tornatore, and A. Pattavina, "Energy-Efficient Baseband Unit Placement in a Fixed/Mobile Converged WDM Aggregation

Network," IEEE Journal on Selected Areas in Communications, vol. 32, pp. 1542–1551, 2014.

[CPLEX] IBM CPLEX Optimizer [Online], Available: http://www-01.ibm.com/software/commerce/optimization/cplex-optimizer/ (accessed March 23, 2017).

[CPRI] Common Public Radio Interface (CPRI) Specification [Online], Available: http://www.cpri.info/spec.html (accessed March 23, 2017).

[Cr16.1] E. Crabbe, I. Minei, S. Sivabalan, and R. Varga, "PCEP Extensions for PCE-Initiated LSP Set Up in a Stateful PCE Model," IETF draft-ietf-pce-pce-initiated-lsp-09, work in progress, 2017. https://www.ietf.org/ (accessed April 18, 2017).

[Cr16.2] E. Crabbe, I. Minei, J. Medved, and R. Varga, "PCEP Extensions for Stateful PCE," IETF draft-ietf-pce-stateful-pce-18, work in progress, 2016. https://www.ietf.org/ (accessed April 18, 2017).

[Cs13] R. Casellas, R. Martínez, R. Muñoz, L. Liu, T. Tsuritani, and I. Morita, "Dynamic Provisioning via a Stateful PCE with Instantiation Capabilities in GMPLS-Controlled Flexi-grid DWDM Networks," in Proceedings of the European Conference on Optical Communication (ECOC), 2013, London, 22–26 September.

[Cu12] F. Cugini, G. Meloni, F. Paolucci, N. Sambo, M. Secondini, L. Gerardi, L. Potí, and P. Castoldi, "Demonstration of Flexible Optical Network based on Path Computation Element," IEEE/OSA Journal of Lightwave Technology (JLT), vol. 30, pp. 727–733, 2012.

[Cu13] F. Cugini, F. Paolucci, G. Meloni, G. Berrettini, M. Secondini, F. Fresi, N. Sambo, L. Potí, and P. Castoldi, "Push-Pull Defragmentation without Traffic Disruption in Flexible Grid Optical Networks," IEEE/OSA Journal of Lightwave Technology (JLT), vol. 31, pp. 125–133, 2013.

[Da14] M. Dallaglio, A. Giorgetti, N. Sambo, F. Cugini, and P. Castoldi, "Impact of Slice-Ability on Dynamic Restoration in GMPLS-based Flexible Optical Networks," in Proceedings of the Optical Fiber Communication Conference, 2014, San Francisco, 9–14 March.

[De05] G. Desaulniers, J. Desrosiers, and M. Solomon, *Column Generation*, Springer, New York, 2005.

[Di17] N. Diamantopoulos, B. Shariati, and I. Tomkos, "On the Power Consumption of MIMO Processing and Its Impact on the Performance of SDM Networks," in Proceedings of the Optical Fiber Communication Conference and Exhibition (OFC), Los Angeles, 19–23 March 2017.

[Di59] E.W. Dijkstra, "A Note on Two Problems in Connection with Graphs," Numerische Mathematik, vol. 1, pp. 269–271, 1959.

[Em08] F. Emmert-Streib and M. Dehmer, *Information Theory and Statistical Learning*, Springer Science & Business Media, New York, 2008.

[Es10] R. Essiambre, G. Kramer, P.J. Winzer, G.J. Foschini, and B. Goebel, "Capacity Limits of Optical Fiber Networks," IEEE/OSA Journal of Lightwave Technology (JLT), vol. 28, pp. 662–701, 2010.

[Fe12] M. Feuer, L.E. Nelson, X. Zhou, S.L. Woodward, R. Isaac, B. Zhu, T.F. Taunay, M. Fishteyn, J.M. Fini, and M.F. Yan, "Joint Digital Signal Processing Receivers for Spatial Superchannels," IEEE Photonics Technology Letters, vol. 24, pp. 1957–1960, 2012.

[Fe15] M. Feuer, "Optical Routing for SDM Networks," in Proceedings of the European Conference and Exhibition on Optical Communications (ECOC), Valencia, 28–30 September 2015.

[Fe95] T. Feo and M. Resende, "Greedy Randomized Adaptive Search Procedures," Springer Journal of Global Optimization, vol. 6, pp. 109–133, 1995.

[Finisar] Finisar, "Programmable Narrow-Band Filtering Using the WaveShaper 1000E and WaveShaper 4000E," white paper, Available: http://www.finisar.com (accessed March 23, 2017).

[Floodlight] Project Floodlight [Online], Available: http://www.projectfloodlight.org/ (accessed March 23, 2017).

[Fo02] A. Forsgren, P.E. Gill, and M.H. Wright, "Interior Methods for Nonlinear Optimization," Society for Industrial and Applied Mathematics Review, vol. 44, no. 4, pp. 525–597, 2002.

[Fo15] N.K. Fontaine, T. Haramaty, R. Ryf, H. Chen, L. Miron, L. Pascar, M. Blau, B. Frenkel, L. Wang, Y. Messaddeq, S. LaRochelle, R.J. Essiambre, Y. Jung, Q. Kang, J.K. Sahu, S.U. Alam, D.J. Richardson, and D.M. Marom, "Heterogeneous Space-Division Multiplexing and Joint Wavelength Switching Demonstration," in Proceedings of the Optical Fiber Communication Conference and Exhibition (OFC), Los Angeles, 22–26 March 2015.

[G694.1] ITU-T SG15, "Spectral Grids for WDM Applications: DWDM Frequency Grid," ITU-T G.694.1, 2012. https://www.itu.int/rec/T-REC-G.694.1/en (accessed April 18, 2017).

[Ga79] M.R. Garey and D.S. Johnson, *Computers and Intractability: A Guide to the Theory of Np-Completeness*, W.H. Freeman & Co., New York, 1979.

[Ge03] A. Gençata and B. Mukherjee, "Virtual-Topology Adaptation for WDM Mesh Networks Under Dynamic Traffic," IEEE Transactions on Networking, vol. 11, pp. 236–247, 2003.

[Ge07] D. Gesbert, M. Kountouris, R. Heath, C.-B. Chae, and T. Salzer, "Shifting the MIMO Paradigm," IEEE Signal Processing Magazine, vol. 24, pp. 36–46, 2007.

[Ge10] M. Gendreau and J.Y. Potvin, *Handbook of Metaheuristics*, International Series in Operations Research & Management Science, Springer, New York, 2010.

[Ge12] O. Gerstel, M. Jinno, A. Lord, and S. Ben Yoo, "Elastic Optical Networking: A New Dawn for the Optical Layer?," IEEE Communications Magazine, vol. 50, pp. s12–s20, 2012.

[Gh12] M. Gharbaoui, B. Martini, and P. Castoldi, "Anycast-based Optimizations for Inter-Data-Center Interconnections," (Invited Paper) IEEE/OSA Journal of Optical Communications and Networking (JOCN), vol. 4, pp. B168–B178, 2012.

[Gi14.1] Ll. Gifre, L. Velasco, N. Navarro, and G. Junyent, "Experimental Assessment of a High Performance Back-End PCE for Flexgrid Optical

Network Re-optimization," in Proceedings of the IEEE/OSA Optical Fiber Communication Conference (OFC), 2014, San Francisco, 9–14 March.

[Gi14.2] Ll. Gifre, F. Paolucci, A. Aguado, R. Casellas, A. Castro, F. Cugini, P. Castoldi, L. Velasco, and V. Lopez, "Experimental Assessment of In-Operation Spectrum Defragmentation," (Invited Paper) Springer Photonic Network Communications, vol. 27, pp. 128–140, 2014.

[Gi15.1] Ll. Gifre, F. Paolucci, O. González de Dios, L. Velasco, L.M. Contreras, F. Cugini, P. Castoldi, and V. López, "Experimental Assessment of ABNO-Driven Multicast Connectivity in Flexgrid Networks," (Invited Paper) IEEE/OSA Journal of Lightwave Technology (JLT), vol. 33, pp. 1549–1556, 2015.

[Gi15.2] Ll. Gifre, R. Martínez, R. Casellas, R. Vilalta, R. Muñoz, and L. Velasco, "Modulation Format-Aware Re-optimization in Flexgrid Optical Networks: Concept and Experimental Assessment," in Proceedings of the European Conference on Optical Communication (ECOC), 2015, Valencia, 28–30 September.

[Gi15.3] Ll. Gifre, F. Paolucci, L. Velasco, A. Aguado, F. Cugini, P. Castoldi, and V. López, "First Experimental Assessment of ABNO-Driven In-operation Flexgrid Network Re-optimization," (Invited Paper) IEEE/OSA Journal of Lightwave Technology (JLT), vol. 33, pp. 618–624, 2015.

[Gi16] Ll. Gifre, M. Tornatore, L.M. Contreras, B. Mukherjee, and L. Velasco, "ABNO-Driven Content Distribution in the Telecom Cloud," Elsevier Optical Switching and Networking (DOI: j.osn.2015.07.006), 2016.

[Gl97] F. Glover and M. Laguna, *Tabu Search*, Kluwer Academic Publishers, Norwell, 1997.

[Go11] J. Gonçalves and M. Resende, "Biased Random-Key Genetic Algorithms for Combinatorial Optimization," Journal of Heuristics, vol. 17, pp. 487–525, 2011.

[Go13] L. Gong, X. Zhou, X. Liu, W. Zhao, W. Lu, and Z. Zhu, "Efficient Resource Allocation for All-Optical Multicasting over Spectrum-Sliced Elastic Optical Networks," IEEE/OSA Journal of Optical Communications and Networking, vol. 5, pp. 836–847, 2013.

[Gr04] W.D. Grover, *Mesh-based Survivable Networks: Options and Strategies for Optical, MPLS, SONET, and ATM Networking*, Prentice Hall, Upper Saddle River, 2004.

[Gr10] S. Gringeri, B. Basch, V. Shukla, R. Egorov, and T.J. Xia, "Flexible Architectures for Optical Transport Nodes and Networks," IEEE Communications Magazine, vol. 48, pp. 40–50, 2010.

[Gu07] L. Guo and L. Li, "A Novel Survivable Routing Algorithm with Partial Shared-Risk Link Groups (SRLG)-Disjoint Protection based on Differentiated Reliability Constraints in WDM Optical Mesh Networks," IEEE/OSA Journal of Lightwave Technology (JLT), vol. 25, pp. 1410–1415, 2007.

[Gur] Gurobi Optimizer 4.5, Available: http://www.gurobi.com (accessed March 23, 2017).

[Ha94] F. Harary, *Graph Theory*, Reading, Menlo park, 1994.

[Hm94] J.D. Hamilton, *Time Series Analysis*, Princeton University Press, Princeton, 1994.

[Hu05] Y. Huang, J.P. Heritage, and B. Mukherjee, "Connection Provisioning with Transmission Impairment Consideration in Optical DWDM Networks with High-Speed Channels," IEEE/OSA Journal of Lightwave Technology (JLT), vol. 23, pp. 982–993, 2005.

[Hu10] S. Huang, M. Xia, C. Martel, and B. Mukherjee, "A Multistate Multipath Provisioning Scheme for Differentiated Failures in Telecom Mesh Networks," IEEE/OSA Journal of Lightwave Technology (JLT), vol. 28, pp. 1585–1596, 2010.

[Hw13] I. Hwang, B. Song, and S. Soliman, "A Holistic View on Hyper-Dense Heterogeneous and Small Cell Networks," IEEE Communications Magazine, vol. 51, pp. 20–27, 2013.

[Ji09] M. Jinno, H. Takara, B. Kozicki, Y. Tsukishima, Y. Sone, and S. Matsuoka, "Spectrum-Efficient and Scalable Elastic Optical Path Network: Architecture, Benefits, and Enabling Technologies," IEEE Communications Magazine (ComMag), vol. 47, pp. 66–73, 2009.

[Ji10] M. Jinno, B. Kozicki, H. Takara, A. Watanabe, Y. Sone, T. Tanaka, and A. Hirano, "Distance-Adaptive Spectrum Resource Allocation in Spectrum-Sliced Elastic Optical Path (SLICE) Network," IEEE Communications Magazine (ComMag), vol. 48, pp. 138–145, 2010.

[Ji12] M. Jinno, H. Takara, Y. Sone, K. Yonenaga, and A. Hirano, "Multiflow Optical Transponder for Efficient Multilayer Optical Networking," IEEE Communications Magazine (ComMag), vol. 50, no. 5, pp. 56–65, 2012.

[Kh16] P. Khodashenas, J.M. Rivas-Moscoso, D. Siracusa, F. Pederzolli, B. Shariati, D. Klonidis, E. Salvadori, and I. Tomkos, "Comparison of Spectral and Spatial Super-Channel Allocation Schemes for SDM Networks", IEEE/OSA Journal of Lightwave Technology (JLT), vol. 34, pp. 2710–2716, 2016.

[Kl11] M. Klinkowski and K. Walkowiak, "Routing and Spectrum Assignment in Spectrum Sliced Elastic Optical Path Network," IEEE Communications Letters, vol. 15, pp. 884–886, 2011.

[Kl13] M. Klinkowski, M. Ruiz, L. Velasco, D. Careglio, V. Lopez, and J. Comellas, "Elastic Spectrum Allocation for Time-Varying Traffic in Flexgrid Optical Networks," IEEE Journal on Selected Areas in Communications (JSAC), vol. 31, pp. 26–38, 2013.

[Kl15] D. Klonidis, F. Cugini, O. Gerstel, M. Jinno, V. Lopez, E. Palkopoulou, M. Sekiya, D. Siracusa, G. Thouénon, and C. Betoule, "Spectrally and Spatially Flexible Optical Network Planning and Operations," IEEE Communications Magazine, vol. 53, pp. 69–78, 2015.

[Ku00] R. Kuehl, *Design of Experiments*, Thomson Learning, Pacific Grove, 2000.

[Le96] K. Lee and V. Li, "A Wavelength Rerouting Algorithm in Wide-Area All-Optical Networks," IEEE/OSA Journal of Lightwave Technology (JLT), vol. 14, pp. 1218–1229, 1996.

[Li13] R. Lin, M. Zukerman, G. Shen, and W. Zhong, "Design of Light-Tree based Optical Inter-Datacenter Networks," IEEE/OSA Journal of Optical Communications and Networking, vol. 5, pp. 1443–1455, 2013.

[Lo14] V. Lopez, B. Cruz, O. González de Dios, O. Gerstel, N. Amaya, G. Zervas, D. Simeonidou, and J.P. Fernandez-Palacios, "Finding the Target Cost for Sliceable Bandwidth Variable Transponders," IEEE/OSA Journal of Optical Communications and Networking (JOCN), vol. 6, pp. 476–485, 2014.

[Lo16] V. López and L. Velasco, *Elastic Optical Networks: Architectures, Technologies, and Control*, in Optical Networks book series, Springer, New York, 2016.

[Lu13] W. Lu and Z. Zhu, "Dynamic Service Provisioning of Advance Reservation Requests in Elastic Optical Networks," IEEE/OSA Journal of Lightwave Technology (JLR), vol. 31, pp. 1621–1627, 2013.

[M2120] "International Multi-operator Paths, Sections and Transmission Systems Fault Detection and Localization Procedures," ITU-T Rec. M.2120, 2002. https://www.itu.int/itudoc/itu-t/aap/sg4aap/history/m2120/m2120.html (accessed April 18, 2017).

[Ma03] E. Martins and M. Pascoal, "A New Implementation of Yen's Ranking Loopless Paths Algorithm," 4OR: Quarterly Journal of the Belgian, French and Italian Operations Research Societies, vol. 1, no. 2, pp. 121–133, 2003.

[Ma14] R. Martínez, Ll. Gifre, R. Casellas, L. Velasco, R. Muñoz, and R. Vilalta, "Experimental Validation of Active Frontend–Backend Stateful PCE Operations in Flexgrid Optical Network Re-optimization," in Proceedings of the European Conference on Optical Communication (ECOC), 2014, Cannes, 22–24 September.

[Ma15] R. Martínez, R. Casellas, R. Vilalta, and R. Muñoz, "Experimental Assessment of GMPLS/PCE-Controlled Multi-Flow Optical Transponders in FlexGrid Networks," in Proceedings of the Optical Fiber Communication Conference (OFC), 2015, Los Angeles, 22–26 March.

[Mo17] F. Morales, M. Ruiz, Ll. Gifre, L.M. Contreras, V. López, and L. Velasco, "Virtual Network Topology Adaptability based on Data Analytics for Traffic Prediction," (Invited Paper) IEEE/OSA Journal of Optical Communications and Networking (JOCN), vol. 9, pp. A35–A45, 2017.

[Mo99] G. Mohan and C. Murthy, "A Time Optimal Wavelength Rerouting Algorithm for Dynamic Traffic in DWDM Networks," IEEE/OSA Journal Lightwave Technology (JLT), vol. 17, pp. 406–417, 1999.

[Mr15] D. Marom and M. Blau, "Switching Solutions for WDM-SDM Optical Networks," IEEE Communications Magazine, vol. 53, pp. 60–68, 2015.

[Mr17] D. Marom, P.D. Colbourne, A. D'Errico, N.K. Fontaine, Y. Ikuma, R. Proietti, L. Zong, J.M. Rivas-Moscoso, and I. Tomkos, "Survey of Photonic Switching Architectures and Technologies in Support of Spatially and Spectrally Flexible Optical Networking," Journal of Optical Communications and Networking, vol. 9, pp. 1–26, 2017.

[Mu16] F. Musumeci, C. Bellanzon, N. Carapellese, M. Tornatore, A. Pattavina, and S. Gosselin, "Optimal BBU Placement for 5G C-RAN Deployment over WDM Aggregation Networks," IEEE/OSA Journal of Lightwave Technology, vol. 34, pp. 1963–1970, 2016.

[Na11] S. Namiki, T. Kurosu, K. Tanizawa, J. Kurumida, T. Hasama, H. Ishikawa, T. Nakatogawa, M. Nakamura, and K. Oyamada, "Ultrahigh-Definition Video Transmission and Extremely Green Optical Networks for Future," IEEE Journal of Selected Topics in Quantum Electronics, vol. 17, pp. 446–457, 2011.

[Na15] A. Napoli, M. Bohn, D. Rafique, A. Stavdas, N. Sambo, L. Poti, M. Nölle, J. Fischer, E. Riccardi, A. Pagano, A. Giglio, M. Svaluto, J. Fabrega, E. Hugues-Salas, G. Zervas, D. Simeonidou, P. Layec, A. D'Errico, T. Rahman, and J. Fernández-Palacios, "Next Generation Elastic Optical Networks: The Vision of the European Research Project IDEALIST," IEEE Communications Magazine, vol. 53, pp. 152–162, 2015.

[Nk15] K. Nakajima, P. Sillard, D. Richardson, M. Li, R. Essiambre, and S. Matsuo, "Transmission Media for an SDM-based Optical Communication System," IEEE Communications Magazine, vol. 53, pp. 44–51, 2015.

[OMNet] OMNet++ [Online], Available: http://www.omnetpp.org/ (accessed March 23, 2015).

[ONF] Open Networking Foundation [Online], Available: https://www.opennetworking.org (accessed March 23, 2015).

[OPENNEBULA] OpenNebula, Cloud management [Online], Available: http://opennebula.org/ (accessed March 23, 2015).

[OpenStack] OpenStack [Online], Available: http://www.openstack.org/ (accessed March 23, 2015).

[Pa10] F. Palmieri, U. Fiore, and S. Ricciardi, "A GRASP-based Network Re-optimization Strategy for Improving RWA in Multi-Constrained Optical Transport Infrastructures," Computer Communications (ComCom), vol. 33, pp. 1809–1822, 2010.

[Pa14] F. Paolucci, A. Castro, F. Cugini, L. Velasco, and P. Castoldi, "Multipath Restoration and Bitrate Squeezing in SDN-based Elastic Optical Networks," (Invited Paper) Springer Photonic Network Communications, vol. 28, pp. 45–57, 2014.

[Pe02] S. Pettie and V. Ramachandran, "An Optimal Minimum Spanning Tree Algorithm," Journal of the ACM, vol. 49, pp. 16–34, 2002.

[Pe10] L. Pessoa, M. Resende, and C. Ribeiro, "A Hybrid Lagrangean Heuristic with GRASP and Path Relinking for Set-k Covering," AT&T Labs Research Technical Report, AT&T Labs, 2010.

[Pe12] O. Pedrola, A. Castro, L. Velasco, M. Ruiz, J.P. Fernández-Palacios, and D. Careglio, "CAPEX Study for Multilayer IP/MPLS over Flexgrid Optical Network," IEEE/OSA Journal of Optical Communications and Networking (JOCN), vol. 4, pp. 639–650, 2012.

[Pi04] M. Pióro and D. Medhi, *Routing, Flow, and Capacity Design in Communication and Computer Networks*, The Morgan Kaufmann Series in Networking, Elsevier, San Francisco, 2004.

[Po08] Y. Pointurier, M. Brandt-Pearce, S. Subramaniam, and B. Xu, "Crosslayer Adaptive Routing and Wavelength Assignment in All-Optical Networks," IEEE Journal on Selected Areas in Communications (JSAC), vol. 26, pp. 1–13, 2008.

[Po13] F. Ponzini, L. Giorgi, A. Bianchi, and R. Sabella, "Centralized Radio Access Networks over Wavelength-Division Multiplexing: A Plug-and-Play Implementation," IEEE Communications Magazine, vol. 51, pp. 94–99, 2013.

[Pv10] C. Pavan, R. Morais, J. Ferreira, and A. Pinto, "Generating Realistic Optical Transport Network Topologies," IEEE/OSA Journal of Optical Communications and Networking, vol. 2, pp. 80–90, 2010.

[Ra87] P. Raghavan and C.D. Tompson, "Randomized Rounding: A Technique for Provably Good Algorithms and Algorithmic Proofs," Combinatorica, vol. 7, no. 4, pp. 365–374, 1987.

[Ra99] B. Ramamurthy, H. Feng, D. Datta, J.P. Heritage, and B. Mukherjee, "Transparent versus Opaque versus Translucent Wavelength-Routed Optical Networks," in Proceedings of the IEEE/OSA Optical Fiber Communication Conference (OFC), 1999, San Diego, February 21–26.

[Re10] M. Resende, R. Martí, M. Gallego, and A. Duarte, "GRASP and Path Relinking for the Max-Min Diversity Problem," Elsevier Computers and Operations Research, vol. 37, pp. 498–508, 2010.

[Re11] R. Reis, M. Ritt, L. Buriol, and M. Resende, "A Biased Random-Key Genetic Algorithm for OSPF and DEFT Routing to Minimize Network Congestion," International Transactions in Operational Research, vol. 18, pp. 401–423, 2011.

[RFC3031] E. Rosen, A. Viswanathan, and R. Callon, "Multiprotocol Label Switching Architecture," IETF RFC 3031, 2001.

[RFC3209] D. Awduche, L. Berger, D. Gan, T. Li, V. Srinivasan, and G. Swallow, "RSVP-TE: Extensions to RSVP for LSP tunnels," IETF RFC 3209, 2001. https://www.ietf.org/ (accessed April 18, 2017).

[RFC4090] P. Pan, G. Swallow, and A. Atlas, "Fast Reroute Extensions to RSVP-TE for LSP Tunnels," IETF RFC 4090, 2005. https://www.ietf.org/ (accessed April 18, 2017).

[RFC4655] A. Farrel, J.-P. Vasseur, and J. Ash, "A Path Computation Element (PCE)-based Architecture," IETF RFC 4655, 2006. https://www.ietf.org/ (accessed April 18, 2017).

[RFC5212] K. Shiomoto, D. Papadimitriou, JL. Le Roux, M. Vigoureux, and D. Brungard, "Requirements for GMPLS-based Multi-Region and Multi-Layer Networks (MRN/MLN)," IETF RFC 5212, 2008. https://www.ietf.org/ (accessed April 18, 2017).

[RFC5521] E. Oki, T. Takeda, and A. Farrel, "Extensions to the Path Computation Element Communication Protocol (PCEP) for Route Exclusions," IETF RFC 5521, 2009. https://www.ietf.org/ (accessed April 18, 2017).

[RFC5557] Y. Lee, JL. Le Roux, D. King, and E. Oki, "Path Computation Element Communication Protocol (PCEP) Requirements and Protocol Extensions in Support of Global Concurrent Optimization," IETF RFC 5557, 2009. https://www.ietf.org/ (accessed April 18, 2017).

[RFC5623] E. Oki, T. Takeda, JL. Le Roux, and A. Farrel, "Framework for PCE-based Inter-Layer MPLS and GMPLS Traffic Engineering," IETF RFC 5623, 2009. https://www.ietf.org/ (accessed April 18, 2017).

[RFC6006] Q. Zhao, D. King, F. Verhaeghe, T. Takeda, Z. Ali, and J. Meuric, "Extensions to the Path Computation Element Communication Protocol (PCEP) for Point-to-Multipoint Traffic Engineering Label Switched Paths," IETF RFC6006, 2010. https://www.ietf.org/ (accessed April 18, 2017).

[RFC6107] K. Shiomoto and A. Farrel, "Procedures for Dynamically Signaled Hierarchical Label Switched Paths," IETF RFC 6107, 2011. https://www.ietf.org/ (accessed April 18, 2017).

[RFC7752] H. Gredler, J. Medved, S. Previdi, A. Farrel, and S. Ray, "North-Bound Distribution of Link-State and TE Information Using BGP," IETF RFC 7752, 2016. https://www.ietf.org/ (accessed April 18, 2017).

[Ri16] J. Rivas-Moscoso, B. Shariati, A. Mastropaolo, D. Klonidis, and I. Tomkos, "Cost Benefit Quantification of SDM Network Implementations based on Spatially Integrated Network Elements," in Proceedings of the European Conference and Exhibition on Optical Communication (ECOC), Düsseldorf, 19–21 September 2016.

[Ri17] J. Rivas-Moscoso, B. Shariati, D. Marom, D. Klonidis, and I. Tomkos, "Comparison of CD(C) ROADM Architectures for Space Division Multiplexed Networks," in Proceedings of the European Conference on Optical Communication (ECOC), 2017, Gothenburg, 17–21 September.

[Ro08] J.P. Roorda and B. Collings, "Evolution to Colorless and Directionless ROADM Architectures," in Proceedings of the Optical Fiber Communication/National Fiber Optic Engineers Conference (OFC/NFOEC), paper NEW2, San Diego, 24–28 February 2008.

[Ru13.1] M. Ruiz, L. Velasco, J. Comellas, and G. Junyent, "A Traffic Intensity Model for Flexgrid Optical Network Planning under Dynamic Traffic Operation," in Proceedings of the IEEE/OSA Optical Fiber Communication Conference (OFC), 2013, Anaheim, 19–21 March.

[Ru13.2] M. Ruiz, M. Pióro, M. Zotkiewicz, M. Klinkowski, and L. Velasco, "Column Generation Algorithm for RSA Problems in Flexgrid Optical Networks," Springer Photonic Network Communications (PNET), vol. 26, pp. 53–64, 2013.

[Ru14.1] M. Ruiz, L. Velasco, A. Lord, D. Fonseca, M. Pióro, R. Wessäly, and J.P. Fernández-Palacios, "Planning Fixed to Flexgrid Gradual Migration: Drivers and Open Issues," IEEE Communications Magazine, vol. 52, pp. 70–76, 2014.

[Ru14.2] M. Ruiz and L. Velasco, "Performance Evaluation of Light-Tree Schemes in Flexgrid Optical Networks," IEEE Communications Letters, vol. 18, pp. 1731–1734, 2014.

[Ru15] M. Ruiz and L. Velasco, "Serving Multicast Requests on Single Layer and Multilayer Flexgrid Networks," IEEE/OSA Journal of Optical Communications and Networking (JOCN), vol. 7, pp. 146–155, 2015.

[Ru16.1] M. Ruiz, M. Germán, L.M. Contreras, and L. Velasco, "Big Data-Backed Video Distribution in the Telecom Cloud," Elsevier Computer Communications, vol. 84, pp. 1–11, 2016.

[Ru16.2] M. Ruiz, F. Fresi, A.P. Vela, G. Meloni, N. Sambo, F. Cugini, L. Poti, L. Velasco, and P. Castoldi, "Service-Triggered Failure Identification/Localization Through Monitoring of Multiple Parameters," in Proceedings of the European Conference on Optical Communication (ECOC), 2016, Düsseldorf, 19–21 September.

[Ry12] R. Ryf, S. Randel, A.H. Gnauck, C. Bolle, A. Sierra, S. Mumtaz, M. Esmaeelpour, E.C. Burrows, R. Essiambre, P.J. Winzer, D.W. Peckham, A.H. McCurdy, and R. Lingle, "Mode-Division Multiplexing over 96 km of Few-Mode Fiber Using Coherent 6×6 MIMO Processing," IEE/OSA Journal of Lightwave Technology, vol. 30, pp. 521–531, 2012.

[Sa11.1] N. Sambo, F. Cugini, G. Bottari, G. Bruno, P. Iovanna, and P. Castoldi, "Lightpath Provisioning in Wavelength Switched Optical Networks with Flexible Grid," in Proceedings of the European Conference on Optical Communication (ECOC), 2011, Geneva, 18–22 September.

[Sa13] N. Sambo, G. Meloni, G. Berrettini, F. Paolucci, A. Malacarne, A. Bogoni, F. Cugini, L. Potì, and P. Castoldi, "Demonstration of Data and Control Plane for Optical Multicast at 100 and 200 Gb/s with and without Frequency Conversion," IEEE/OSA Journal of Optical Communications and Networking, vol. 5, pp. 667–676, 2013.

[Sa15] N. Sambo, P. Castoldi, A. D'Errico, E. Riccardi, A. Pagano, M. Svaluto, J. Fàbrega, D. Rafique, A. Napoli, S. Frigerio, E. Hugues-Salas, G. Zervas, M. Nolle, J. Fischer, A. Lord, and J. Fernandez-Palacios, "Next Generation Sliceable Bandwidth Variable Transponders," IEEE Communications Magazine, vol. 53, pp. 163–171, 2015.

[Sh02] G. Shen, W. Grover, T.H. Cheng, and S.K. Bose, "Sparse Placement of Electronic Switching Node for Low Blocking in Translucent Optical Networks," OSA Journal of Optical Networking (JON), vol. 1, pp. 424–441, 2002.

[Sh07] G. Shen, and R.S. Tucker, "Translucent Optical Networks: The Way Forward," IEEE Communication Magazine (ComMag), vol. 45, pp. 48–54, 2007.

[Sh11] G. Shen, Q. Yang, S. You, and W. Shao, "Maximizing Time-Dependent Spectrum Sharing between Neighboring Channels in CO-OFDM Optical Networks," in Proceedings of the IEEE International Conference on Transparent Optical Networks (ICTON), Stockholm, 26–30 June 2011.

[Sh16.1] B. Shariati, P.S. Khodashenas, J.M. Rivas-Moscoso, S. Ben-Ezra, D. Klonidis, F. Jimenez, L. Velasco, and I. Tomkos, "Investigation of Mid-term Network Migration Scenarios Comparing Multi-band and Multi-fiber Deployments," in Proceedings of the Optical Fiber Communication Conference and Exhibition (OFC), Anaheim, 20–24 March 2016.

[Sh16.2] B. Shariati, P.S. Khodashenas, J.M. Rivas-Moscoso, S. Ben-Ezra, D. Klonidis, F. Jimenez, L. Velasco, and I. Tomkos, "Evaluation of the Impact of Different SDM Switching Strategies in a Network Planning Scenario," in Proceedings of the Optical Fiber Communication Conference and Exhibition (OFC), Anaheim, 20–24 March 2016.

[Sh16.3] B. Shariati, J.M. Rivas-Moscoso, D. Klonidis, S. Ben-Ezra, F. Jiménez, D.M. Marom, P.S. Khodashenas, J. Comellas, L. Velasco, and I. Tomkos, "Options for Cost-Effective Capacity Upgrades in Backbone Optical Networks," in Proceedings of the European Conference on Network and Optical Communication (ECOC), Düsseldorf, 19–21 September 2016.

[Sh16.4] B. Shariati, D. Klonidis, D. Siracusa, F. Pederzolli, J.M. Rivas-Moscoso, L. Velasco, and I. Tomkos, "Impact of Traffic Profile on the Performance of Spatial Superchannel Switching in SDM Networks," in Proceedings of the European Conference on Optical Communication (ECOC), Düsseldorf, 19–21 September 2016.

[SND] SNDlib, Available: http://sndlib.zib.de/ (accessed March 23, 2017).

[So11] Y. Sone, A. Watanabe, W. Imajuku, Y. Tsukishima, B. Kozicki, H. Takara, and M. Jinno, "Bandwidth Squeezed Restoration in Spectrum-Sliced Elastic Optical Path Networks (SLICE)," IEEE/OSA Journal of Optical Communications and Networking (JOCN), vol. 3, pp. 223–233, 2011.

[SPARK] Apache Spark, Available: http://spark.apache.org/ (accessed March 23, 2017).

[Ta11] T. Takagi, H. Hasegawa, K. Sato, Y. Sone, A. Hirano, and M. Jinno, "Disruption Minimized Spectrum Defragmentation in Elastic Optical Path Networks That Adopt Distance Adaptive Modulation," in Proceedings of the European Conference on Optical Communication (ECOC), 2011, Geneva, 18–22 September.

[Ta15] J. Tapolcai, P. Ho, P. Babarczi, and L. Rónyai, "Neighborhood Failure Localization in All-Optical Networks via Monitoring Trails," IEEE/ACM Transactions on Networking, vol. 23, pp. 1719–1728, 2015.

[To17] H. Tode and Y. Hirota, "Routing, Spectrum, and Core and/or Mode Assignment on Space-Division Multiplexing Optical Networks," Journal of Optical Communications and Networking, vol. 9, pp. A99–A113, 2017.

[Ts06] J. Tsai, S. Huang, D. Hah, and M. Wu, "$1 \times N^2$ Wavelength-Selective Switch with Two Cross-Scanning One-Axis Analog Micromirror Arrays in a 4-f Optical System," IEEE/OSA Journal of Lightwave Technology (JLT), vol. 24, pp. 897–903, 2006.

[TS23.401] ETSI, "General Packet Radio Service (GPRS) Enhancements for Evolved Universal Terrestrial Radio Access Network (E-UTRAN) Access," 3GPP TS 23.401, 2015. https://portal.3gpp.org/desktopmodules/Specifications/SpecificationDetails.aspx?specificationId=849 (accessed April 18, 2017).

[Tz13] A. Tzanakaki, M.P. Anastasopoulos, and K. Georgakilas, "Dynamic Virtual Optical Networks Supporting Uncertain Traffic Demands,"

IEEE/OSA Journal of Optical Communications and Networking, vol. 5, pp. A76–A85, 2013.

[Va16] A.P. Vela, A. Via, M. Ruiz, and L. Velasco, "Bringing Data Analytics to the Network Nodes," in Proceedings of the European Conference on Optical Communication (ECOC), 2016, Düsseldorf, 19–21 September.

[Ve09] L. Velasco, S. Spadaro, J. Comellas, and G. Junyent, "Shared-Path Protection with Extra-Traffic in ASON/GMPLS Ring Networks," OSA Journal of Optical Networking (JON), vol. 8, pp. 130–145, 2009.

[Ve10] L. Velasco, F. Agraz, R. Martínez, R. Casellas, S. Spadaro, R. Muñoz, and G. Junyent, "GMPLS-based Multi-Domain Restoration: Analysis, Strategies, Policies and Experimental Assessment," IEEE/OSA Journal of Optical Communications and Networking (JOCN), vol. 2, pp. 427–441, 2010.

[Ve12.1] L. Velasco, M. Klinkowski, M. Ruiz, and J. Comellas, "Modeling the Routing and Spectrum Allocation Problem for Flexgrid Optical Networks," Springer Photonic Network Communications (PNET), vol. 24, pp. 177–186, 2012.

[Ve12.2] L. Velasco, M. Klinkowski, M. Ruiz, V. López, and G. Junyent, "Elastic Spectrum Allocation for Variable Traffic in Flexible-Grid Optical Networks," in Proceedings of the IEEE/OSA Optical Fiber Communication Conference (OFC), 2012, Los Angeles, March 4–8.

[Ve13.1] L. Velasco, P. Wright, A. Lord, and G. Junyent, "Saving CAPEX by Extending Flexgrid-based Core Optical Networks towards the Edges," IEEE/OSA Journal of Optical Communications and Networking (JOCN), vol. 5, pp. A171–A183, 2013.

[Ve13.2] L. Velasco, A. Asensio, J.Ll. Berral, V. López, D. Carrera, A. Castro, and J.P. Fernández-Palacios, "Cross-Stratum Orchestration and Flexgrid Optical Networks for Datacenter Federations," IEEE Network Magazine, vol. 27, pp. 23–30, 2013.

[Ve14.1] L. Velasco, A. Castro, M. Ruiz, and G. Junyent, "Solving Routing and Spectrum Allocation Related Optimization Problems: From Off-Line to In-Operation Flexgrid Network Planning," (Invited Tutorial) IEEE/OSA Journal of Lightwave Technology (JLT), vol. 32, pp. 2780–2795, 2014.

[Ve14.2] L. Velasco, D. King, O. Gerstel, R. Casellas, A. Castro, and V. López, "In-Operation Network Planning," IEEE Communications Magazine (ComMag), vol. 52, pp. 52–60, 2014.

[Ve14.3] L. Velasco, A. Asensio, J.Ll. Berral, A. Castro, and V. López, "Towards a Carrier SDN: An Example for Elastic Inter-Datacenter Connectivity," (Invited Paper) OSA Optics Express, vol. 22, pp. 55–61, 2014.

[Ve15] L. Velasco, L.M. Contreras, G. Ferraris, A. Stavdas, F. Cugini, M. Wiegand, and J.P. Fernández-Palacios, "A Service-Oriented Hybrid Access Network and Cloud Architecture," IEEE Communications Magazine, vol. 53, pp. 159–165, 2015.

[Ve17] L. Velasco, A.P. Vela, F. Morales, and M. Ruiz, "Designing, Operating and Re-Optimizing Elastic Optical Networks," (Invited Tutorial) IEEE/OSA Journal of Lightwave Technology (JLT), vol. 35, pp. 515–526, 2017.

[Vr05] S. Verbrugge, D. Colle, P. Demeester, R. Huelsermann, and M. Jaeger, "General Availability Model for Multilayer Transport Networks," in Proceedings of Design of Reliable Communication Networks (DRCN), October 2005, Island of Ischia, October 16–19.

[Wa11.1] X. Wang, Q. Zhang, I. Kim, P. Palacharla, and M. Sekiya, "Blocking Performance in Dynamic Flexible Grid Optical Networks—What Is the Ideal Spectrum Granularity?" in Proceedings of the European Conference on Optical Communication (ECOC), 2011, Geneva, 18–22 September.

[Wa11.2] X. Wan, L. Wang, N. Hua, H. Zhang, and X. Zheng, "Dynamic Routing and Spectrum Assignment in Flexible Optical Path Networks," in Proceedings of the IEEE/OSA Optical Fiber Communication Conference (OFC), 2011, Los Angeles, March 6–10.

[Wa11.3] Y. Wang, X. Cao, and Y. Pan, "A Study of the Routing and Spectrum Allocation in Spectrum-Sliced Elastic Optical Path Networks," in Proceedings of the IEEE International Conference on Computer Communications (INFOCOM), Shanghai, 10–15 April 2011.

[Wa12] Q. Wang and L. Chen, "Performance Analysis of Multicast Traffic over Spectrum Elastic Optical Networks," in Proceedings of the Optical Fiber Communication (OFC), 2012, Los Angeles, March 4–8.

[Wa14] K. Walkowiak, A. Kasprzak, and M. Klinkowski, "Dynamic Routing of Anycast and Unicast Traffic in Elastic Optical Networks," in Proceedings of the IEEE International Conference on Communications (ICC), Sydney, 10–14 June 2014.

[We11] K. Wen, Y. Yin, D. Geisler, S. Chang, and S. Yoo, "Dynamic On-demand Lightpath Provisioning Using Spectral Defragmentation in Flexible Bandwidth Networks," in Proceedings of the European Conference on Optical Communication (ECOC), 2011, Geneva, 18–22 September.

[Wi01] P. Willis, *Carrier-Scale IP Networks: Designing and Operating Internet Networks*, IET Digital Library, Stevenage, 2001.

[Wi06] P.J. Winzer and R. Essiambre, "Advanced Optical Modulation Formats," Proceedings of the IEEE, vol. 94, pp. 952–985, 2006.

[Wi12] P.J. Winzer, "Optical Networking Beyond WDM," IEEE Journal Photonics, vol. 4, pp. 647–651, 2012.

[WIKISIZE] Wikipedia, Available: http://en.wikipedia.org/wiki/Wikipedia_talk:Size_of_Wikipedia#size_in_GB (accessed March 23, 2017).

[Wo13] S. Woodward, W. Zhang, B. Bathula, G. Choudhury, R. Sinha, M. Feuer, J. Strand, and A. Chiu, "Asymmetric Optical Connections for Improved Network Efficiency," IEEE/OSA Journal of Optical Communications and Networking, vol. 5, pp. 1195–1201, 2013.

[Ya05.1] X. Yang and B. Ramamurthy, "Dynamic Routing in Translucent DWDM Optical Networks: The Intradomain Case," IEEE/OSA Journal of Lightwave Technology (JLT), vol. 23, pp. 955–971, 2005.

[Ya05.2] X. Yang, L. Shen, and B. Ramamurthy, "Survivable Lightpath Provisioning in DWDM Mesh Networks under Shared Path Protection and Signal Quality Constraints," IEEE/OSA Journal of Lightwave Technology (JLT), vol. 23, pp. 1556–1567, 2005.

[Ya08] W. Yao and B. Ramamurthy, "Rerouting Schemes for Dynamic Traffic Grooming in Optical DWDM Networks," Elsevier Computer Networks (ComNet), vol. 52, pp. 1891–1904, 2008.

[Ye71] J. Yen, "Finding the k Shortest Loopless Paths in a Network," Management Science, vol. 17, pp. 712–716, 1971.

[Yo14] M. Yoshinari, Y. Ohsita, and M. Murata, "Virtual Network Reconfiguration with Adaptability to Traffic Changes," IEEE/OSA Journal of Optical Communications and Networking, vol. 6, pp. 523–535, 2014.

[Yu05] S. Yuan and J. Jue, "Dynamic Lightpath Protection in DWDM Mesh Networks under Wavelength-Continuity and Risk-Disjoint Constraints," Elsevier Computer Networks (ComNet), vol. 48, pp. 91–112, 2005.

[Zh00] Y. Zhu, G. Rouskas, and H. Perros, "A Comparison of Allocation Policies in Wavelength Routing Networks," Springer Photonic Network Communications (PNET), vol. 2, pp. 265–293, 2000.

[Zo15] M. Zotkiewicz, M. Ruiz, M. Klinkowski, M. Pióro, and L. Velasco, "Reoptimization of Dynamic Flexgrid Optical Networks After Link Failure Repairs," IEEE/OSA Journal of Optical Communications and Networking (JOCN), vol. 7, pp. 49–61, 2015.

Index

Provisioning, Recovery, and In-operation Planning in Elastic Optical Networks,
First Edition. Luis Velasco and Marc Ruiz.
© 2017 John Wiley & Sons, Inc. Published 2017 by John Wiley & Sons, Inc.